Progress in Vaccinology

Contraception Research for Today and the Nineties
Progress in Birth Control Vaccines

Progress in Vaccinology

Progress in Vaccinology

Volume 2

G.P. Talwar, Editor

Progress in Vaccinology

With 77 Illustrations

Springer-Verlag
New York Berlin Heidelberg
London Paris Tokyo

Professor G.P. Talwar
National Institute of Immunology
JNU Complex, Shahid Jeet Singh Marg
New Delhi – 110 067, India

Library of Congress Cataloging-in-Publication Data
Progress in vaccinology.
   (Progress in vaccinology ; v. 2)
   Includes bibliographies and index.
   1. Vaccines. I. Talwar, G.P. II. Series. [DNLM:
1. Communicable disease control. 2. Immunotherapy.
3. Vaccines. W1 PR685 v.2 / QW 805 P9638]
QR189.P74 1989   615'.372   88-33629
ISBN 0-387-96734-6

Copyright is not claimed for U.S. Government employees. The papers in this volume were presented at a Symposium on Progress in Vaccinology, 1–5 December 1986, New Delhi, India.

Typeset by David E. Seham Associates, Inc., Metuchen, New Jersey.
Printed and bound by Arcata Graphics/Halliday, West Hanover, Massachusetts.
Printed in the United States of America.

9 8 7 6 5 4 3 2 1

ISBN 0-387-96734-6 Springer-Verlag New York Berlin Heidelberg
ISBN 3-540-96734-6 Springer-Verlag Berlin Heidelberg New York

Series Preface

Vaccines have historically been considered to be the most cost-effective method for preventing communicable diseases. It was a vaccine that enabled global eradication of the dreaded disease smallpox. Mass immunization of children forms the anchor of the strategy of the World Health Organization (WHO) to attain "health for all" status by the year 2000.

Vaccinology is undergoing a dimensional change with the advances that have taken place in immunology and genetic engineering. Vaccines that confer short or inadequate immunity or that have side effects are being replaced by better vaccines. New vaccines are being developed for a variety of maladies. Monoclonal antibodies and T cell clones have been employed to delineate the immunodeterminants on microbes, an approach elegantly complemented by computer graphics and molecular imaging techniques. Possibilities have opened for obtaining hitherto scarce antigens of parasites by the DNA recombinant route. Better appreciation of the idiotypic network has aroused research on anti-idiotypic vaccines. Solid-phase synthesis of peptides is leading to an array of synthetic vaccines, an approach that is expected to attain its full potential once the sequences activating suppressor cells are discovered and the rules for presentation of antigens to T and B cells are better worked out.

A new breed of vaccines is on the horizon that seeks to control fertility. Originally conceived to intercept a step in the reproductive process, they are conceptual models for developing approaches to regulate the body's internal processes. The importance of lymphokines and monokines in the induction of the immune response and in killing parasites is realized, and specific or nonspecific routes are employed to elicit their formation. Interleukins and interferons have been produced by DNA recombinant methods and experimental approaches initiated to coexpress the genes for such regulators with microbial antigens. The old smallpox vaccine, vaccinia, is appearing in a new garb with genetically engineered foreign genes. The technology for manufacture of vaccines, especially for cell culture-based organisms, is undergoing changes, with new cell lines, promoters for better expression, and automation.

Contemporary vaccinology is a multidisciplinary science (and technology) that is developing rapidly. Findings are reported in disparate journals. Periodical reviews by experts assimilating relevant progress in a given field would be of immense value to investigators, funding agencies, manufacturers, and users of the vaccines, the public health authorities. This series aims to provide comprehensive reviews on topics relating to various aspects of vaccinology by leading investigators.

G.P. Talwar

Preface

The series began with a volume reviewing progress on birth control vaccines. The present volume is a "curtain raiser" to vaccines against communicable diseases. Several key issues are discussed. Five chapters deal with the public health usage of vaccines for universal immunization, and the problems associated with adequate supply of vaccines and their maintenance in active form. New technologies for manufacture of vaccines are described. Every vaccine is not equally effective everywhere. Carefully planned trials in South India by the ICMR-WHO with a hitherto accepted vaccine, BCG, did not produce protection against pulmonary tuberculosis in children after a 5-year observation period, and results of this important study are reviewed by the investigator in charge. Drs. Lagrange and Hurtrel discuss immunity in tuberculosis. The oral polio vaccine has also not given universal immunization. Genetic and environmental factors appear to influence the efficacy of vaccines.

A vaccine for hepatitis B, either prepared from plasma-derived material or obtained by DNA recombinant techniques, is on the market. The nature of immunity induced by the HB surface protein and the biological significance of antibody to the pre-S region are new observations reviewed in Part IV. Dr. Pillot and co-workers describe the characteristics of the virus they isolated from fecal samples collected during an African epidemic. The virus is apparently the elusive causative agent of non-A non-B hepatitis.

Eleven chapters are devoted to vaccines (and related issues) for gastrointestinal tract infections. The section includes the last report written by Dr. R. Germanier, the discoverer of the oral attenuated typhoid vaccine. This vaccine has given encouraging results in large-scale trials in Egypt. Drs. Jan Holmgren and Ann-Mari Svennerholm and their colleagues have contributed two remarkable chapters on their work on oral cholera vaccine and on mucosal immunity in relation to ETEC vaccine. *Shigella* vaccines are reviewed by Dr. Sam Formal and his colleagues and rotavirus vaccine by Dr. Kapikian and coworkers. A number of leading investigators from

India have written on their findings and thoughts on amebiasis, giardiasis, and cholera.

Drs. Almond and Parkhouse have reviewed their important work on helminth infections, and Dr. Renu Lal describes her work on immuno-diagnosis of filariasis. Veterinary vaccines are not forgotten. Advances in the development and testing of leprosy vaccines are reviewed, as are the immunodiagnostic approaches to discovering the transmitters of the infection. The antisporozoite vaccine for malaria is discussed by the NIH group, and the first trial of the vaccine by the Walter Reed Institute is reported by Dr. Ballou. The impact that water management approaches can have on the vector is discussed by the pioneer of this approach, Dr. V. P. Sharma. Dr. Dyann Wirth reviews new approaches to epidemio-logical studies on malaria and leishmaniasis.

The last section of the book has five chapters. Two address the use of vaccinia as a vector for vaccines for hepatitis, rabies, and virus-borne tumors. Engineered vaccines of vaccinia have great potential, demanding only a single administration. They induce both cell-mediated and humoral immunity and are affordable for mass-scale use. The chapter by Drs. M. P. Kieny and R. Lathe demonstrates the efficacy of the vaccinia–rabies vaccine by the oral route. Dr. Carl Alving discusses the use of liposomes as carrier of vaccines. Dr. Eli Sercarz analyzes lessons of interest to vac-cinology based on their elegant work on suppressor and helper determi-nants in lysozymes. Finally, Dr. Jonas Salk deliberates on the way vac-cinology is progressing.

G.P. Talwar

Contents

PART IX MALARIA AND LEISHMANIASIS

PART X NEW APPROACHES TO VACCINOLOGY

Contributors

S.C. Adlakha, Indian Council of Agricultural Research, Krishi Bhavan, New Delhi-110 001 India.

Christina Åhrén, Department of Medical Microbiology, University of Göteborg, S-413 46 Göteborg, Sweden.

Bernadino Albuquerque, Instituto de Medicina, Tropical de Manaus, CEP: 69000, Manaus, Amazonas, Brazil.

N.M. Almond, Collaborative Centre, National Institute for Biological Standards and Control, South Mimms, Potters Bar, Herts EN6 3QG England.

Carl R. Alving, Department of Membrane Biochemistry, Walter Reed Army Institute of Research, Washington, DC 20307-5100 USA.

Edwin L. Anderson, Marshall University, Huntington, WV, USA.

J. Armand, Institut Merieux, 58 Avenue Leclerc, bp 7046, 69348 Lyon Cedex 07 France.

F. Arminjon, Institut Merieux, 58 Avenue Leclerc, bp 7046, 69348 Lyon Cedex 07 France.

W. Ripley Ballou, Department of Immunology, Walter Reed Army Institute of Research, Washington, DC 20307-5100 USA.

Robert Barker, Harvard School of Public Health, Department of Tropical Public Health, 665 Huntington Avenue, Boston, MA 02115 USA.

Robert B. Belshe, Marshall University, Huntington, WV, USA.

V.P. Bhardwaj, Central JALMA Institute for Leprosy, Taj Ganj, Agra-282 001 India.

Alok Bhattacharya, Laboratory of Parasitic Diseases, National Institute of Allergy and Infectious Diseases, National Institutes of Health, Bethesda, MD 20892 USA.

Sudha Bhattacharya, Laboratory of Parasitic Diseases, National Institute of Allergy and Infectious Diseases, National Institutes of Health, Bethesda, MD 20892 USA.

Robert E. Black, Center for Vaccine Development, Division of Geographic Medicine, University of Maryland School of Medicine, Baltimore, MD 21201 USA.

Barry R. Bloom, Department of Microbiology and Immunology, Albert Einstein College of Medicine, Bronx, NY 10461 USA.

A. Budkowska, Unite d 'Immunologie Microbienne, WHO Centre of Reference and Research on Viral Hepatitis, Institut Pasteur 75724, Paris, France.

Jerry M. Buysse, Department of Bacterial Immunology, Division of Communicable Diseases and Immunology, Walter Reed Army Institute of Research, Washington, DC 20307-5100 USA.

M. Cadoz, Institut Merieux, 58 Avenue Leclerc, bp 7056, 69348 Lyon Cedex 07 France.

Robert M. Chanock, Laboratory of Infectious Diseases, National Institutes of Allergy and Infectious Diseases, National Institutes of Health, Bethesda, MD 20892 USA.

Cynthia Christy, University of Rochester, Rochester, NY, USA.

John Clemens, International Centre for Diarrhoeal Disease Research Bangladesh, Dhaka, Bangladesh.

Mary Lou Clements, Johns Hopkins University, Baltimore, MD, USA.

Jacinto Convit, Institute de Biomedicine, Caracas 101 Venezuela.

S.J. Cryz, Jr., Swiss Serum and Vaccine Institute, CH-3001 Berne, Switzerland.

Chiranjib Dasgupta, Division of Immunology, National Institute of Cholera and Enteric Diseases, Calcutta 700 010 India.

C. Dauguet, Department de Virologie, Institut Pasteur 75724 Paris, France.

Vidal F. de la Cruz, Molecular Vaccines Inc., 19 Firstfield Road, Gaithersburg, MD 20878 USA.

Louis S. Diamond, Laboratory of Parasitic Diseases, National Institute of Allergy and Infectious Diseases, National Institutes of Health, Bethesda, MD 20892 USA.

Raphael Dolin, University of Rochester, Rochester, NY, USA.

Heitor Dourado, Instituto de Medicina, Tropical de Manaus, CEP: 69000, Manaus, Amazonas, Brazil.

P. Dubreuil, INSERM U 131, 32, rue des Carnets 92141 Clamart, France.

Jorge Flores, Laboratory of Infectious Diseases, National Institute of Allergy and Infectious Diseases, National Institutes of Health, Bethesda, MD 20892 USA.

Samuel B. Formal, Department of Enteric Infections, Walter Reed Army Institute of Research, Washington, DC 20307-5100 USA.

M. Galimand, Unite du Cholera, Institut Pasteur 75724, Paris, France.

N.K. Ganguly, Department of Experimental, Medicine and Parasitology, Postgraduate Institute of Medical Education and Research, Chandigarh-160 012 India.

R. Germanier, deceased.

Asoke C. Ghose, Division of Immunology, National Institute of Cholera and Enteric Diseases, Calcutta 700 010 India.

B.K. Girdhar, Central JALMA Institute for Leprosy, Taj Ganj, Agra-282 001 India.

Roger I. Glass, Centers for Disease Control, Atlanta, GA, USA.

Marino Gonzalez, Central University of Venezuela, Caracas, Venezuela.

Mario Gorziglia, Laboratory of Infectious Diseases, National Institute of Allergy and Infectious Diseases, National Institutes of Health, Bethesda, MD 20892 USA.

Leif Gothefors, University of Umea, Umea, Sweden.

Kim Y. Green, Laboratory of Infectious Diseases, National Institute of Allergy and Infectious Diseases, National Institutes of Health, Bethesda, MD 20892 USA.

Thomas L. Hale, Department of Enteric Infections, Walter Reed Army Institute of Research, Washington, DC 20307-5100 USA.

Neal A. Halsey, Johns Hopkins University, Baltimore, MD, USA.

Scott B. Halstead, Health Sciences Division, The Rockefeller Foundation, 1133 Avenue of the Americas, New York NY 10036 USA.

Jan Holmgren, Department of Medical Microbiology, University of Göteborg, S-413 46 Göteborg, Sweden.

Yasutaka Hoshno, Laboratory of Infectious Diseases, National Institute of Allergy and Infectious Diseases, National Institutes of Health, Bethesda, MD 20892 USA.

B. Hurtrel, Institut Pasteur, 25 rue du Docteur Roux, 75724 Paris Cedex 15 France.

Becky M. Itty, National Institute of Immunology, New Delhi 110 067 India.

Alice Jacob, Microbiology Division, Central Drug Research Institute, Lucknow 226 001 India.

William R. Jacobs, Department of Microbiology and Immunology, Albert Einstein College of Medicine, Bronx, NY 10461 USA.

T. Jacob John, ICMR Centre of Advanced Research in Virology, Christian Medical College Hospital, Vellore, Tamil Nadu 632 004 India.

V.R. Kalyanaraman, Pasteur Institute of India, Coonoor, Nilgiris, Tamil Nadu 643 103 India.

Christine Kapfer, Department of Enteric Infections, Walter Reed Army Institute of Research, Washington, DC 20307-5100 USA.

Albert Z. Kapikian, Laboratory of Infectious Diseases, National Institute of Allergy and Infectious Diseases, National Institutes of Health, Bethesda, MD 20892 USA.

H.K. Kar, Department of Dermatology, Ram Manohar Lohia Hospital, New Delhi 110 011 India.

Kiran Katoch, Central JALMA Institute for Leprosy, Taj Ganj, Agra 282 001 India.

M.P. Kieny, Transgene S.A., 11 rue de Molsheim, 67000 Strasbourg, France.

Dennis J. Kopecko, Department of Bacterial Immunology, Division of Communicable Diseases and Immunology, Walter Reed Army Institute of Research, Washington, DC 20307-5100 USA.

P.H. Lagrange, Institut Pasteur, 25 rue du Docteur Roux, 75724 Paris Cedex 15 France.

Altaf A. Lal, Laboratory of Parasitic Diseases, National Institute of Allergy and Infectious Diseases, National Institutes of Health, Bethesda, MD 20892 USA.

Renu B. Lal, Laboratory of Parasitic Diseases, National Institute of Allergy and Infectious Diseases, National Institutes of Health, Bethesda, MD 20892 USA.

R. Lathe, CNRS-LGME and U184-INSERM, 11 rue Humann, 67085 Strasbourg Cedex, France.

Y. Lazizi, INSERM U 131, 32 rue des Carnets 92141 Clamart, France.

Myron M. Levine, Center for Vaccine Development, Division of Geographic Medicine, University of Maryland School of Medicine, Baltimore, MD 21201 USA.

Yolanda Lopez-Vidal, Department of Medical Microbiology, University of Göteborg, 5-41346 Göteborg, Sweden.

Genevieve A. Losonsky, University of Maryland School of Medicine, Baltimore, MD 21201 USA.

R.C. Mahajan, Department of Experimental Medicine and Parasitology, Postgraduate Institute of Medical Education and Research, Chandigarh-160 012 India.

Thomas F. McCutchan, Laboratory of Parasitic Diseases, National Institute of Allergy and Infectious Diseases, National Institutes of Health, Bethesda, MD 20892 USA.

Vijay Mehra, Department of Microbiology and Immunology, Albert Einstein College of Medicine, Bronx, NY 10461 USA.

Karen Midthun, Laboratory of Infectious Diseases, National Institute of Allergy and Infectious Diseases, National Institutes of Health, Bethesda, MD 20892 USA.

R.S. Misra, Department of Dermatology, Safdarjang Hospital, New Delhi 110 029 India.

Robert L. Modlin, Department of Medicine, University of Southern California, School of Medicine, Los Angeles, CA 90033 USA.

B. Montagnon, Institut Merieux, 58 Avenue Leclerc, bp 7046, 69348 Lyon Cedex 07 France.

Bernard Moss, Laboratory of Viral Diseases, National Institute of Allergy and Infectious Diseases, National Institutes of Health, Bethesda, MD 20892 USA.

K.D. Moudgil, Department of Biochemistry, All India Institute of Medical Sciences, New Delhi 110 029 India.

A. Mukherjee, Institute of Pathology, Indian Council of Medical Research, Safdarjang Hospital, New Delhi 110 029 India.

R. Mukherjee, National Institute of Immunology, New Delhi 110 067 India.

Abu Salim Mustafa, Laboratory for Immunology, Department of Pathology, Norwegian Radium Hospital, Montebello, N-0310 Oslo 3, Norway.

R.M.E. Parkhouse, National Institute for Medical Research, The Ridgeway, Mill Hill, London NW7 1AA England.

S.A. Patil, Central JALMA Institute for Leprosy, Taj Ganj, Agra-282 001 India.

Irene Perez-Schael, Central University of Venezuela, Caracas, Venezuela.

J. Pillot, Unite d'Immunologie Microbienne, WHO Centre of Reference and Research on Viral Hepatitis, Institut Pasteur 75724 Paris, France.

Louis Potash, Flow Laboratories, Inc, McLean, VA, USA.

Robert H. Purcell, Laboratory of Infectious Diseases, National Institute of Allergy and Infectious Diseases, National Institutes of Health, Bethesda, MD 20892 USA.

V. Ramalingaswami, UNICEF, 3 United Nations Plaza, New York, NY 10017 USA.

G. Ramu, Central JALMA Institute of Leprosy, Taj Ganj, Agra-282 001 India.

B.U. Rao, Southern Regional Station, Indian Veterinary Research Institute, Bangalore 560 024 India.

Thomas H. Rea, Department of Medicine, University of Southern California, School of Medicine, Los Angeles, CA 90033 USA.

Margaret B. Rennels, University of Maryland School of Medicine, Baltimore, MD 21201 USA.

William O. Rogers, Harvard School of Public Health, Department of Tropical Public Health, 665 Huntington Avenue, Boston, MA 02115 USA.

M. Roumiantzeff, Institut Merieux, 58 Avenue Leclerc, bp 7046, 69348 Lyon Cedex 07 France.

Soumitra Roy, Transfusion Medicine Research Program, Department of Laboratory Medicine, University of California, San Francisco, CA 94143 USA.

David Sack, International Centre for Diarrhoeal Disease Research, Bangladesh, Dhaka, Bangladesh.

Jonas Salk, The Salk Institute for Biological Studies, PO Box 85800, San Diego, CA 92138 USA.

Joaquin Sanchez, Department of Medical Microbiology, University of Göteborg, S-413 46 Göteborg, Sweden.

Mathuram Santosham, Johns Hopkins University, Infectious Diseases Research Center, Whiteriver, AZ, USA.

Stephen D. Sears, Johns Hopkins University, Baltimore, MD, USA.

U. Sengupta, Central JALMA Institute for Leprosy, Taj Ganj, Agra 282 001 India.

Eli E. Sercarz, Department of Microbiology, University of California, Los Angeles, CA 90024 USA.

A.K. Sharma, Department of Dermatology, Ram Manohar Lohia Hospital, New Delhi 110 011 India.

M.D. Sharma, Unite d'Immunologie Microbienne, WHO Centre of Reference and Research on Viral Hepatitis, Institut Pasteur 75724 Paris, France.

R.C. Sharma, Malaria Research Centre, 22 Sham Nath Marg, Delhi 110 054 India.

R.D. Sharma, Department of Veterinary Medicine, Haryana Agricultural University, Hisar, Haryana 125 004 India.

V.P. Sharma, Malaria Research Centre, 22 Sham Nath Marg, Delhi 110 054 India.

Sudhir Sinha, Central JALMA Institute for Leprosy, Taj Ganj, Agra-282 001 India.

Jotna Sokhey, Ministry of Health and Family Welfare, Nirman Bhawan, New Delhi 110 001 India.

Scott Snapper, Department of Microbiology and Immunology, Albert Einstein College of Medicine, Bronx, NY 10461 USA.

Brahm S. Srivastava, Centre for Biotechnology, Jawaharlal Nehru University, New Delhi 110 067 India.

Ranjana Srivastava, Microbiology Division, Central Drug Research Institute, Lucknow 226 001 India.

Marc C. Steinhoff, Johns Hopkins University, Baltimore, MD, USA.

Laksami Suesebang, Malaria Division, Ministry of Public Health, Bangkok 10200 Thailand.

Ann-Mari Svennerholm, Department of Medical Microbiology, University of Göteborg, S-413 46 Göteborg, Sweden.

G.P. Talwar, National Institute of Immunology, New Delhi 110 067 India.

Malabi M. Venkatesan, Department of Bacterial Immunology, Division of Communicable Diseases and Immunology, Walter Reed Army Institute of Research, Washington, DC 20307-5100 USA.

Timo Vesikari, University of Tampere, Tampere, Finland.

V.K. Vinayak, Department of Experimental Medicine, Postgraduate Institute of Medical Education and Research, Chandigarh 160 012 India.

Girish N. Vyas, Transfusion Medicine Research Program, Department of Laboratory Medicine, University of California, San Francisco, CA 94143 USA.

Goran Wadell, University of Umea, Umea, Sweden.

Kenneth S. Warren, Health Sciences Division, The Rockefeller Foundation, 1133 Avenue of the Americas, New York, NY 10036 USA.

Judith A. Welsh, Laboratory of Parasitic Diseases, National Institute of Allergy and Infectious Diseases, National Institutes of Health, Bethesda, MD 20892 USA.

Dyann F. Wirth, Harvard School of Public Health, Department of Tropical Public Health, 665 Huntington Avenue, Boston, MA 02115 USA.

Peter F. Wright, Vanderbilt University, Nashville, TN, USA.

S.A. Zaheer, National Institute of Immunology, New Delhi 110 067 India.

Maurizio Zanetti, The Division of Immunology, Medical Biology Institute, La Jolla, CA 92037 USA.

Part I
Opening Remarks

CHAPTER 1

Vaccinology: The Two Revolutions

Kenneth S. Warren

The two revolutions in vaccinology consist of the biotechnology revolution, which is providing an unprecedented opportunity to produce new and better vaccines, and the children's revolution of UNICEF, which administers the vaccines now available throughout the world. These two revolutions clearly synergize: the promise of new and better vaccines has galvanized the application of available vaccines and the widespread use of vaccines encourages investigators and funders to develop new and better ones.

Since Jenner discovered the smallpox vaccine in 1790, approximately 20 vaccines have been developed for human use, although only nine of them are generally available at the present time. Some are relatively ineffective, and side effects may range from the unpleasant to the lethal. The new-found power of the biotechnology revolution, however, suggests that from a pace of one vaccine per decade the rate may increase by at least an order of magnitude over the next 20 years. At a workshop on vaccine innovation and supply convened by the Institute of Medicine of the National Academy of Sciences of the United States, it was reported that the Department of Defense has 42 vaccines under development, the National Institutes of Health 28 and the Rockefeller Foundation, with its relatively meager resources, 6. The latter six vaccines were all directed toward diseases of the developing world, such as malaria, schistosomiasis, filariasis, and hookworm.

People today think of biotechnology in terms of the recombinant DNA technique through which bacterial recipients of genetic material coding for a particular protein (putative vaccine) produce it in large amounts. However, biotechnology is far more than that. It is a fusion of three major sciences—immunology, cell biology, and molecular biology—which provides a multiplicity of methods by which vaccines can be developed.

1. Biotechnology enables the identification of protective antigens, usually proteins, through the use of monoclonal antibodies and genetic probes.
2. Genetic engineering produces large amounts of a given protein not only in bacteria but in yeasts and mammalian cells as well.

3. Genetic engineering also permits deletion of genes that are responsible for virulence, replication, and toxin production.
4. It can insert genes for protective antigens into vectors including viruses (e.g., vaccinia), bacteria (e.g., *Salmonella typhi*), and mycobacteria.
5. Finally, the structure of the protein antigens may be determined not only directly but also from the structure of the DNA coding for the protein, which is easier to determine. Small pieces of the large protein molecules (epitopes) may be the active immunizing sites, which may lead to the simple chemical production of "synthetic vaccines."

Some vaccines are now undergoing improvement. For example, a crucial step in replacing the controversial BCG vaccine for tuberculosis has been the cloning of the entire genome of myobacterial tuberculosis. With respect to pertussis, an acellular vaccine prepared in Japan is now being tested in Sweden. More advanced work is being done involving the use of a single protein from *Bordetella pertussis,* which has been made into a tox-oid-like vaccine. With respect to polio, the three-dimensional structure of the entire virus has been determined including the exact location of every single atom. Furthermore, protective genes for polio 3 have been inserted into polio 1, and polio 2 should soon be added, thereby making three vaccines into one. For typhoid, a series of modified bacteria have been made, one of which is unable to metabolize galactose, leading to the production of toxic intermediates that gradually kill the bacillus. Other strains require aromatic amino acids that are not present in the tissues, causing the organisms to slowly starve to death. Thus living typhoid bacilli that cannot multiply within the tissues may serve as efficient oral vaccines. A wide variety of cholera vaccines have been developed and are undergoing testing. Among them are a vaccine containing the B (binding) subunit of the cholera toxin and killed bacteria; another involves merely the deletion of the A subunit of the toxin, which is responsible for the severe watery diarrhea.

With respect to new vaccines, enormous progress is being made on the three greatest killers of children: the diarrheas, respiratory infections, and malaria. Not only are vaccines being tested for rotavirus (attenuated animal viruses) and enterotoxigenic *Escherichia coli* (a synthetic vaccine combining both the heat-stable and the heat-labile toxins), but protective genes for *Shigella* have been added to the suicidal *Salmonella typhi* bacteria. It is possible that protective antigens for all of the major diarrheal diseases will be added to the typhoid vector. For the respiratory infections, the protective carbohydrate antigens of *Streptococcus pneumoniae* (pneumococcus) are being fused to protein carriers to render them effective in children below 2 years of age. Respiratory syncytial virus has been cloned and at least one of its proteins sequenced. For falciparum malaria, two sporozoite (infective stages injected by mosquitoes) vaccines are in clinical trial, one genetically engineered and the other synthetic and consisting of

12 amino acids. Blood-stage vaccines are in advanced stages of development in Sweden and Australia, with one group focusing on synthetic vaccines and the other on genetic engineering and vaccinia vectors.

With respect to the complex helminth organisms, many laboratories have produced protective monoclonal antibodies against schistosomiasis. Using them, several groups have extracted protective antigens, and at least two have cloned these antigens. A protective antigen has been isolated and cloned for filariasis, and a putatively protective antigen has been cloned for hookworm.

Work is proceeding on hepatitis A, parainfluenza, dengue, influenza, herpes, rabies, *Mycobacterium leprae,* leishmaniasis, amebiasis, and many others. There is a plethora of vaccines for hepatitis B, including those isolated from infected human sera and those genetically engineered. One of the latter, produced in yeast, is the first genetically engineered vaccine approved for human use. Because hepatitis B virus is a major cause of hepatocellular carcinoma, it constitutes the first anticancer vaccine. Finally, antifertility vaccines have been studied for some time, with pioneering work being done in India. An anti-beta human chorionic gonadotropin vaccine is now being tested in humans in Australia.

All of this work would be useless if the vaccines were not being used. In December 1982, UNICEF announced a revolution for children, emphasizing four cost-effective means for rapidly decreasing childhood morbidity and mortality in the developing world: growth charts, oral rehydration, breast feeding, and immunization. Jonas Salk and Robert McNamara then suggested a major emphasis on immunization, which led to a meeting in Bellagio, Italy in 1984 entitled "Protecting the World's Children: Vaccines and Immunization Within Primary Health Care." A task force for child survival was organized by the five sponsoring agencies—UNICEF, WHO, United Nations Development Programme, the World Bank, and the Rockefeller Foundation—to coordinate a global effort. In 1985 a second meeting was held in Cartagena, Colombia to report on progress. Colombia has achieved 80% coverage, and major campaigns were underway in El Salvador, Burkina Faso, Senegal, Nigeria, and Turkey. India has begun to immunize all of its children as a memorial to Indira Gandhi, and China has pledged to increase its coverage from 50% to virtually all of its children. In March 1988 a third meeting held in Talloires, France reported that global immunization had reached 50%, that the goal of 80% could be attained in 1990, and that a campaign to eradicate polio from the face of the earth by the year 2,000 had been initiated.

Thus the feedback loop between the two revolutions has been established. The children of the world are being immunized; and if the power of biotechnology to produce new and better vaccines is fostered, the well-being of children throughout the world, both North and South, will be remarkably improved.

Part II
Public Health Perspective—
The Present Scene

CHAPTER 2

Vaccinology and the Goal of Health for All

V. Ramalingaswami

Journey Through History

Edward Jenner, who can truly be described as the first immunologist, coined the term *vaccine* from the Latin *vaccinus,* meaning "of the Cows" in 1798, almost two centuries ago (13). In 1796 Jenner inoculated an 8-year-old boy with material from a cowpox lesion for prevention of smallpox. The last case of naturally occurring smallpox was reported from Somalia in October 1977. Thus it took 181 years to rid the world of smallpox after the discovery of an effective vaccine. Smallpox is now a finished story.

Rabies—An Unfinished Story

Louis Pasteur continued to use the term vaccine in honor of Jenner, thus giving it a permanent place in history. Pasteur introduced inactivated microbes as immunizing agents. In 1885, after testing a vaccine made of inactivated rabies virus in animals, he administered it to a 9-year-old boy severely bitten by a rabid dog and facing certain death. Fourteen doses of the vaccine were given, and the boy did not develop rabies. By 1890 antirabies vaccination centers were established in the major cities of the world, and some of these centers were designated Pasteur Institutes.

The world has now celebrated the 100th anniversary of vaccination against rabies, appropriately at the Pasteur Institute in Paris. Alas, however, effective control of rabies in the world as a whole is not yet within our grasp. The threat of rabies to public health continues in the developing countries despite advances in antirabies vaccine since the time of Pasteur. At a Fogarty International Center symposium on controlling rabies in wild and domestic animals as the goal of human rabies control, promising results were reported with an oral attenuated, antirabies vaccine in controlling rabies in foxes and raccoons. A DNA recombinant antirabies vaccine has been made using the vaccinia virus as vector with the possibility of producing the vaccine economically and on a mass scale. Studies are being planned on an international collaborative basis to immunize domestic and wildlife animals with this recombinant vaccine.

The problem of rabies affects both developed and developing countries, but with a difference. In developing countries the principal source of human rabies is the rabid dog. In India alone, for example, 25,000 persons die each year from this disease, and nearly half a million receive postexposure antirabies immunizations. Neuroparalytic accidents occur with the use of sheep brain vaccine to the extent of 1/5,000 to 1/11,000 cases. The Pan American Health Organization has taken the initiative to bring urban rabies under control in Latin America by 1990. In the industrialized world, although human rabies has been largely eliminated and dogs are no longer a source of the disease, the infection has become established in a number of wildlife species, posing a constant threat to man and domestic animals even though the number of rabies cases in humans acquired from this source is low.

BCG—The Enigma Continues

Continuing our historical journey, we move from Louis Pasteur to Albert Calmette and Camille Guérin, who in 1921 produced an attenuated strain of *Mycobacterium bovis* that causes tuberculosis, thus leading to the BCG vaccine. Mass vaccination with BCG against tuberculosis was quickly started in many countries using different strains but all originating from the strain developed by Calmette and Guérin. It is now part of the Expanded Program of Immunization (EPI) established by WHO and UNICEF. Under this program, 57% of children in the developing world have been immunized with BCG by age 1 by 1987. The efficacy of BCG in protection against tuberculosis remains a debatable issue, the uncertainty being compounded by the results of the South Indian trial, often erroneously described as "the Madras Monumental Failure Trial." Initially reported as showing no protection against adult pulmonary tuberculosis (18) further follow-up of children from 5 to 12½ years demonstrated up to 50% protection against pulmonary tuberculosis (19). Despite the doubts about adult tuberculosis, there is a consensus that BCG protects against childhood forms of tuberculosis, which are serious with permanent sequelae.

It is of interest that studies in Karonga district of Northern Malawi indicate that BCG is perhaps more effective in the control of leprosy than tuberculosis (7), and it is said that BCG is "sufficiently protective" against leprosy in East and Central Africa. In the Madras study, too, some degree of protection was observed against leprosy (19). There is thinking that BCG should be considered a serious candidate for leprosy control in these regions.

Randomized clinical trials with BCG over the years revealed protection against tuberculosis ranging from 0 to 80% and against leprosy ranging from 20 to 80% at different times. Vaccine strains and their immunogenic potencies, ethnic, racial, ecological, and climatic factors, host factors such as host genetics and nutritional status, strain variability, and prevalence of atypical mycobacteria in the environment may be determinants of efficiency of BCG in a given geographical location.

Diphtheria, Tetanus, Pertussis—Still Some Distance To Go

In 1923 Gaston Ramos discovered that bacterial toxins responsible for diphtheria and tetanus could be inactivated with formaldehyde, giving rise to *toxoids,* which can then be used to protect against these diseases. Later, in 1925, a vaccine against pertussis was developed that was a suspension of killed *Bordetella pertussis* organisms. The DPT triple vaccine is a component of EPI, and up to now approximately 52% of the target group of children in the developing world have been reached with the full immunization schedule.

In Japan the extensive use of whole cell pertussis vaccine brought about a remarkable decrease in the incidence of pertussis, from approximately 160/100,000 during the late 1940's to negligible levels by the mid-1960 (14). Similar protection with the vaccine was observed in other countries, e.g., Britain. However, despite the success, public concern over reactions to whole cell vaccines including a few cases of brain damage in Japan and some other developed countries led to lower utilization rates, a *rise in pertussis incidence,* and a demand for purer, less reactogenic subunit vaccines. This goal has now been accomplished in Japan. Two basic types of acellular pertussis vaccine, both with reduced endotoxin content but one containing more filamentous hemagglutinin (FHA) than the other, have been licensed for routine use in Japan from late 1981. More than 20 million doses of this vaccine have been administered there to date. A two-component acellular vaccine and a highly purified acellular toxoid-vaccine have been tested in Sweden in a multicentric, randomized, double-blind, placebo-controlled trial. The results reported at a meeting in Stockholm in December 1987 show that both the acellular vaccines are effective in preventing severe, culture positive disease in 6-11 month old children but leave several questions unanswered such as optimum composition of acellular pertussis vaccine and safety and efficacy in children under 5 months of age.

Tetanus is an eminently preventable disease but remains uncontrolled in many developing countries. In India deaths from neonatal tetanus range from 1/1000 to 15/1000 live births in urban areas and 5/1000 to 67/1000 live births in rural areas. It is estimated that 500,000 deaths occur from neonatal tetanus annually in the countries of Southeast Asia and the eastern Mediterranean (8). For the elimination of tetanus the strategy is clear and comprises four components (17): (a) to produce and maintain protective levels of antibody during pregnancy; (b) the practice of scientific midwifery using sterile instruments kits: (c) continuous multiple immunizations during infancy and childhood; and (d) administration of routine booster dose to all those with accidental wounds.

The triad of diseases—diphtheria, tetanus, pertussis—is still a long way from being brought fully under control in the Third World. The necessity of administering multiple doses and the side reactions, although mild for the large part, are impediments. For the diphtheria and tetanus components, a two-dose schedule would be more acceptable, and therein lie future developments.

Yellow Fever and Polio Vaccines—Fidelity of Vaccine Strains

The yellow fever vaccine, which appeared during the 1930s, used Max Theiler's 17D strain. The continued fidelity of the 17D strain is a matter of gratification. The 1950s saw the discovery of vaccines against poliomyelitis: first the inactivated injectable vaccine by Jonas Salk followed by the live attenuated oral vaccine by Albert Sabin. The oral vaccine is used in the EPI. It is easy to administer, is inexpensive, and has a good track record in controlling epidemics and in routine national immunization programs. More recently the killed Salk vaccine's advantages are being increasingly recognized: It is a better immunogen, it requires only two doses compared with the three to five doses of oral vaccines that must be administered in developing country settings, it is becoming comparable in cost because of the mass production technologies in continuous cell culture systems, and it is easy to administer along with DPT—hence there is better patient compliance.

The era of live attenuated virus vaccines continued with the arrival of a vaccine against measles during the early 1960s. With further attenuation there is now a safe and highly effective vaccine against measles, capable of producing long-lasting immunity with a single dose. 53% of children under age 1 are now being immunized against poliomyelitis and about 43% against measles under the EPI in the developing countries as a whole. Temperature vulnerability of oral polio vaccine is a problem. There is room for further improvement in the cost and stability of measles vaccine. Encouraging results are being obtained with the use of the Edmonston-Zagreb strain in infants below 9 months of age.

Impact of Modern Biology on Vaccinology

A-vaccine-a-decade was the pace of progress since Pasteur until the new biology and biotechnology revolution appeared in recent years (11). Of 200 infectious diseases of man caused by bacteria, viruses, and parasites, there are today highly effective vaccines against nine diseases of childhood. Thirteen other vaccines of varying quality and efficacy are being used in high-risk groups in different countries. The availability of biotechnologies such as genetic engineering to design attenuated organisms or the expression or synthesis of pure antigens has led to the prediction by Jordan (11) that new or improved vaccines can be developed for 28 diseases within the next decade, resulting in a several fold increase in research and development of vaccines.

Biotechnology enables one to obtain insights into the precise nature of antigens that are recognized by the immune system, the specific nature of microbial surface proteins, characterization of virus surface components, bacterial toxins, stage-specific parasite surface proteins, determination of three-dimensional structures of proteins in relation to their immunological domains, the capacity to clone the genes coding for critical regions of these proteins, and understanding the molecular nature of the pathogenicity of microorganisms. Once a candidate antigen is identified, several pathways are open: genetic engineering of cell types such as bacteria, yeasts, or mammalian cells to induce the expression of antigens; the insertion of gene(s) into virus vectors such as vaccinia or herpes; chemical synthesis of small peptide segments of the protein; manipulating live organisms to reduce virulence yet retain the capacity to induce immunity to yield nonvirulent mutants that are immunogenic; and mimicking the antigenic structure of certain parasites and viruses using anti-idiotype antibodies resulting in anti-antibodies with a configuration resembling the immunogenic protein anti-idiotypes that can prime or actually stimulate an effective immune response (6).

Low-cost, safe, effective, thermally stable, pure vaccines are needed. Fewer doses to achieve protection and immunization at birth provide advantages. Problems of thermal instability, the necessity of repeated multiple vaccinations, low immunogenicity, and genetic instability are challenges to be addressed by the new biology.

Vaccinology for the Tropics: Three Special Missions

Vaccinology for the tropics has three special missions: (a) to deal with the diarrhea–pneumonia complex, which together with malnutrition is a critical determinant of the high infant mortality rates in the tropics; (b) to fight some common cancers of the tropics; and (c) to stem population growth through antifertility vaccines. Only the first two are considered here.

Oral rehydration therapy is cutting down mortality rates from acute diarrheal disease in infants and young growing children, but morbidity persists. Clean water supply and sanitation schemes are making progress, yet the task of getting oral rehydration therapy into the hands of mothers and matters of personal hygiene are impeding progress in the control of acute diarrheal diseases.

Diarrhea–Pneumonia Complex

There is need for vaccines against the major pathogenic agents of diarrhea and dysentery. Oral vaccines are best, as they stimulate local mucosal secretory immunity.

A combined whole cell/toxin B subunit oral vaccine has just been reported to be safe and moderately effective against cholera in a field study in Bangladesh (2), but a key question remains as to its long-term efficacy, as a rapid decrease in immunity is a major drawback of the time-honored parenteral killed whole cell vaccine (4). A genetically engineered, live oral cholera–typhoid vaccine based on Ty21a is currently undergoing trials in Adelaide, Australia. The era of oral cholera vaccines is with us. Likewise, there is much interest in an all-purpose, oral, attenuated, antidysentery vaccine consisting of protective antigens of *Shigella* type 1 and other relevant *Shigella* types in the form of a multivalent cocktail vaccine. The devastating emergence of *Shigella* type 1 dysentery in India and Bangladesh in recent years imparts a sense of urgency to this effort. Rotaviruses are the single most important agents of acute diarrhea in children in developing countries. Vaccines based on bovine and rhesus rotavirus strains are currently being evaluated.

Although quantification is difficult, there are believed to be close to 7.0 million cases of typhoid fever occurring annually in Southeast Asia with a population of 1.4 billion people, 750,000 cases in West Asia with a population of 94 million; 4.4 million cases in Africa with a population of 369 million; and 23,000 cases in the industrialized world with a population of 1.1 billion. There are estimated to be about 15.5 million cases of typhoid occurring in the world today annually, excluding China (3). School-age children in developing countries are particularly at risk. Results of studies with the use of Ty21a vaccine originally in Egypt and now in Chile and Indonesia and the new trials with vi polysaccharide capsular surface antigen in Nepal, which are presented in this volume, provide hope for the control of typhoid fever.

Another intestinal pathogen, *Entoaemeba histolytica,* is a tissue-lysing protozoan parasite that is known to infect 500 million people worldwide excluding China, with 8 to 10% of them manifesting clinical disease and 0.2% of the illnesses proving to be fatal (9). This ubiquitous and defiant parasite may now be ready to yield to molecular biology. Host defenses against this infection are beginning to be recognized as is the key role played by the macrophage. The possibility of developing virulence or adherence factors as antigens, of live, attenuated strains, and prevention of transmission of infection to prevent encystation or excystation are being explored. It may be recalled that it was at the School of Tropical Medicine, Calcutta, in 1912 that Sir Leonard Rogers described the efficacy of the root of *Caphaelos ipecacuanha* for the treatment of amebic dysentery and hepatitis (15).

Immunization, oral rehydration, breast feeding, adequate weaning foods, nutrient fortification, growth monitoring, sanitation and safe water, and education are the strategies to deal with the diarrhea–pneumonia–malnutrition complex in the primary health care setting (12).

Acute Respiratory Infections: Major Factor in Child Survival

Several studies have revealed the importance of acute respiratory infections to child mortality and morbidity of equal, if not greater, magnitude than acute diarrheal disease. The control of these infections through chemotherapy and immunoprophylaxis in the primary care setting is a chapter of biomedical endeavor that is opening up. The demonstration of almost 60% efficiency in preventing death from lower respiratory tract infections as the sole cause of death in children under age 5 by administering a polyvalent pneumococcal capsular polysaccharide vaccine in Papua, New Guinea (16) coupled with the high immunogenicity of a new conjugate vaccine composed of capsular polysaccharide of *Hemophilus influenzae* type b linked to an *Neisseria meningitidis* outer membrane protein found in infants 2 to 6 months of age (5) are indications of a forthcoming immunological attack on respiratory infections during childhood.

Molecular Biology and Tropical Cancers

Developments in molecular biology and immunology have led to visions of making immunizing agents against some of the common cancers of the tropics. Hepatitis B virus and primary carcinoma of the liver, the human papilloma virus and cancer of the uterine cervix, and the Epstein-Barr virus and Burkitt's lymphoma and nasopharyngeal carcinoma are examples in which the virus genome has been found to be integrated into the host cell genome. Approximately 500,000 new cases of cervical cancer occur each year, the bulk of them in the developing countries. Of these cases, 100,000 are seen in India alone. About 250,000 new cases of hepatocellular carcinoma occur each year, and again most occur in the developing world: China, Southeast Asia, and Africa. The fact that there are estimated to be 200 million carriers of hepatitis B virus in the world with the attendant risks of developing chronic liver disease and liver cancer is mind-boggling (1). About 50,000 new cases of nasopharyngeal carcinoma occur every year, most of them in Southern China and Southeast Asia; and 5,000 to 10,000 fresh cases of Burkitt's lymphoma arise annually, mainly in the malarial areas of Africa.

There are providential gaps in cancer progression that can be exploited in cancer-control programs (1). For example, there is a gap of 2 to 3 months from perinatal infection to persistence of hepatitis B virus infection in blood, providing an opportunity to intervene through vaccination and prevent the carrier state. Unlike animal cancers caused by viruses, it appears that human cancer viruses act more slowly, and other factors may have to impinge on the altered cells in a multistep process before clinical cancer sets in.

While the horizons of vaccine development have widened immensely as a result of advances in modern biology and biotechnology, there are still a number of diseases, as pointed out by Jordan (11), for which a vaccine may not be feasible within the next decade or so. This lack may be due to a variety of factors: multiplicity of serotypes, defiance of the organism to be cultured in vitro (as with leprosy), multiple life cycles with multiple antigens, the onset of infection early in infancy, a long incubation period (as with AIDS), insufficient knowledge of pathogenesis and immunity, instability and reversion to virulence of attenuated organisms, lack of suitable adjuvants, and so on. Jordan cited AIDS, trachoma, onchocerciasis, lymphatic filariasis, schistosomiasis, amebiasis, leishmaniasis, non-A non-B hepatitis, and viral encephalitis as examples in this category. This list makes obvious the need to press on with existing tools such as vector control measures including biological methods, source reduction through environmental management, chemotherapy, and chemoprophylaxis as vital tools for the control of tropical diseases.

Concluding Thoughts: Voyage of a Vaccine

Molecular biology, cell biology, immunology, and genetics have placed in our hands immense possibilities for improving the safety and efficacy of existing vaccines and discovering new vaccines against intractable diseases. Vaccination is the single most cost-effective procedure in the health history of mankind. Yet the goal of immunizing all children against serious illnesses is still eluding us. We have today a number of effective and safe vaccines, and the list is lengthening. We need, in addition to these tools, a strategy for delivery. The EPI is making progress. India's reported success in the Intensive District Program provides some hope that the goal of immunizing all eligible children against the six EPI diseases may yet be achievable by 1990, but it calls for supreme efforts orchestrated toward a clearly defined goal, performance orientation, and persuasive political and administrative commitment from all sections of society. New vaccines are not of much use if they cannot be used. Having a vaccine and an effective immunization program are not the same thing.

Controversies about vertical and horizontal programs in primary health care should not lead to slackening of our efforts to achieve universal immunization. Equally, whether the smallpox miracle can be repeated need not detain us. What we wish to emulate, as Hopkins said, is not the smallpox mold but the smallpox model (10).

The discovery of new vaccines or improvement of existing ones is an essential but not sufficient step in the fulfillment of the goal of health for all. The endeavor must involve the voyage of the vaccine from the factory to the village, resulting in immunization of the entire target population. I hope that the concept of vaccinology as visualized in these volumes encompasses this wider view and is not restricted to the molecular biology of vaccine discovery and the technology of manufacture.

Vaccines are complete technologies dealing with primary prevention such as sanitation and safe water in the conception of Lewis Thomas. They break the dangerous partnership between infection and malnutrition. Immunization services can create real assets for health services in developing countries. A comprehensive global childhood immunization program can serve as a catalyst for primary health care. Immunization can be the entering wedge of primary health care movement leading to the goal of health for all.

References

1. Blumberg BS: Overview: hepatitis B virus and primary hepatocellular carcinoma, in *Hepatitis Viruses and Hepatocellular Carcinoma*. Tokyo, Academic Press, 1985, pp 3–13.
2. Clemens JD, Harris JR, Khan MR, et al: Field trial of oral cholera vaccines in Bangladesh. *Lancet* 1986;2:124–127.
3. Edelman R, Levine MM: Summary of an international workshop on typhoid fever. *Rev Infect Dis* 1986;8:329–349.
4. Editorial: Oral cholera vaccines. *Lancet* 1986;2:722–723.
5. Einhorn MS, Anderson EL, Winberg GA, et al: Immunogenicity in infants of Haemophilus influenzae type B polysaccharide in a conjugate vaccine with Neisseria meningitidis outer membrane protein. *Lancet* 1986;2:299–302.
6. Fields BH: What biotechnology has to offer vaccine development in *International Symposium on Vaccine Development and Utilization*. Washington, DC, Agency for International Development, 1986.
7. Fine PEM, Ponnighaus JM, Maine N, et al: Protection efficiency of BCG against leprosy in Northern Malawi. *Lancet* 1986;2:499–54.
8. Galazka, A: *Immunization of Pregnant Women*. EPI/GEN/83/5. Geneva, WHO, 1985.
9. Guerrant RL: Amoebiasis: introduction, current status and research questions. *Rev Infect Dis* 1986;8:218–227.
10. Hopkins, DR. Beyond smallpox eradication: assignment children. 1985;69/72:235–242.
11. Jordan WS Jr: Impediments to the development of additional vaccines, vaccines against important diseases will not be available in the next decade in *International Symposium on Vaccine Development and Utilization*. Washington, DC, Agency for International Development, 1986.
12. Keusch GT, Scrimshaw NS: Selective primary health care strategies for control of disease in the developing world. XXIII. Control of infection to reduce the prevalence of infantile and childhood malnutrition. *Rev Infect Dis* 1986;8:273–287.

13. Population Reports: *Issues in World Health, Immunizing the World's Children.* Series 1, No. 5, March-April 1986.
14. Report of the U.S. Public Health Service Interagency Group to Monitor Vaccine development, production and usage: *Pertussis and Pertussis Vaccines in Japan.* Washington, DC, U.S. Department of Health and Human Services, 1986.
15. Rogers L: The rapid cure of amoebic dysentery and hepatitis by hypodermic injection of soluble salts of emetine. *Br Med J* 1912;1:1424–1425.
16. Riley ID, Lehman D, Alpers MP, et al: Pneumococcal vaccine prevents death from acute Lower respiratory tract infections in Papua New Guinea children. *Lancet* 1986;2:877–880.
17. Shofield F: Selective primary care: strategies for control of disease in the developing world. XXII. Tetanus—a preventable problem. *Rev. Infect Dis.* 1986;8:144–156.
18. Tuberculosis Prevention Trial, Madras. Trial of BCG vaccines in South India for tuberculosis prevention. *Ind. J. Med. Res.,* 1980;72(Suppl):1–74.
19. Tripathy SP: The case for BCG. *Ann. Nat Acad Med Sci India* 1983;19:11–21.

CHAPTER 3

Toward Universal Immunization: 1990

Jotna Sokhey

Disease Burden

The disease burden manifests as morbidity, mortality, and serious complications following illnesses, including lifelong physical disabilities. Many diseases indirectly increase the disease burden by debilitating children, leading to poor general health status and growth. The health problems facing a large country such as India are essentially linked to poverty and overpopulation, which in turn lead to less satisfactory health facilities, poor sanitary and environmental conditions, and poor nutrition. The women and children are more vulnerable to all these factors.

Diphtheria, whooping cough, tetanus, poliomyelitis, tuberculosis, measles, and typhoid fever are caused by different microorganisms and have varying clinical symptoms. All these diseases, however, have three things in common: They are serious diseases, often leading to severe lifelong complications and death; they are common during early childhood; and all of them are preventable by immunization.

Based on sample surveys conducted in 1981 and 1982, it has been estimated that in the absence of an immunization program in India nearly 0.3 million infants would die annually due to neonatal tetanus, and more than 0.15 million children would develop paralytic poliomyelitis every year. The neonatal tetanus mortality rate was 13.3/1000 live births in India in the rural areas and 3.2/1000 in the urban areas in 1981 and 1982. The incidence rate of paralytic poliomyelitis was estimated to be 1.5/1000 to 1.7/1000 children 0 to 4 years of age (around 7/1000 live births) (2,3,5,10).

The estimated incidence of the other vaccine-preventable diseases is equally large. Rough estimates indicate that the total number of the cases of all the diseases together would be nearly 40 million if there was no immunizaton program in the country. The expected deaths are around 1.5 million in the absence of an immunization program. The figures are based on the average morbidity and mortality rates.

The estimates for diphtheria, pertussis, measles, and tuberculosis are based on the assumptions that 1% of the infants will get diphtheria, 80%

pertussis, 90% measles, and 0.1% tuberculosis. The case fatality rate has been presumed to be 3% for all diseases. For diphtheria and tuberculosis it is estimated to be 10% and 90%, respectively. It has been presumed that 22.9 million children were born in India in 1985.

Although the incidence of the diseases has decreased as a result of the immunization program, they continue to be major public health problems. Poliomyelitis, for example, is the major cause of lameness in children, and neonatal tetanus accounts for a large number of neonatal deaths.

The infant mortality rate (IMR) in India is also still high. Several combined measures have been adopted in line with the National Health Policy to reduce the IMR to levels below 87/1000 live births by 1990 and 60/1000 by the year 2000 (6). Rapid acceleration of the immunization coverage is expected to lead to a fall in the infant mortality rates. Already positive trends are evident with the IMR recorded at 95/1000 in 1985. It is the first time that the levels have fallen to less than 100/1000 in the country. Infant mortality decreased from 200/1000 live births in 1901 to 129/1000 in 1970. After stagnating at around 125/1000 through the 1970s, it dipped to (per 1000) 120 in 1979, 114 in 1980, 110 in 1981, 105 in 1982, and 104 in 1983 and 1984 (12).

Expanded Program on Immunization

The expanded program on immunization (EPI) was started in 1978 with the objective of reducing morbidity and mortality due to diphtheria, pertussis, tetanus, poliomyelitis, tuberculosis, and typhoid fever by making vaccination services àvailable to all eligible children and pregnant women

Table 3.1. Phased targets, by year, under the EPI program.

Vaccine[a]	1985–1986	1986–1987	1987–1988	1988–1989	1989–1990
TT(PW) 2\b					
E	258.0	253.0	248.0	243.0	239.0
T	129.0	152.0	186.0	219.0	239.0
%	50.0	60.0	75.0	90.0	100.0
DPT 3/Polio 3/BCG 1					
E	233.0	228.0	225.0	221.0	215.0
T	140.0	153.0	169.0	177.0	183.0
%	60.0	67.0	75.0	80.0	85.0
Measles 1					
E	233.0	228.0	225.0	221.0	215.0
T	23.0	57.0	100.0	142.0	183.0
%	10.0	25.0	45.0	65.0	85.0

All figures relating to targets and performance are the number of pregnant women with two doses or a booster dose of TT and infants with three doses of DPT and polio vaccines and one dose of BCG and measles vaccines. Results are given in millions.
[a]E = estimated eligibles. T = annual vaccination targets. % = percent coverage of the eligibles.

Table 3.2. Districts included in the program.

Year	No. of districts	Total No.	% Age of districts under UIP
1985–1986	30	30	7
1986–1987	60	90	22
1987–1988	90	180	44
1988–1989	120	300	73
1989–1990	123 +	423 +	100

by 1990. Measles vaccine was included in the program in 1985–1986. It was designed to achieve self-sufficiency in the production of vaccines required for the program (4,7–9,11,12). The phased targets, by year, are shown in Table 3.1.

Universal Immunization Program

A major shift in strategies was adopted in 1985–1986 with the launching of the universal immunization program (UIP) to significantly enhance the immunization coverage of the eligible population and to improve the quality of the services. Districts have been identified selectively with the objective of achieving at least 85% coverage of the infants with three doses each of DPT and OPV and one dose each of BCG and measles vaccines. The pregnant women are given two doses of TT (8). A district covers a population of 1.5 to 2.0 million. The number of districts in each state varies considerably from three districts in Mizoram to 56 in Uttar Pradesh.

Thirty districts were taken up in 1985–1986 and another 62 districts in 1986–87. An additional 90 districts have since been included in the program, bringing the total number of districts to 182. The balance of the districts were to be covered in the next two years so that by the end of 1990 all the districts are covered under UIP (Table 3.2). The population currently covered under UIP and the number of infants and pregnant women are shown in Table 3.3.

Table 3.3. Populations in the program.

Year	Districts	UIP figures in millions				
		Pop. of India	Pop.	%Pop.	Preg. women	Infants
1985	30	754.71	65.6	8.7	2.17	1.93
1986	92	769.98	186.4	24.2	6.10	5.30
1987	182	785.22	382.5	48.7	12.16	11.12

Magnitude of the Task

The UIP is one of the largest immunization programs in the world. The magnitude of the work involved is staggering. By 1990 the services are to be provided to more than 23 million pregnant women and 18 million infants annually. During the 5-year period from 1985 to 1990, it is proposed that more than 92 million pregnant women and 82 million infants are to be immunized. More than 600 million injections must be administered and more than 1000 million doses of various vaccines utilized.

The magnitude of the task is evident by the fact that all the pregnant women and infants in India need to be contacted in order to provide the services. When such contacts have to be maintained at least five times for the infants and twice for the pregnant women, there is no scope for any default in the services. The total number of contacts required are estimated to be 140 million annually and more than 600 million during the 5-year period.

Prerequisites for Success

The successful implementation of a program of the above magnitude in a large country such as India requires careful planning, effective management, and optimal use of available resources. The program is complex, requiring close coordination between personnel engaged in different tasks at various levels. A sense of commitment and urgency can take us toward our goal. Equally obviously, a program for the health of children must have the support and cooperation of the community, the voluntary sector, and professional bodies.

Additional inputs are being provided under the program to make it operationally feasible. The cold chain is being strengthened by providing storage and transportation equipment. During the 5-year period 275 walk-in coolers, 28,000 refrigerators, 300,000 vaccine carriers, 15.2 million syringes, and 60 million needles, among other items, are expected to be supplied under the program.

Training and upgrading the skills of the workers is receiving the highest priority under the program. A 5-day training course is held for the medical officers in charge of the program at the district level. So far, 206 officers have attended the training courses. In addition, a 4-day course is held for the medical officers at the primary health center level. A total of 2117 medical officers have participated in the training programs. The health workers and supervisors attend a 2-day task-oriented course; 27,989 multipurpose workers have been trained; and 23,882 grass root level workers of other categories attended 1-day briefing sessions. Special training material for the medical officers and a manual for the health workers have been printed. The manual is available in the regional languages.

Publicity material has been developed to enhance public demand. Radio and television are being made use of for the first time at the national level. Thrust is being given to improved field monitoring and reporting. A post of District Immunization Officer has been created for better planning and management and increased field supervision; 182 such posts have so far been created.

The government of India has allocated Rs 2400 million to meet the expenditure on the program over the 5-year period, including recurring expenses (12).

Problems

One of the major difficulties in the rapid expansion of the program is the storage and distribution of vaccines at the recommended temperatures. Regular supplies must be ensured to a large number of immunization centers over huge distances from the manufacturing institutions. When one takes into consideration the erratic power supply and the unavailability of ice in many places, the task becomes stupendous.

Some the other major problems that adversely affect the immunization program are inadequate development of health infrastructure, poor communication facilities, and inadequate facilities for mobility. A primary health center, on average, covers an area of more than 257 km^2 and a subcenter of 36 km^2. A large proportion of the productive time of the health workers is spent on traveling because of the large distances to be covered and the relatively poor communication and transport facilities.

India has a large network of health infrastructure with more than 655 community health centers (CHCs), 11,029 primary health centers (PHCs), and 83,000 subcenters in the rural areas in addition to the hospitals, dispensaries, MCH clinics, and postpartum centers in the urban areas. There is still, however, a large balance of infrastructural development with 4650 CHCs, 9352 PHCs, and more than 40,000 subcenters to be established (12). The shortfalls in filling the vacancies is considerable. The central government is providing full support to the state health authorities for speedier development of infrastructure, filling the vacancies, and training the staff.

Community Participation

A major reason for low coverage in many areas is that many people do not realize the serious nature of the diseases or the services that are available for their prevention. Ignorance regarding the need for completing the schedule is one of the reasons for high dropout rates. Coverage levels would increase substantially if the dropout rates were minimized. The voluntary organizations and professional bodies are being requested to support the program by creating an increased awareness and enhancing demand for the services.

Field monitoring has been stepped up to maintain the high quality of services. Potency testing of field samples of oral polio vaccine has been started under the program to monitor the quality of the cold chain for vaccines.

Reported Performance

The immunization coverage of pregnant women and infants has increased considerably since the program was started in 1978. During the 5 years from 1980 to 1985, 37.2 million pregnant women received two doses or a booster dose of TT; 49.8 million and 26.5 million children received three doses of DPT and polio vaccines, respectively; and 66.7 million children were given one dose of BCG vaccine.

In 1986–1987, it was estimated the 46% of the pregnant women were given two doses of Tetanus Toxoid. The coverage of infants with three doses each of DPT and OPV was estimated to be 57 and 49 percent respectively. Coverage with BCG vaccine and measles vaccine was reported to be 49 and 17 percent. The above coverages are for children under one year of age.

During the period 1986–1990 it is proposed to cover more than 82 million infants and more than 93 million pregnant women. The increase in the total numbers to be covered is substantial. The task is even more difficult, as the age group to be covered has been reduced from 0 to 2 years to under 1 year only.

By the end of 1990 more than 23 million pregnant women and 18 million infants are expected to be immunized annually. At the end of 1979–1980 and 1985–1986 the corresponding figures were 4.8 million and 9.2 million pregnant women and 6.9 million and 12.8 million infants, respectively.

The success of the program will be evaluated in terms of the reduction in the number of patients with vaccine-preventable diseases. The surveillance system is being geared to document the impact of the services.

An increased awareness, a sense of urgency, and dedication among the medical professionals, voluntary sector, and the community can make the difficult task of universal immunization coverage achievable within the short time span of 5 years. We are at the threshold of making a major breakthrough in the program resulting in a substantial decrease in the number of cases. The services, however, would need to be accelerated rapidly in order the achieve the goal of universal immunization coverage by 1990.

References

1. *Annual Reports of the Department of Family Welfare.* New Delhi, Ministry of Health and Family Welfare, Government of India.
2. Basu RN, Sokhey J: A baseline study on neonatal tetanus in India. *Pak Pediatr J* 1982;6:184–197.
3. Basu RN, Sokhey J: Prevalence of poliomyelitis in India. *Indian J Pediatr* 1984;51:515–519.
4. Bhargava I, Sokhey J: Immunization programme in India. *Indian Pediatr* 1985;22:97–106.
5. Bhargava I, Sokhey J (eds): *The Control of Neonatal Tetanus in India.* New Delhi, Ministry of Health and Family Welfare, Government of India, 1983.
6. *National Health Policy.* New Delhi, Ministry of Health and Family Welfare, Government of India, 1983.
7. Sokhey J: The national programme for the control of poliomyelitis. *Indian J Public Health* 1985;25:168–174.
8. Sokhey J: *Universal Immunization Programme 1985–1986: Data at a Glance.* New Delhi, Ministry of Health and Family Welfare, Government of India, 1985.
9. Sokhey J, Bhargava I: Management of the immunization programme in India. *Indian Pediatr* 1986;23:669–675.
10. Sokhey J, Bhargava I: The control of neonatal tetanus in India. *Indian Pediatr* 1984;21:515–519.
11. Sokhey J, Bhargava I, Basu RN: *Handbook on the Immunization Programme in India.* Ministry of Health & Family Welfare, Government of India, 1984.
12. *Towards Universal Immunization—1990.* New Delhi, Ministry of Health and Family Welfare, Government of India, 1985.

CHAPTER 4

New and Improved Techniques for Vaccine Production

M. Roumiantzeff, J Armand, F. Arminjon, B. Montagnon, and M. Cadoz

Vaccines represent the most spectacular success of applied immunology. Since the discoveries of Jenner and Pasteur, large numbers of vaccines have been developed, and today more than 30 products are regularly and widely used in humans. Over the past decade discoveries and technological progress have, however, transformed the production possibilities of bacterial and viral vaccines, and the aim of this presentation is to illustrate these new and improved vaccine production techniques using a few examples. These limited examples are drawn mainly from our experience at the Institut Mérieux and cover the fields of both bacterial and viral vaccines.

Bacterial Vaccines

The preparation of bacterial vaccines has a long tradition, but today several of these "traditional" vaccines have been put into question for reasons of safety or efficacy. Crude bacterial preparations, consisting of inactivated bacterial cells, have disappeared or are soon to do so, replaced by new vaccines prepared from one or several antigenic components purified from the original infectious organism. Most of these new preparations are inactivated vaccines, which offer the best safety and protective efficacy.

For certain products the antigenic vaccine components must be extracted, whereas for others the bacteria produce a toxin excreted into the culture medium, which must be separated from the culture supernatant and then purified. The tetanus and diphtheria vaccines proposed by Ramon in 1923 serve as models. The toxin and toxoid purification and concentration stages have, however, been improved with modern technology (8).

Capsular Polysaccharide Vaccines

Meningococcal Vaccine

The meningococcal vaccine is the prototype for biochemically well defined polysaccharide vaccines. Gotschlich et al (12) developed the first vaccine

Fig. 4.1. Industrial biofermentor used for mass culture of bacteria.

against group C meningococcus, since which time group A and group C meningococcal polysaccharides have been produced on an industrial scale (3). Mass cultures of the bacteria are routinely performed using large industrial biofermentors (Fig. 4.1). The extraction of capsular surface polysaccharides is clearly defined in order to obtain high yields and to protect the immunogenicity of the polysaccharides. Stabilization procedures using the addition of lactose have been developed (38). The industrial development of meningococcus vaccine is reviewed elsewhere (27).

Pneumococcus Vaccine

The pneumococcus vaccine was developed some years later (2). The problems encountered in its development were linked to the multiplicity of serological types required in the final preparation in order to obtain sufficient epidemiological coverage. The first generation of pneumococcal vaccines comprised 12 to 15 valencies; the Institut Mérieux vaccine contained 14 types, each of which represented by 50 μg of purified polysaccharide. Some years later a WHO-sponsored standardization was proposed and accepted (25). Most producers then adopted a formulation with 23 valencies (Table 4.1), in which the total polysaccharide dose is in fact reduced by comparison with the first generation vaccines, as each of the 23 components is limited to 25 μg of purified polysaccharide (4).

Table 4.1. Pneumococcal polysaccharide vaccine comprising 23 serotypes.

1	7F	12F	19A
2	8	14	19F
3	9N	15B	20
4	9V	17F	22F
5	10A	18C	23F
6B	11A		33F

The serotype numbers in boldface are the new serotypes introduced into the 23-component formula in addition to those present in the initial 14-component formula. Two serotypes were replaced: 6B replaced 6A, and 33F replaced 25.

Hemophilus influenzae Type B Vaccine

The *Hemophilus influenzae* type B vaccine constitutes the most recent addition to the family of polysaccharide vaccines. The difficulties encountered in its development were not linked to the extraction techniques because the immune property depends on an unique capsular polysaccharide, polyribosylribitolphosphate (PRP) (1). This unique PRP component corresponds to a simple formulation. In contrast, it rapidly became apparent from the early clinical studies that this vaccine had immunological limitations (23) and enabled obtention of good protective immunity only in children more than 2 years of age; an increase in the polysaccharide dose not improving immunization in young children. The immunizing capacities of the preparation were transformed by a method of coupling the polysaccharide to a carrier protein, which was described by Schneerson et al. (31). This transformation enabled good immunity with two vaccine doses. At the Institut Mérieux we chose the tetanus toxoid as the carrier protein, coupled with the polysaccharide, using the classic method of cyanogen bromide.

Toxoid and Subunit Toxoid Vaccines

Our increasing knowledge of the role of bacterial exotoxins provides a promising approach to the prevention of various bacterial diseases. During the last year a new generation of vaccines, using toxoid or subunit toxoid components as protective antigen, have been designed and are presently being developed. We describe two examples that could be of great importance in future applications.

Cholera Vaccine

Current injectable cholera vaccines, widely used throughout the world, are still recommended today, although the protective effect obtained is recognized to be mediocre. Fundamental studies on the structure and mechanism of action of cholera toxin, however, have transformed our approach to vaccination against this disease. Holmgren et al. (14) demonstrated that the cholera enterotoxin plays a major role in pathogenesis and protection. This toxin consists of two subunits, A and B: The A subunit is responsible for the pathogenic effect of the toxin through adenylate cyclase activation and thus for all symptoms of the disease. The B subunit is involved in binding the toxin to the ganglioside receptors of the gut mucosa. This B subunit is devoid of pathogenic effects but can stimulate a solid intestinal immunity. The formulation of a new oral cholera vaccine has been proposed that comprises both killed cholera whole cells (WC) and the B subunit (BS) component of cholera toxin (33). The whole cell components are produced by culture in 1000-L biofermentors using a syncase medium. Bacterial cells removed by centrifugation are inactivated by heat or formalin treatment.

The same culture of *Vibrio cholerae,* Inaba strain, that serves as the source for cholera toxins is used to obtain the crude culture filtrate. The main technical problem concerns production and extraction of the cholera toxin, and in particular the industrial preparation of a purified B subunit, the principal component of the new oral vaccine. Industrial separation and purification techniques developed by the Institut Mérieux using ion-exchange (35) and immunoaffinity (34) chromatography with Spherosil DEAE dextran have made it possible to resolve the difficulties of large-scale cholera toxin preparation. The preparation and attachment of the ganglioside receptor, in this case the lyso-GM_1 ganglioside on porous silica beads, has facilitated large-scale purification of cholera toxin (11). The GM_1 receptor ganglioside, hydrolyzed to lyso-GM_1, is covalently coupled to the Spherosil DEAE dextran. A glass column of these porous silica beads (Spherosil), onto which a layer of DEAE dextran has been adsorbed, enables binding of the cholera toxin, after which the toxin can be eluted using an acid citrate buffer (36) Adaptation to an industrial scale of this affinity chromatography has been achieved (Fig. 4.2). A l-kg column enables treatment of 1000 L of *Vibrio cholerae* culture and the isolation of 20 g of cholera toxin per cycle (Fig. 4.2). The B subunit is then purified by gel filtration (37).

An oral vaccine composed of B subunit and killed whole cells (BS-WC) has been formulated: 1 mg of BS plus 1×10^{11} killed WCs, the latter comprising 2.5×10^{10} cells of each of four preparations: heat-killed *Vibrio cholerae* 01 Inaba; heat-killed *V. cholerae* 01 Ogawa; formalin-killed *V. cholerae* 01 Inaba, El Tor bio type: and formalin-killed *V. cholerae* 01 Ogawa, classical biotype. This BS-WC vaccine is currently under inves-

Fig. 4.2. Column for immunoaffinity chromatography using Spherosil DEAE dextran with ganglioside receptor for large-scale purification of cholera toxin.

tigation in a major field trial in Bangladesh (7), which has been described and discussed elsewhere.

Acellular Pertussis Vaccine

Killed whole cell pertussis vaccine is one of the most widely used vaccines in the world, particularly in its triple combination with diphtheria and tetanus (DPT), which forms the basis for the Expanded Program on Immunization. It has, however, been criticized for the limits to its efficacy and, above all, the occasionally serious side effects it engenders. A total reassessment is presently under way, following pressure from health authorities in several countries and from the WHO (13,16).

A first generation of acellular pertussis vaccines has been developed in Japan (30). The nationwide use of this vaccine, produced by several manufacturers, has highlighted the prominent role of two antigens in the pathogenicity of the disease, as well as in the protective efficacy and side effects of the vaccines. These two antigens are the pertussis toxin (also known as lymphocytosis-promoting factor, or LPF) and filamentous hemagglutinin (FHA).

Several attempts are currently being made to obtain a well standardized, highly purified, stable nontoxic vaccine compatible with mass production. Institut Mérieux has developed original procedures including the following steps (Fig. 4.3):

1. *Bordetella pertussis* culture in a liquid medium, enabling industrial production of a vaccine suitable for mass immunization program.
2. Separate purification of two antigens by the application of chromatographic processes. This separation enables rational determination of the necessary proportion of each antigen in the final vaccine.
3. An original detoxification process of the toxin using glutaraldehyde, which enables obtention of a perfectly stable anatoxin free of all toxicity.

A vaccine containing detoxified pertussis toxin alone and a bivalent vaccine containing this toxoid and FHA antigen are presently undergoing systematic clinical trials (5). The Swedish trials presently under way (22) using Japanese vaccines are attempting to determine the protective role of each of the two antigens and to evaluate the tolerance of these new formulations in children less than 2 years of age. The Institut Mérieux vaccine is under evaluation in 3-month-old children in France and other countries.

Viral Vaccines

Viral vaccines for human use have undergone considerable development; they are prepared as live vaccines using attenuated viral strains, or as inactivated vaccines using viral preparations inactivated by appropriate physical and chemical processes. However, the industrial development, particularly of inactivated vaccines, has been hindered by two major factors: the origin of the animal cells providing the support for viral multiplication, and the culture limitations of these mother cells. Today the utilization of cell lines has removed the multiplication limitations of primary cells, and microcarrier culture techniques have enabled large-scale industrial production. Two examples of inactivated vaccine production—polio and rabies vaccines—are presented to illustrate the major improvements achieved in the preparation of viral vaccines; and the quality control techniques necessitated by the development of heteroploid cell lines, in particular Vero cells, are described.

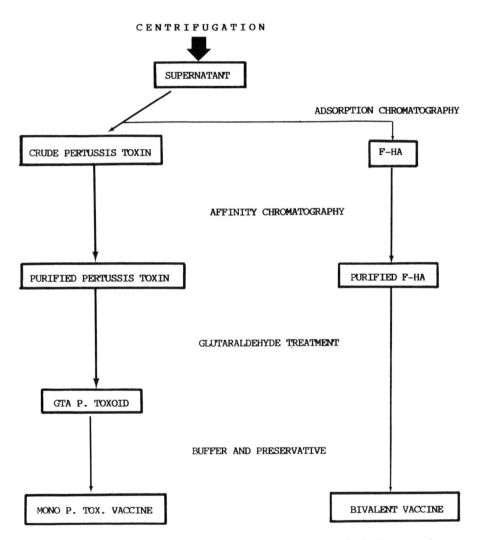

Fig. 4.3. Acellular pertussis vaccines. Production procedures for both monovalent toxoid vaccine and bivalent toxoid and FHA vaccine.

Production of Inactivated Polio and Rabies Vaccines

Inactivated Polio Vaccine Produced on Vero Cells

Inactivated polio vaccine, developed by Salk before Sabin's attenuated vaccine, was produced in several European countries, particularly France, The Netherlands, and Scandinavia. However, production levels were limited because of the limited supply of monkeys whose sacrifice provided the primary kidney cells necessary for poliomyelitis virus multiplication. The major revolution in terms of cell and viral culture came about with the publication of Van Wezel's microcarrier technique (40). The second stage, enabling industrialization of polio virus culture, was the adaptation of the Vero cell line to microcarriers, thereby enabling the culture of Vero cells and polio virus in large biofermentors (17). This technique was rapidly developed to reach the scale of 1000-L microcarrier cultures of Vero cells and polio virus. Production and control techniques have been reported in detail by Montagnon et al. (18,19).

Inactivated Rabies Vaccine Produced on Vero Cells

Cell culture vaccines marked a watershed in the long history of rabies vaccine development. The adaptation and culture of rabies virus on human diploid cells by Wiktor et al (41) marked the first fundamental step of this evolution. The production, control, and clinical results regarding the various rabies vaccines produced in cell culture have been extensively reviewed (26). The first cell culture techniques, and particularly vaccine production on human diploid cells, gave vaccines with remarkable safety and efficacy but did not resolve the problems of mass production that would enable a sufficiently economical vaccine that could be applied regularly in the developing countries subject to a high level of endemic rabies (28). For this reason, in 1983 the Institut Mérieux started developing a purified vaccine produced on Vero cells, known as purified Vero rabies vaccine (PVRV), based on the model of the new inactivated polio vaccine produced also on Vero cells.

The main production and control stages of PVRV were inspired by those used for inactivated polio vaccine. We do not go into detail here regarding these stages, as they have been reported elsewhere (10,21,28). Figure 4.4 illustrates the principal stages of biofermentor culture of Vero cells, rabies virus culture, concentration, inactivation, and purification in the production of PVRV vaccine. Furthermore, the results obtained with this new vaccine have been reported for preexposure vaccination (9,29) and postexposure treatment (6,32) under various epidemiological conditions.

Vero Cell Vaccine: Quality Control

The technological developments regarding vaccine production (design of biofermentors, culture techniques, and the transfer of microcarriers and

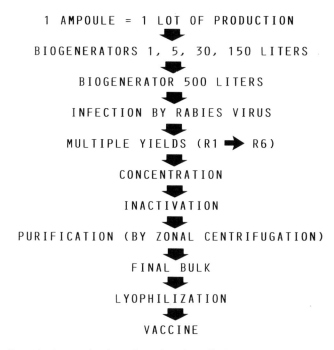

Fig. 4.4. Steps in the production of one lot of purified Vero rabies vaccine (PVRV).

cells) and the various regulatory processes of culture parameters for cells and viruses were rapidly achieved (19). It took longer, however, to develop suitable control techniques, norms applicable to these techniques, and above all norms enabling the use of heteroploid cell lines.

The first major step was taken in 1981, with the definition of new requirements for poliomyelitis vaccine (inactivated) by a WHO Expert Committee on Biological Standardization (42). This document laid the foundations for rules of good manufacturing practice and production controls at the various stages of cell culture, virus culture, inactivation, and vaccine preparation.

Two main control techniques were thus introduced to guarantee the quality of cell substrates and the purity of the final product. The first techniques aimed at demonstrating the absence of tumorigenicity in the cells. The test for tumorigenicity, first described by Van Steenis and Van Wezel (39), gauges the capacity of local multiplication and invasion of live cells inoculated subcutaneously in the ventral region of newborn rats immunosuppressed with anti-rat thymocyte globulin. Twenty–one days after injection tumorigenic control cell lines (HeLa, Hep-2, and KB) produce a local tumorigenic effect manifested by the development of subcutaneous tumor nodules at the inoculation site and a general effect manifested by the development of pulmonary metastases. The test has been

Table 4.2. Comparative tumorigenicity of various cell lines: Vero cells compared with reference tumorigenic cell lines (HeLa, Hep-2, KB).

Cell line	Passages	Subcutaneous tumor nodules		Pulmonary metastases	
		Frequency[a]	%	Frequency[a]	%
Vero	137 to 159	0/195[a]	0	0/195	0
HeLa	195 to 253	206/206	100	125/206	60,7
Hep-2	60 to 133	111/114	97	62/114	54,3
KB	26 to 70	22/40	55	18/40	45

[a]Number of local tumors or metastases reported per number of inoculated animals 21 days after subcutaneous inoculation with 10^6 cells per animal. Clinical observations were confirmed by histopathology.

adapted to Vero line passages (19). The cumulative results obtained to date for Vero cells are reported in Table 4.2. It should be pointed out that an absence of tumorigenicity has been observed for Vero cells between the 137th and 159th passages; the Working Cell Bank, which serves as a base for cell production, is at the 137th passage, and virus production is carried out at the 142nd passage.

A second technique essential to the control of viral products based on the culture of heteroploid cells was developed to guarantee the purity of the final product. In 1981 the WHO Expert Committee indicated the general direction, stating that the validity of purification processes should be ensured by a reduction in the level of cellular DNA between the initial cell harvest and the final vaccine. The measurement test for residual vero cell DNA by DNA–DNA molecular hybridization (Fig. 4.5) has been described in detail by Montagnon et al, as have the results obtained for the purification and preparation stages of polio vaccine (20). The figures reported by the Institut Mérieux team and those by several other laboratories have been used in the definition of new international norms, published by WHO in 1987 (44).

The broad security coefficient provided by purification processes has been discussed by Petricciani (24). The value in terms of safety of viral production performed on cell lines when compared with the risks of cultures on primary cells must be emphasized. The short period between the sacrifice of an animal serving as cell donor and the viral culture performed on these cells does not permit sufficient time to carry out the slow and difficult controls that could enable demonstration of viral contaminants of the oncovirus or retrovirus type (15,43), whereas establishment of a diploid or heteroploid cell line provides ample time for the systematic detection of markers for any potential contaminant. At the levels of the Working Cell Bank and viral production for Vero cells, the accumulation since 1982 of increasingly perfected tests has demonstrated a total absence of any virus, reverse transcriptase activity, any DNA retrovirus, or protein

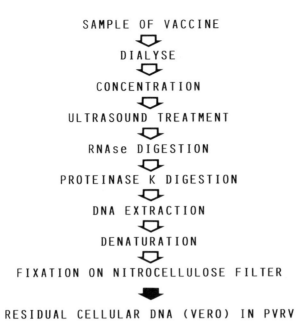

SAMPLE OF VACCINE

DIALYSE

CONCENTRATION

ULTRASOUND TREATMENT

RNAse DIGESTION

PROTEINASE K DIGESTION

DNA EXTRACTION

DENATURATION

FIXATION ON NITROCELLULOSE FILTER

RESIDUAL CELLULAR DNA (VERO) IN PVRV

Fig. 4.5. Main stages in the test for measuring residual Vero cell DNA by DNA–DNA molecular hybridization.

markers for HTLV 1, HTLV 3, HTLV 4, STLV 1, and STLV, 3 (P.J. Kanki, personal communication).

Conclusion

New and improved production techniques are available for bacterial and viral vaccines. These vaccines represent a wide range of products with highly varying antigenic components. Thus new techniques for their production must be adapted to each situation with regard to extraction, concentration, and purification. Increasingly sophisticated quality control methods are required for all these products, which are adapted to eliminate any pathogenic risk. These improvements must be suited to large-scale industrial manufacturing practice, as it is the sole means to overcome the major problem of economical vaccine production.

Acknowledgments. We gratefully acknowledge all our colleagues who participated in the research and development of the vaccines described in this publication, and especially J.L. Tayot. We also thank Miss D. Salle for her skilled secretarial assistance.

References

1. Anderson P, Smith DH: Isolation of the capsular polysaccharide from the culture supernatant of Haemophilus influenzae b. *Infect Immun* 1977;15:472–477.
2. Austrian R, Douglas RM, Schiffman G, et al: Prevention of pneumococcal pneumonia by vaccination. *Trans Assoc Am Physician* 1976;89:184–194.
3. Ayme G, Donikian R, Mynard MC, Lagrandeur G: Production and controls of serogroup A Neisseria meningitidis polysaccharide vaccine, in *Table Ronde sur l'Immunoprophylaxie de la Méningite Cérébrospinale, Saint-Paul de Vence, France.* Lyon, Fondation Mérieux, 1973, pp 4–30.
4. Cadoz M, Armand J, Arminjon F, et al: A new 23 valent pneumococcal vaccine: immunogenicity and reactogenicity in adults. *J Biol Stand* 1985;13:261–265.
5. Cadoz M, Arminjon F, Quentin-Millet MJ, Armand J: Safety and immunogenicity of Mérieux acellular pertussis vaccines in adult volunteers. Presented at: Workshop on Acellular Pertussis Vaccines Organized by the Interagency Group to Monitor Vaccine Development, CDC, FDA, NIH. Bethesda, 1986.
6. Chadli A, Merieux C, Arrouji A, Ajjan N: Study of the efficacy of a vaccine produced from rabies virus cultivated on Vero cells, in Vodopija I, et al (eds): *Improvements in Rabies Post-exposure Treatment* Zagreb, Institute of Public Health, 1985, pp 129–136.
7. Clemens JD, Sack DA, Harris JR, et al: Field trial of oral cholera vaccines in Bangladesh. *Lancet* 1986;2:124–127.
8. Court G, Nguyen C, Tayot JL: Purification of tetanus toxin by affinity chromatography on porous silica beads derivatized with polysialogangliosides, in *Sixth International Conference on Tetanus.* Lyon, Fondation Mérieux, 1982, pp 331–345.
9. Dureux B, Canton PH, Gerard A, et al: Rabies vaccine for human use, cultivated on Vero cells. *Lancet* 1986;2:98.
10. Fournier P, Montagnon B, Vincent-Falquet JC, et al: A new vaccine produced from rabies virus cultivated on Vero cells, in Vodopija I, et al (eds): *Improvements in Rabies Postexposure Treatment.* Zagreb, Institute of Public Health, 1985, pp 115–121.
11. Fredman P, Nilsson O, Tayot JL, Svennerholm L: Separation of ganglioside on a new type of anion-exchange resin. *Biochim Biophys Acta* 1980;618:42–52.
12. Gotschlich EC, Liu TY, Artenstein MS: Human immunity to the meningococcus. III. Preparation and immunochemical properties of the group A, group B, and group C meningococcal polysaccharides. *J Exp Med* 1969;129:1349–1365.
13. Granström M, Granström G, Gillenius P, Askelöf P: Neutralizing antibodies to pertussis toxin in whooping cough. *J Infect Dis* 1985;151:646–649.
14. Holmgren J, Svennerholm AM, Lönnroth I, et al: Development of improved cholera vaccine based on subunit toxoid. *Nature* 1977;269:602–604.
15. Homma T, Kanki PJ, King NW, et al: Lymphoma in macaques: association with virus of human T lymphotrophic family. *Science* 1984;225:716–718.
16. Manclark CR: Proceedings of the Fourth International Symposium on Pertussis. *Dev Biol Stand* 1986;61:3–576.
17. Montagnon BJ, Fanget B, Nicolas AJ: The large scale cultivation of Vero cells in microcarrier culture for virus vaccine production—preliminary results for killed poliovirus vaccine. *Dev Biol Stand* 1981;47:55–64.

18. Montagnon B, Vincent-Falquet JC, Fanget B: Thousand litre scale microcarrier culture of Vero cells for killed polio virus vaccine: promising results. *Dev Biol Stand* 1984;55:37–42.
19. Montagnon B, Fanget B, Vincent-Falquet JC: Industrial scale production of inactivated poliovirus vaccine prepared by culture of Vero cells on microcarrier. *Rev Infect Dis* 1984;6(suppl 2):S341–344.
20. Montagnon BJ, Martin R, Barraud B, et al: Residual Vero cell DNA and inactivated polio vaccine: detection by DNA-DNA molecular hybridization, in Hope. HE, Petricciani JC (eds): *Abnormal Cells, New Products and Risk.* Gaithersburg, Maryland, Tissue Culture Association, 1985, pp 82–89.
21. Montagnon B, Fanget B, Fournier P, et al: Vaccin rage PVRV: pureté et stabilité, in *La Rage et la Brucellose dans le Bassin Méditerranéen et la Péninsule Arabe, Montpellier, France.* Lyon, Fondation Mérieux, 1986, pp 147–152.
22. Olin P: Clinical trials of acellular pertussis vaccines in Sweden: an attempt to solve the problem of pertussis vaccination. Presented at the Sclavo International Conference on Bacterial Vaccines and Local Immunity, Sienna, Italy, November 17–19, 1986.
23. Peltola H, Käythy H, Sivonen A: Haemophilus influenzae type b capsular polysaccharide vaccine in children: a double-blind field study of 100,000 vaccinees 3 months to 5 years of age in Finland. *Pediatrics* 1977;60:730–737.
24. Petricciani JC: Old issues confront biotechnology and new products. *Vaccine* 1984;2:235–236.
25. Robbins JB, Austrian R, Lee CJ, et al: Considerations for formulating the second generation pneumococcal capsular polysaccharide vaccine with emphasis on the cross reactive types within groups. *J Infect Dis* 1983;148:1136–1159.
26. Roumiantzeff M, Ajjan N, Branche R, et al: Rabies vaccine produced in cell culture: production, control and clinical results, in Kurstak E, Al-Nakib W, Kurstak C (eds): *Applied Virology I.* New York, Academic Press, 1984, pp 241–296.
27. Roumiantzeff M, Armand J, Thoinet M, Taylor JL: Production and use of vaccines: technology of vaccine production, in Bell R, Torrigiani G, (eds): *New Approaches to Vaccine Development.* Basel, Schwabe, 1984, pp 477–493.
28. Roumiantzeff M, Ajjan N, Montagnon B, Vincent-Falquet JC: Rabies vaccines produced in cell culture. *Ann Inst Pasteur Virol* 1985;136E:413–424.
29. Roumiantzeff M, Ajjan N, Vincent-Falquet JC: Experience in pre-exposure vaccination, in *Rev Infect Dis* 1988;4.
30. Sato Y, Kimura M, Fukumi H: Development of a pertussis component vaccine in Japan. *Lancet* 1984;1:122–126.
31. Schneerson R, Barrera O, Sutton A, Robbins JB: Preparation, characterization and immunogenicity of Haemophilus influenzae type b polysaccharide-protein conjugates. *J Exp Med* 1980;152:361–376.
32. Suntharasamai P, Warrell MJ, Warrell DA, et al: New purified Vero-cell vaccine prevents rabies in patients bitten by rabid animals. *Lancet* 1986;2:129–131.
33. Svennerholm AM, Jertborn M, Gothefors L, et al: Mucosal antitoxic and antibacterial immunity after cholera disease and after immunization with a combined B subunit-whole cell vaccine. *J Infect Dis* 1984;149:884–893.
34. Tardy M, Tayot JL, Roumiantzeff M, Plan R: Immunoaffinity chromatography

on derivatives of porous silica beads: industrial extraction of antitetanus antibodies from placental blood and plasma, in Epton R (ed): *Chromatography of Synthetic and Biological Polymers*. Chichester, Ellis Horwood, 1978, vol 2, pp 298–313.

35. Tayot JL, Tardy M, Gattel P, et al: Industrial ion exchange chromatography of proteins on DEAE-Dextran derivates of porous silica beads, in Epton R (ed): *Chromatography of Synthetic and Biological Polymers*. Chichester, Ellis Horwood, 1978, vol 2, pp 95–110.

36. Tayot JL, Tardy M: Isolation of cholera toxin by affinity chromatography on porous silica beads covalently coupled ganglioside GM_1, in Svennerholm L, Dreyfus H, Urban PF (eds): *Structure and Function of Gangliosides*. New York, Plenum 1980, pp 471–478.

37. Tayot JL, Holmgren J, Svennerholm L, et al: Receptor-specific large-scale purification of cholera toxin on silica beads derivatized with lyso GM_1 ganglioside. *Eur J Biochem* 1981;113:249–258.

38. Tiesjema RH, Beuvery EC, Tepas BJ: Enhanced stability of meningococcal polysaccharide vaccines by using lactose as a menstrum for lyophilisation. *Bull WHO* 1977;55:43–48.

39. Van Steenis G, Van Wezel AL: Use of ATG-treated new-born rat for in vivo tumorigenicity testing of cell substrates. *Dev Biol Stand* 1982;50:37–46.

40. Van Wezel AL: Growth of cell strains and primary cells on microcarrier in homogeneous culture. *Nature* 1967;216:64–65.

41. Wiktor TJ, Fernandes MV, Koprowski H: Cultivation of rabies virus in human diploid cell strain WI 38. *J Immunol* 1964;93:353–356.

42. WHO Expert Committee on Biological Standardization: Requirements for poliomyelitis vaccine (inactivated). *WHO Tech Rep Ser* 1982;673:annex 2.

43. WHO: T-Lymphotropic retroviruses of non-human primates. *Weekly Epidemiol Rec* 1985;60:269–270.

44. WHO Expert Committee on Biological Standardization: Requirements for rabies vaccine (inactivated) for human use produced in continuous all lines. *WHO Tech Rep Ser* 1987;760:annex 9.

CHAPTER 5

Transfer of Vaccine Production to the Developing World: Rabies Vaccine

Scott B. Halstead

Ours has been termed "the age of new vaccines" (1). From the standpoint of rabies in the developing world, a more apt characterization might be "the age of old vaccines." Semple-type vaccines for postbite treatment of humans continue to be manufactured and used in more than 50 developing countries where rabies is a significant health problem (2). Perhaps we should call this era "the age of no vaccines." Far too many developing countries fail to manufacture sufficient veterinary rabies vaccine to permit serious planning for the control of dog and cat rabies, let alone the implementation of nationwide control programs. It is estimated that at least 280 million dogs in enzootic areas should be vaccinated. At present fewer than 10% of these dogs receive vaccine (2).

It is difficult to avoid the conclusion that stagnation in programs to prevent rabies in the developing world is linked to four key problems: (a) the relatively low priority accorded the prevention of human rabies; (b) the need for effective interministerial cooperation for successful rabies control programs; (c) the need for national self-sufficiency in vaccine production; and (d) cost.

With no major international donor agencies supporting rabies control, the greatest constraint to the design and implementation of effective programs is cost. It is the relatively low cost and small investment in facilities and manpower that keep toxic, neuroparalytic Semple-type rabies vaccines on the market in many countries. The cost factor seems to have inhibited investment in the production of anti-rabies serum in horses, let alone in humans. Cost discourages many countries from contemplating the production of human and veterinary vaccines in tissue cultures. Without low-cost, efficacious vaccines, national planning for rabies control is paralyzed. The World Health Organization (WHO) and the Rockefeller Foundation have initiated a program to relieve this gridlock. It is proposed to supplant the generic vaccine developed by Pasteur and modified by Semple with a tissue culture-based vaccine. The same process and the same facility can be used to manufacture both veterinary and human rabies vaccines. This program should serve to expand national competence to produce other tissue culture-based vaccines.

In October 1984 the Veterinary Public Health Unit of the Communicable Disease Division of the WHO sponsored a consultation on the transfer of technology for the production of rabies vaccine (2). The strong consensus of participants—virologists, public health officials, veterinarians, and vaccine manufacturers from the developed as well as the developing world—was that rabies vaccines produced in adult brain tissue should be displaced as quickly as possible. It was recommended that countries with annual vaccine requirements of more than 100,000 doses should consider local production of rabies vaccine in tissue cultures.

Two months later WHO, with Rockefeller Foundation support, convened a group of experts in Geneva (3): George Baer, Centers for Disease Control, Atlanta; Geoffrey Letchworth, University of Wisconsin School of Veterinary Medicine; Holger Lundbeck, a Swedish rabies authority; Pierre Reculard, Pasteur Vaccins; Geoffrey Schild, U.K. National Institute for Biological Standards and Control; Dr. A.L. Van Wezel, Rijksinstituut voor de Volksgezondheid; and myself as chair. The Group decided that the continuous African green monkey kidney cell line, Vero, was likely to be the most trouble-free substrate for commercial-scale production of rabies vaccines in developing countries. A vero cell vaccine is produced commercially and licensed for human use in France (4). With minimal or no virus concentration, infected supernatant fluids can be inactivated for use in animals. After removal of DNA, inactivated virus is used in humans. It was recommended that many countries with a population of 20 million persons and more might have the scientific and economic infrastructure to support indigenous production of vaccines using modern methods. Countries with a lower population more likely need to rely on importation; groups of smaller countries might form consortia for regional vaccine production.

When the Group of Experts met again early in 1986, assessment visits had been made to Burma, Thailand, Indonesia, Egypt, Mexico, Colombia, India, and China. More recently visits have been made to Cameroon and Morocco. Based on these visits, the Group agreed that the optimal site for an initial technology transfer program should be the semigovernmental Veterinary Products Company of Colombia (VECOL) located in Bogota. VECOL is a well financed veterinary biologicals manufacturing firm with a broad line of successful products and progressive management. Furthermore, a memorandum between the Pan American Health Organization, the Ministry of Health, the Ministry of Agriculture of Colombia, and VECOL provided the legal umbrella for the importation of equipment plus an agreement by the Ministry of Health to allow VECOL to manufacture and market rabies vaccine for use in humans.

Despite the recognition by the Group of Experts that many developing countries now have the technical and economic base required to support modern rabies vaccine production, a critical problem remained to be solved: how to find a rabies virus strain adapted to grow in Vero cells that was in the public domain? The answer was supplied by Dr. Pierre Sureau, Director WHO Collaborating Centre and Research in Rabies. Rabies Unit, Institut Pasteur, Paris, France. The Pasteur strain, which is generic by virtue of its origin and subsequent adaptation in publically supported laboratories, has been made available to the WHO for use in the technology transfer program.

At the Institute Armand Frappier (IAF) in Montreal, Canada, the Pasteur strain was quickly shown to replicate well in Vero, and acceptable antigen yields were obtained in several batches of experimental inactivated vaccine. Funds have been provided to IAF to assemble training staff and to purchase the supplies and equipment to furnish a production facility at VECOL.

Had all gone as planned, the vaccine production technicians plus quality control staff from the National Institute of Health in Colombia would have begun training in November 1986. It is with a deep sense of personal loss as well as a great loss to the vaccine production community at large that I note the unexpected death of Dr. Anton Van Wezel, which removed the experience, leadership, and dedication to technology transfer that have been essential for this program to date. In sorrow, but out of respect for his ideals, we will go on.

The successful transfer to key developing countries of commercial-scale tissue culture-based vaccine production—and the VECOL program will use a microcarrier fermentation process—will provide a manufacturing base for other human and veterinary vaccines including gene constructs that use mammalian cells as a production substrate. If this technology transfer program does succeed, we will be at the dawn of a new age, an age when affordable, high quality rabies vaccines for man as well as animals will be the basis for improved treatment of dog bite victims and for preventing the spread of rabies virus in domestic animals.

References

1. Schmeck HM Jr: The new age of vaccines. *New York Times Magazine* April 29, 1984.
2. WHO: *Report of a WHO Consultation on Transfer of Technology for Production of Rabies Vaccine*. WHO/Rab/85.197. Geneva, WHO 1984.
3. WHO: *WHO Group of Experts on Rabies Vaccine Production Techniques*. Geneva, WHO, 1984.
4. Fournier P, Montagnon B, Vincent-Falquet JC, et al: A new vaccine produced from rabies virus cultivated in vero cells, in Vodopija I, Nicholson KG, Smerdel S, Bijok U (eds): *Improvements in Rabies Post-exposure Treatment*. Zagreb, Institute of Public Health, 1985, pp 123–127.

CHAPTER 6

Policy for Developing Countries for Storage, Distribution, and Use of Essential Vaccines for Immunoprophylaxis

S.C. Adlakha and V.R. Kalyanaraman

The importance of veterinary vaccines lies not only in their value as significant tools for immunoprophylaxis of animal diseases but also for evaluating the newer technologies for vaccine production for the control of human diseases. The first large-scale vaccine produced using genetic engineering was against diarrhea of piglets in 1982 in The Netherlands by the Dutch firm Akzo. There is great potential for the development of a vaccine against the animal forms of sleeping sickness, which is transmitted by tse–tse flies; the control of this disease could permit breeding of cattle in vast tracts of Africa, which has not been possible hitherto because of this disease. Significant progress has been made in the development of a vaccine for foot-and-mouth disease. The gene that codes for one of the virus's coat proteins, VPI, has already been cloned and inserted into *Escherichia coli*. Once purified, this protein is likely to act as an effective vaccine against the foot-and-mouth disease virus. The VPI vaccines produced by genetic engineering could offer substantial advantage in terms of cost as well. Among the advances in this area is the cloning of a rabies virus antigen.

Genetic engineering not only promises to provide protection against diseases that cannot be fought by conventional vaccines, it also provides safer substitutes of existing antivirus vaccines. Viruses themselves are not used in these vaccines; however one of the proteins made by the bacteria is used, thereby eliminating the risk that the vaccine could be contaminated with live viruses. If this goal is achieved, it would no longer be necessary to develop and maintain, at considerable expense, high-security laboratories for vaccine production.

There are a number of serious virus diseases for which inexpensive and effective vaccines could be developed using the latest technology. The vital economic importance of eradicating, or at least controlling, viral diseases in animals and the role genetic engineering can play in such endeavors has been recognized by the United Nations Industrial Development Organization. This body has assigned a high priority to the development of genetically engineered vaccines against several diseases, including foot-and-mouth disease, rabies, African horse sickness, and blue tongue of sheep.

In most of the developing countries, outdated production technology is being used for many vaccines. Developing countries must take cognizance of the following facts.

1. Progressive use of advanced technology for large-scale, economical manufacture of good-quality vaccines.
   a. Use of biofermenters in the large-scale production of bacterial vaccines with a reliable monitoring system to control and correct the main culture parameters, temperature, pH, and aeration, which will achieve better and economical production.
   b. New techniques for the production of viral vaccines such as the availability of new synthetic and recombinant DNA technology leading to better production standardization, control, and stability of the vaccine. The cloned vaccines are on trial; e.g., the protective coat of the foot-and-mouth disease virus has been cloned and expressed. In addition to cost, the advantages of cloned protein vaccines are for their safety, lack of risk of undesirable mutation, thermal stability, and better shelf life.
2. Requirements of the various vaccines depending on the national disease control policy.

Considering the economy of the scale of production, it would be desirable for large countries to set up or modernize production centers of their own; it would be advantageous for the small countries of the region to obtain their vaccine requirements from a country in that same region which has facilities for large-scale vaccine production rather than venture on its own vaccine production or continue with its outdated manufacturing unit. Moreover, vaccine banks can be developed on a regional basis. As a matter of fact, for veterinary vaccines this idea has already been taken up by the Animal Production and Health Commission for Asia and the Pacific (APHCA) of the Food and Agriculture Organization. It is a good example of technical cooperation among developing countries that could be expanded.

The need for testing the safety and potency of vaccines has been recognized by all countries despite the fact that financing of quality control schemes is an overwhelming burden on the vaccine manufacturer. In this connection, standardization of vaccine production protocols by the developing countries is essential if vaccines of dependable quality are to be produced.

Although it is necessary for the manufacturing units to follow good manufacturing practices, it is equally important that a remunerative price is paid for the products manufactured by such units. The prerequisite for good manufacturing practices is provision of adequate modern facilities, including modern sophisticated equipment plus on uninterrupted supply of power and water. Optimal utilization of the installed capacity is essential; and if at certain point of time it is not possible, recourse may be taken to diversification of products using the same essential facility. An example is the utilization of the presently surplus cell cultures of foot-and-mouth disease vaccine production capacity in India for the production of inactivated cell culture rabies vaccine.

Regular foolproof cold chain would have to be developed to replace the improvised unsatisfactory procedures of storage and transport of vaccines. The loss of biological activity in transit could be avoided by seeking co-operation of the airlines and railways to accord top priority to shipment of vaccines. They could also be motivated to develop cold-storage facilities to stock vaccines at booking centers. Incentives should be given to private transporters to use refrigerated vans for the transport of vaccines. The current practice of shipping vaccine by ordinary transport without proper refrigeration should be banned until the time that such vaccines are developed that can withstand the higher temperatures that develop in containers in transit.

Immunoprophylaxis

When suitable strategies for control of communicable diseases are planned for a developing country, the twin problems of large population and low level of hygiene and nutrition deserve primary consideration. However, it takes considerable time to implement effective programs to combat these problems. On the other hand, an effective program involving immuno-prophylactics for prevention of communicable diseases as an accepted public health measure in the promotion of primary health care could be more easily organized and implemented with little delay.

Manufacture of Immunoprophylactics

The manufacture of immunoprophylactics in a developing country takes into consideration the following factors.

1. Vaccines required for the Expanded Program of Immunization (EPI): DTP group of vaccines, BCG, polio, measles
2. Vaccines required on large scale but not under EPI, e.g., rabies vaccine.
3. Vaccines for local use or for selected group of beneficiaries, e.g., JE, KFD vaccines
4. Vaccines that are available in developed countries which are not cost-effective in a developing country, thereby limiting their use, e.g., hepatitis vaccine, tissue culture rabies vaccine
5. Vaccines in various stages of research and development in developed countries, e.g., recombinant vaccines, acellular pertussis vaccine, oral bait vaccines, improved vaccines for cholera and typhoid, anti-*Shigella* vaccine

Hence any policy formulated must be clearly based on a situation analysis and a realistic view of priorities, resource assessment, proper manpower development, planning, adequate funding, and of course utilization of modern technologies in the field of biotechnology. The molecular biology design must be sufficient to produce large quantities of safe, potent immunoprophylactics to cater to the entire country's need and at the same time be cost-effective. These modern systems of biotechnology lend themselves to making multiple vaccines through the same systems and thus enhance their potential usefulness. In regard to manpower development, one must plan for augmentation of trained manpower at different levels: production, quality control, distribution, and administration.

If the policy is to be evolved taking all the points mentioned into consideration, some questions must be answered.

1. Is the developing country to have many small units duplicating the work or to have large, centralized units with concentrated facilities?
2. If the requirement of the various vaccines is huge, is it advisable to put all the "eggs in one basket"?
3. Being an important aspect of national health, should the program be in the public or the private sector or both?

A special advisory group of experts should be constituted to help develop the specific infrastructure and to suggest the most appropriate linkages within as well as outside the country so that the scientific talent can be accumulated faster and the "take off" time reduced considerably. Once a considered decision and recommendation is made in this regard, the "action program" should be instituted on a time- and target-oriented schedule without delay.

Despite the fact that discussions regarding transfer of technology go on at various levels between developed and developing countries, and international agencies try their best to play an active role in this regard, little progress seems to be made owing to various constraints. Thus action on this front should become a fast reality without strings attached—economic or otherwise. How can one expect "health for all in 2000 AD" to become a reality otherwise?

Quality Control of Immunoprophylactics

The special nature of immunobiologicals, different from conventional drugs, calls for creation of a special organization or authority, completely independent and comprehensive, to supervise and control the quality of these products at the national level as well as to improve and insist on good manufacturing practices. At present there is a big lacuna in this regard in developing countries.

This national authority should also utilize the services of various scientific institutions and laboratories in the country for testing potency, immune response, serosurveys, and field efficacy; preparation of national standards; monitoring the cold chain system; and special quality control tests such as DNA analysis, karyology, tumorigenicity, isozyme studies, monoclonal assays, etc. Such a system avoids unnecessary duplication of facilities and reduces the cost. Of course, apart from the national level of quality control, it should be ensured that every manufacturing unit has a competent and fully equipped quality control laboratory to screen their products without bias.

Storage and Distribution

One of the major difficulties in the rapid expansion of an immunization program is the storage and distribution of immunoprophylactics at the recommended temperature without loss of potency at the consumer level. The cold storage facilities in a developing country vary significantly from urban to rural areas.

The cold chain essentially consists of three vital elements: cold storage equipment, transportation by the shortest and quickest route, and motivated and trained manpower. There should be no breakdown at any point of this cold chain. The service and repair facilities should be excellent and rapid, and a power supply of constant voltage should be ensured. Because more than one agency is involved in maintaining the cold chain, coordination must be established between the agencies for effective functioning of the cold chain. Cold chain maintenance may be the most strenuous task to be accomplished in a developing country, and any policy formulated must devote ample attention and high priority to the cold chain and proper delivery system.

Use of Immunoprophylactics

There are a few points that must be noted with regard to the use of immunoprophylactics.

1. The use of an immunoprophylactic should cover the eligible population and thereby result in reduction of the disease incidence in the country.

2. The coverage should be given at the right time and by the right technique.

3. The most important point of any immunization program is to deliver suitable vaccine in adequately potent form to those who need it most. Strategies for distribution and administration therefore must be formulated as a single or combined approach of fixed immunization centers, outreach operations, and/or mass campaigns. The system of immunization should be such that it produces the desired result.

4. The dosage schedules should be formulated in such manner as to give effective immunity in a developing country, e.g., dosage of OPV.

5. Integration of vaccination sources should be organized so that all the

vaccines are available at each vaccination session in order to reduce the number of contacts to a minimum. This procedure helps reduce the dropout rate.

6. Utmost care should be taken to formulate policies for contraindications for immunoprophylactics. The list should be realistic, keeping in mind the health status prevailing in the developing countries.

7. The proper use of any immunoprophylactic depends on:
   a. Procurement and distribution of adequate quantity of essential supplies and equipment
   b. Training of health personnel
   c. Reliable surveillance system
   d. Evaluation and monitoring
   e. Operational research
   f. Community participation and health education

Part III
Environmental Factors Modulating the Efficacy of the Vaccine

CHAPTER 7

Immunity in Tuberculosis: Environmental Versus Intrinsic Factors Modulating the Immune Responsiveness to Mycobacteria

P.H. Lagrange and B. Hurtrel

Infections caused by mycobacteria produce a wide range of immunological reactions. Tuberculosis and leprosy are often considered the prototype of chronic intracellular infections, and the immunological investigation of these diseases has provided insights into the basic mechanisms of both cellular and humoral immunity (30). However, there are few diseases that so effectively express the spectrum of immune responsiveness as leprosy and, to a lesser extent, tuberculosis. Moreover, results from many BCG vaccine trials against tuberculosis and leprosy showed wide variations in protection (8). Two main hypotheses were put forward to explain the observed variations in the efficacy of vaccination: (a) variability in the potency of the strain and preparations used; and (b) the influence of previous sensitization with atypical mycobacteria. These hypotheses are favored no more, however and it is thought that the protective effect of BCG may depend on epidemiological, environment, and immunological factors affecting the infecting agents as well as the host.

Many investigative efforts are directed toward the explanation of the nature of the immune responsiveness variability seen in leprosy (3) and tuberculosis (55). Some progress has been made in elucidating the role of suppressive or helper influences of T cells or their mediators and in reconstituting the immune defects. However, many parameters are involved concerning dependent and independent variables in the various assays used, and they need to be clearly identified in order to clarify the exact role they play in induction, expression, and regulation of the immune protection induced after a natural or experimental infection.

Immune Responses After Mycobacterial Infections

In susceptible hosts, inoculation of pathogenic myobacteria through its own multiplication in macrophages induces specific and nonspecific immune responses. Although most researchers have primarily addressed the question of cellular immunity in tuberculosis or leprosy, few scientists have focused on aspects of the humoral immune response (54). Undoubtedly, specific antibodies are produced in response to natural and experimental mycobacterial infections, but there is evidence that they play no role in host protection (42). On the other hand, some reports show enhancing effects of specific antibodies. Passive immunization with specific anti-mycobacteria sera from BCG-infected mice to a certain extent promote the growth of bacilli in the spleen (17). The mechanism of such an enhancement has not been fully explored but does not seem to follow high bacterial uptake of the spleen. However, because it occurred only after a low inoculum, it might involve a particular distribution of the mycobacteria in a subpopulation of macrophages that is either less able to kill the mycobacteria or more able to promote growth of the bacilli in the absence of the T cell-dependent response, as measured by the delayed-type hypersensitivity (DTH) reaction (18). Moreover, no data were given concerning the effects of such passive transfer on the acquired resistance of the infected mice.

Thus most of our knowledge about immunological mechanisms controlling mycobacterial diseases concerns only cell-mediated immunity (CMI) in experimental or human tuberculosis and leprosy. The specific responses are well characterized by mechanisms that prevent further multiplication and spreading of mycobacteria. These events, called *macrophage activation* and *granuloma formation,* are specific in their induction but are nonspecific in their expression, as they are locally able to inhibit the growth and dissemination of other obligate or facultative unrelated intracellular multiplying microorganisms (20). Several reviews have appeared during the past few years on the immunology of mycobacterial infections (7,11,12,45); from them and other relevant literature it can be concluded that the specificity of the immune response to pathological or nonpathological mycobacteria has been shown to be mediated by thymus-dependent lymphocytes (T lymphocytes). Specific, uncommitted T cells become stimulated in the T cell areas of the draining lymphoid tissues during the induction phase, which occur soon after inoculation of the living bacteria (21). This T cell stimulation can be evaluated in vivo qualitatively and quantitatively by measuring the kinetics of the cells incorporating radiolabeled precursors of DNA (i.e., tritiated thymidine). However, this global evaluation only partially reflects the clonal expansion of specifically engaged T cells. It has been shown that, after intravenous injection of BCG, among the radiolabeling present in the spleen suspension 10 or 20 days after BCG inoculation only a small (25%) but significant proportion of [^{3}H]thymidine uptake was due to T cells (37); and among these radiolabeled T cells, only one-tenth may represent specific committed T lymphocytes. In order to obtain a more precise analysis of the number of cells engaged in protection, it is necessary to resort to passive transfer of these cells in naïve animals using dilution analysis methods. However, passive protection against tuberculosis needs to be evaluated in age- and sex-matched syngeneic recipient mice that had been rendered T cell-deficient prior to transfer by exposure to 500 rad of ionizing radiation (32).

As soon as these specific committed T cells are stimulated, some of them leave the draining T cell areas via the lymphoid ducts through the systemic circulation and reach the inflammatory foci at the inoculation site or everywhere bacteria are present, and then the host develops a granulomatous reaction and macrophage activation. These reactions are elicited by specific recognition of antigens on the surface of infected macrophages by the circulating committed lymphocytes. These events stop further multiplication and dissemination of the mycobacteria. In addition, the host may express a DTH reaction to mycobacterial antigens. Usually in the normal host the development of this cellular hypersensitivity to mycobacterial antigens is associated with the occurrence of an acquired resistance to subsequent challenge of the same or related mycobacteria (33). Nonspecific immunity that has increased during mycobacterial infection can also be detected in vivo and in vitro by measuring the increase in resistance of the mononuclear phagocyte system against unrelated pathogen (i.e., *Listeria monocytogenes*). This nonspecific immunity can mask or modulate the real efficacy of the specific acquired resistance. However, such nonspecific resistance is not directly transferable in naïve recipients without the co-injection of mycobacterial antigens (32). All of these independent variables can be measured in experimental animals in order to evaluate factors involved in the induction or the expression of the immune response after mycobacterial infections. However, many of these variables are correlates of protective immunity, and thus only the transfer of immune T lymphocytes in challenged animals or in vitro microbicidal assays may give the final results on protection.

Activated blood-borne macrophages accumulating in the newly formed tubercle seem to be the major cells involved in acquired resistance against intracellular multiplying mycobacteria. However, the virulent tubercle bacilli may replicate within these cells or persist in a dormant state for a prolonged period. It has been shown that not all macrophages in a tubercle granuloma have the same function (45,53). The subcellular mechanisms that allow mycobacteria to resist intracellular killing in normal macrophages and the nature of the bacteriostatic (rather than bactericidal) mechanisms observed within the phagolysosomes of activated macrophages are still unknown (47). There is evidence of involvement of reactive oxygen, i.e., hydrogen peroxide (H_2O_2) in killing mycobacteria, and a good correlation was demonstrated between the capacity of host macrophages to produce H_2O_2 spontaneously after mycobacterial infection and the natural capacity, tested in vitro, of these macrophages to inhibit the growth of BCG (50). Experiments with both human and mouse cells, in which the criteria for activation were the enhancing macrophage capacity to release H_2O_2 and to kill protozoal and bacterial intracellular multiplying pathogens, give evidence that γ-interferon (γ-IFN) is the principal factor activating macrophages in infectious immunity (24,35). Douvas et al (14) showed that γ-IFN stimulates human macrophages to become leishman-icidal, but *M. tuberculosis* grew better in these stimulated macrophages. It is clearly important to discover which functional macrophage subset is involved in mycobacterial killing and how mycobacterial products, anti-bodies, and T cell products influence the distribution of microorganisms among subsets.

Thus an important question remains: What are the nature and functions of the macrophage subsets in which the clinically relevant mycobacteria are present during infection and their relation with immune T cells (29)? Furthermore, activation of macrophages alone may represent only a part of the acquired resistance mechanisms. Observations in the guinea pig, mouse, and man show that distinct effector T cell functions are not nec-essarily mediated by the same cells (45), and evidence shows that in *Lis-teria* infection distinct murine T cell subsets were involved in promoting immune granuloma formation associated with systemic protection and those involved in locally induced protection and γ-IFN production (34). Thus increasing knowledge of the heterogeneity of purified antigens, lym-phocytes, lymphokines, and macrophages forces us to reevaluate all ex-perimental models used to explain the classical pathway of CMI to intra-cellular bacteria, in order to develop better immunoprophylaxis tools. A careful analysis of factors involved in the induction of protective immunity to mycobacterial infection might help in such a way.

Factors Involved in Induction of Protective Immunity

Numerous models in various experimental animals have been used to evaluate protective immunity in mycobacterial diseases (6). At present, in vivo assays are more numerous and less discussed than in vitro assays. Mortality rate, evaluation of the kinetics of the growth of mycobacteria in the target organs, and immunopathological reactions are the tests used most often.

The only natural "experiments" showing protection performed in man are the BCG vaccination trials against tuberculosis or leprosy. Results from such trials against tuberculosis, as well as against leprosy, showed wide variations in protection, ranging from 0 to 80% (28). However, the design inconsistencies and the epidemiological bias of such trials have been so numerous that interpretation is difficult (8). Nevertheless, im-portant points have emerged from these trials and from subsequent ex-perimental studies: BCG vaccination has both succeeded and failed to protect simultaneously against tuberculosis and leprosy in different trails (43). BCG might fail to protect against secondary tuberculosis in areas where it does protect against both the disseminated phase of primary tu-berculosis and leprosy (other than multibacillary cases). As already discussed, the first point is in agreement with the fact that protection obtained by a mycobacterial strain can also be achieved by immunization with a different strain, as shown in experimental models (36).

Such results are in direct opposition concerning the hypothesis of the immunomodulating effect of various strains of atypical mycobacteria offered by Stanford et al. (52). It was considered that preexisting *Listeria* type of cellular hypersensitivity reaction normally elicited after *Mycobacterium nonchromogicum, M. vaccae,* and *M. Leprae* may be markedly boosted by BCG vaccination, leading to higher protection, whereas the *M. kansasii, M. scrofulaceum,* and *M. bovis* (strain Pasteur BCG) induced a Koch-type cellular hypersensitivity that may lead to a reduced level of protective immunity following BCG vaccination. This hypothesis was challenged using experimental models in guinea pigs (16) and mice (36). These studies showed that prior sensitization with almost all atypicaly mycobacteria, being responsible in the eight BCG trials, was able to actively protect the host after aerogenic challenge (38). The only alteration induced by prior sensitization with such atypical mycobacteria was in some case to mask the protective efficacy of the BCG, as little or no potentiation was observed when BCG was inoculated after preimmunization with atypical mycobacteria. Such masking effect was described by Palmer and Long two decades ago (39).

Moreover, such results support the hypothesis that common antigenic determinants possessed by different mycobacterial strains are relevant to the generation of protective cellular immunity and are either identical or closely cross-reactive in various mycobacterial strains. However, the acquired resistance to mycobacteria or other intracellular multiplying bacteria appears to be critically dependent on immunizing with live rather than dead organisms (38). The basis for the requirement of living organisms for the induction of specific acquired antimycobacterial immunity is not yet understood. It may depend on the characteristics of the bacteria and host factors, as it is speculated that parasitization of host macrophages by live proliferative organisms may allow the presentation of so-called replication bacterial antigens to host T lymphocytes in a fashion that most efficiently induces immune protection (48).

Mycobacterial Factors in the Induction of Protective Immunity

Live BCG induces an effective antituberculosis immunity whether introduced into tissues by the intravenous, subcutaneous, intraperitoneal, aerogenic, or oral route. Usually introduction of 10^5 to 10^6 viable BCG organisms intravenously induces a maximal degree of protection against a virulent *M. tuberculosis* challenge given by the same route 3 weeks after vaccination; but the presence of specific immune T cells can be detected as soon as 1 week *after* immunization when higher doses of vaccine are given (45). No protection was achieved when dead BCG in saline, whatever the dose given, was inoculated into mice (9,31). Not only do live micro-organisms need to be injected, but the bacteria must be able to proliferate in the host macrophages after inoculation in order to induce protection. This evidence was demonstrated by a comparison between two BCG strains; one being streptomycin-susceptible and the other streptomycin-resistant (SMY-res) (10). Only the former was able to induce antituberculous resistance. The inability of the latter to immunize mice after intravenous injection was shown to be associated with rapid inactivation of the in vivo injected population, so that viable counts in both liver and spleen decreased to subthreshold numbers before significant degrees of antituberculous resistance could be induced. Subcutaneous injection of BCG-SMY-res did not induce any lymphoproliferative response in the draining lymph node. Thus it seems that only live and proliferative BCG strains, which are able to resist nonspecific resistance, can induce acquired protection. This finding was further demonstrated using the same BCG vaccine in different strains of mice known to be naturally resistant or naturally susceptible to BCG (4). It was shown that the kinetics and the quality of the acquired immunity were directly related to the level of BCG load in the lymphoid tissues, which in turn is the result of the infecting dose and the host's genetic control against BCG. Human- or Armadillo-derived irradiated and/or heat-killed *M. leprae* inoculations in mice have been shown to induce immune protection against homologous organisms (46) or against *M. tuberculosis* (40) or *M. bovis,* strain (27) BCG. Cloned T cell lines were also obtained after in vivo immunization with irradiated *M. leprae* without any adjuvant (25). They exhibited helper-type activity, producing interleukin-2, macrophage-activating factor, and γ-IFN, and they were further characterized in terms of cross-reactivities with other species of mycobacteria. However, induction of cross-protection against BCG infections by irradiated *M. leprae* lymph node cells occurred only in particular strains of mice (i.e., in C57Bl/6 but not C3H mice) (27), indicating that the immune response pattern evoked after mycobacterial infections is determined partly by the organism used and partly by the genetic factors of the host.

Host Factors in Induction of Protective Immunity

Experimental studies indicate clearly that protection developed after a local infection differs from protection induced after a systemic challenge, depending on the capacity of mycobacterial strains to disseminate and on the properties of the mouse strains used. Moreover, differences in the clinical course of tuberculosis and leprosy in the human population and individuals have long been recognized (30). There are several lines of evidence that suggest that host genes do exist in humans in controlling the innate resistance and susceptibility to tuberculosis (48). Associations with certain HLA antigens have suggested a close association with the expression of leprosy rather than susceptibility or resistance to the infective organism (15). It is clear from experimental studies that many genes interact to determine the level of resistance in an individual. Experimental infectious murine models with intracellular multiplying organisms using outbred, inbred, congenic, recombinant, and mutant strains have been used to analyze such genetic control (6,45). Hence studies with mycobacteria have shown that inbred and outbred mouse strains can be classified into those that are naturally resistant and those that are naturally susceptible. It was of interest to note about the results of such studies that strains previously designated Ity^r (*Salmonella typhimurium* resistant) and Lsh^r (*Leishmania donovani* resistant) were also resistant to intravenous challenge with *M. bovis* (BCG) (19) or *M. lepraemurium* (5,26) and that strains designated Ity^s or Lsh^s were susceptible. Resistance was shown to be dominant and was expressed through control of a gene (or group of genes) located on chromosome 1. Although this gene(s) is not mapped, its expression seems to influence both early multiplication of the organisms within the macrophages (51) and later development of acquired resistance mechanisms (4,23).

In the case of the BCG study (19), differences in resistance were mostly revealed in the spleen after 3 weeks of a low inoculum (2×10^4 viable BCG/mouse) of well dispersed bacilli (strain Montreal) inoculated intravenously. More modest differences were observed in other organs or when higher doses were used. When BCG was inoculated subcutaneously (SC) with various doses (10^4 to 10^6), growth curve kinetics showed no significant differences in resistance in the draining node or at the inoculum site, except at day 1 after challenge (23). Differences in resistance were also noted when various BCG vaccine strains or atypical mycobacteria were used, being found to be related to the natural ability of such strains to replicate in the mouse (13). Usually low inocula of virulent, well dispersed BCG strain injected intravenously into Bcg^r mice showed little multiplication, whereas in Bcg^s mice progressive growth occurred, peaking at 3 weeks and then decreasing. Subsequently, lower numbers of viable bacilli were detected for several weeks in Bcg^s than in Bcg^r strains, a difference that was most marked in liver and lungs and was dose-dependent (29).

These results seem to indicate that the nonspecific natural resistance in Bcg^r mice is able to prevent multiplication and persistence of a low inoculum of an attenuated living strain of *M. bovis* without help from a T cell response. However, when a higher inoculum of BCG is given to Bcg^r mice (i.e., C3H), the mice were less able to prevent late multiplication and dissemination (in lungs). Classically, it was shown that mice that had recovered from the primary infection developed DTH to tuberculin and showed strong resistance to further challenge with BCG or virulent tubercle bacilli (9). However, when mice were inoculated intravenously with a low inoculum of BCG, only the Bcg^s mice developed a protracted DTH reaction to tuberculin (23,41). Furthermore, when several strains of mouse were inoculated subcutaneously with 4×10^6 BCG Pasteur and tested with tuberculin at varying times after sensitization, the time course patterns of such DTH reactions can be clearly separated into three groups, independent of the time after sensitization (23,29).

For all mouse strains tested, except Balb/c, a good correlation existed between natural susceptibility (Bcg^s strains) and the protracted DTH type. Early and intermediate types were observed in Bcg^r strains.

Acquired resistance after such immunization was then evaluated in Bcg^s and Bcg^r strains. Fourteen days after immunization, all mice, including controls, were inoculated intravenously with 10^5, 10^6, or 10^7 viable BCG, and 14 days later the number of viable BCG were counted in spleens. The resistance index was calculated for each mouse strain and compared to the tuberculin DTH level measured 18 or 42 hr after elicitation that has been given on the challenge day in other group of mice used only for DTH measurements without any challenge. A higher resistance index was found only in strains presenting a protracted DTH type. Morever, a linear correlation was found between the resistance index and the DTH levels measured 42 hr after elicitation. On the other hand, no such correlation was found in strains of mouse presenting an 18 hr reaction. This finding might be correlated with the fact that Bcg^r mice had only one-tenth the number of granulomas in liver and spleen compared to the Bcg^s mice (41).

The Balb/c mice constitute an exception to the observed correlation between Bcg gene expression (classified as Bcg^s) and DTH expression, developing the intermediate DTH reaction (1). When viable BCG were recovered from spleen in C57B1/6 and Balb/c mice at varying intervals after an intravenous injection of a low inoculum of BCG Pasteur, no major difference between strains was observed during the first 3 weeks after challenge, but later Balb/c mice were less able than C57B1/6 mice to clear the infection progressively (29).

Such impairment of the late clearance rate was shown to be radiosensitive, as increased clearance occurred in Balb/c mice that were sublethally irradiated (500 rads) 24 hours before the intravenous injection of 1×10^4 viable BCG. Moreover, such infected preirradiated Balb/c mice expressed higher levels of the 42-hour DTH reaction to tuberculin, compared to infected normal Balb/c mice, and it occurred as they recovered from the primary infection. No such radiosensitive phenomena were observed in other Bcg^s or Bcg^r strains of mouse tested.

Thus the disappearance of the 42-hour DTH reaction in Balb-c mice associated with a low clearance rate after the primary infection might be due to induction of an active particular subset of T cells with regulatory functions, observed here after a low inoculum of nondisseminating mycobacterial strain such as BCG. Similar results were also found after *M. lepraemurium* (2) and *Leishmania tropica* (22) infections.

A nylon-wool-adherent radiosensitive suppressor cell population in spleen was demonstrated in the case of *M. lepraemurium* infection, which was able to decrease resistance when injected into infected syngeneic mice

(2). Thy-1$^+$, L3T4$^+$, and I-J phenotype cells, capable of inhibiting the induction and expression of DTH as well as reversing the healing of lesions, were demonstrated in leishmanial infections (22). The susceptibility to irradiation of this suppressor T cell population in those infections, including the BCG, was not found in other Bcg^r strains of mouse. These suppressor T cells are thus different from those induced after a higher dose of BCG or after subcutaneous inoculation of *M. lepraemurium* (44). They might be triggered by different mechanisms, either directly or indirectly, through primary defects in macrophages or in a particular subpopulation of macrophage that is unable to present the protective antigen(s) to the right population of T lymphocytes. Thus depending on the time, the multiplication, and the persistence of the microorganisms, L3T4 lymphocytes accumulate and function as suppressor cells, explicating nonhealing lesions, dissemination, or low clearance rate after primary infection. The immune response may then vary from upgrading reactions (if L3T4 cells are prevented from functioning) to downgrading reactions (if mycobacteria continue to multiply and disseminate). Such variations in the immune response may resemble those observed in human borderline leprosy or in nonreactive intermediate tuberculosis. Other regulatory mechanisms have been described in Bcg^s and Bcg^r mouse strains, trying to explain acquired anergy, in vitro unresponsiveness, and dissemination of microorganisms (6,29,45). Usually they were produced after intravenous injection of high doses of nondisseminating mycobacteria (i.e., BCG) or after subcutaneous injection of disseminating mycobacteria (*M. lepraemurium, M. avium,* or *M. intracellulare*) in mice.

Lymphocyte trapping and induction of at least two types of regulatory suppressor cell (T lymphocytes and macrophages) may contribute to the nonspecific component of the defect, but these cells have never been shown to have a detrimental effect on the course of the infection, even if they were associated with the regulation of the immunopathological response. Therefore they may be irrelevant to the immune protection mechanisms.

Figure 7.1 shows the pathways of anergic and protective responses after mycobacterial infections in different strains of mice according to the route, the inoculum dose, and the possible host genetic control against BCG.

Conclusions

Conclusions drawn from experimental models of mycobacterial infections have been useful in the past for evaluating bacterial factors involved in the induction of CMI in general and particularly about protective immunity. Results obtained from BCG vaccine human trials on tuberculosis and leprosy that show geographic variations in the acquired protection have led to many hypotheses.

Two main hypotheses were developed during the 1960s concerning the

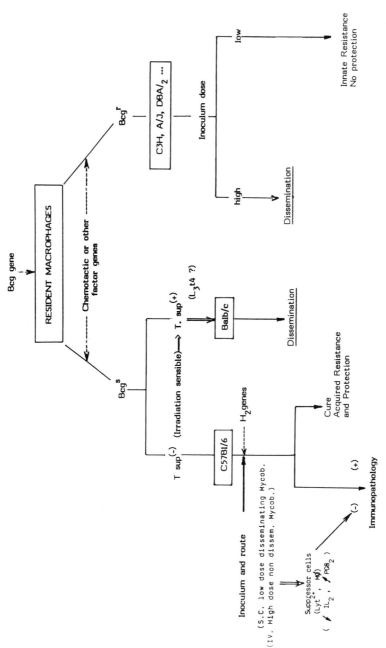

Fig. 7.1. Pathways of anergic, protective and immunopathologic responses after mycobacterial infections in different inbred strains of mice according to the route, the inoculum dose and the possible host genetic control against BCG.

potency of various daughter BCG strains and the influence of previous sensitization with atypical mycobacteria. However, careful analysis of such trials and experimental animal models showed that such hypotheses might not be involved in the results of the last trial, realized in South India (28), as the two vaccines for this trial were the best selected strains and the only effect of atypical mycobacteria might be to immunize partially and then mask differences among a vaccine- and placebo-inoculated population, both being sensitized to prior vaccination. Thus other factors must be studied to explain differences in protection against tuberculosis and leprosy after BCG vaccination. Among such factors, environmental and genetic factors might be involved in the resistance and susceptibility to mycobacterial diseases.

Apart from the fact that nutritional and socioeconomic factors might be co-agents in depressing immune responses and increasing infective contacts with pathogenic mycobacteria, interest in the genetic factors controlling immune responses after mycobacterial infections has increased in recent years (48). However, confusion arose in the literature concerning mostly the definition of resistance or susceptibility in mouse models. In general, the confusion results from the use of different species and strains of mycobacteria (without any definition of their virulence: their ability to disseminate in the host), various inoculum doses, different routes of administration, different expressions of resistance, and different patterns of CMI responsiveness or unresponsiveness. During the 1970s, Smith gave examples of such situations. When several laboratories used animal test systems for evaluating vaccines against tuberculosis, no two laboratories were found to use the same system (49). More controlled studies of this situation showed that each major variable in the animal test system (e.g., species, vaccination route and dose, interval between vaccination and challenge, challenge route and dose) contributed to inducing a difference in ranking order of the effectiveness of a serie of live attenuated vaccines. If one adds variables concerning the host genetic factors, increased confusion is not surprising. Thus there is an urgent need for further in vivo and in vitro studies in several areas.

Acknowledgments. This work was supported partly by the Immunology of Tuberculosis (IMMTUB) of the UNDP/World Bank/WHO Special Program for Research and Training in Tropical Diseases. The authors thank Mrs. J. Lortholary for typing the manuscript.

References

1. Adu HO, Curtis J, Turk JL: Differences in cell-mediated immune responses of "high-resistance" and "low-resistance" mice to a non pathogenic mycobacterium. *Scand J Immunol* 1981;14:467–480.

2. Alexander J: Adoptive transfer of immunity and suppression by cells and serum in early Mycobacterium lepraemurium infections in mice. *Parasitol Immunol* 1979;1:159–166.

3. Bloom BR: Learning from leprosy: a perspective on immunology and the Third World. *J Immunol* 1986;137:1–10.

4. Bourassa D, Forget A, Pelletier M, et al: Cellular immune responses to Mycobacterium bovis (BCG) in genetically susceptible and resistant congenic mouse strains. *Clin Exp Immunol* 1985;62:31–38.

5. Brown IN: Animal models and immune mechanisms in mycobacterial infections, in Ratledge C, Stanford JL (eds): *Biology of Mycobacteria.* London, Academic Press, 1983; vol 2, pp. 173–234.

6. Brown IN, Glynn AA, Plant JE: Inbred mouse strain resistance to Mycobacterium lepraemurium follows the Ity/Lsh pattern. *Immunology* 1982;47:149–156.

7. Chaparas SD: The immunology of mycobacterial infections. *CRC Crit Rev Microbiol* 1982;9:139–197.

8. Clemens JD, Chuong JJH, Feinstein AR: The BCG controversy: a methodological and statistical reappraisal. *JAMA* 1983;249:2362–2369.

9. Collins FM: Acquired resistance to mycobacterial infections. *Adv Tuberc Res* 1972;18:1–30.

10. Collins FM, Montalbine V: Relative immunogenicity of streptomycin-susceptible and resistant strains of BCG. II. Effect of the route of immunization on growth and immunogenicity. *Am Rev Respir Dis* 1975;111:43–51.

11. Collins FM: The immunology of tuberculosis. *Am Rev Respir Dis* 1982;25(suppl 3):42–49.

12. Daniel TM: The immunology of tuberculosis. *Clin Chest Med* 1980;1:189–201.

13. Denis M, Forget A, Pelletier M, et al: Control of the BCG gene of early resistance in mice to infection with BCG substrains and atypical mycobacteria. *Clin Exp Immunol* 1986;63:517–525.

14. Douvas GS, Lookes DL, Vatter AE, Crowle AJ: Gamma interferon activates human macrophages to become tumoricidal and leishmanicidal but enhances replication of macrophages associated mycobacteria. *Infect Immun.* 1985;50:1–8.

15. Edden W van, Gonzalez NM, De Vries RRP, et al: HLA linked control of predisposition to lepromatous leprosy. *J Infect Dis* 1985;151:9–14.

16. Edwards ML, Goodrich JM, Muller A, et al: Infection with Mycobacterium avium-intracellulare and the protective effect of bacille Calmette-Guérin. *J Infect Dis* 1982;145:733–741.

17. Forget A, Benoit JC, Turcotte R, Gusew-Chartrand N: Enhancement activity of antimycobacterial sera in experimental Mycobacterium bovis (BCG) infection in mice. *Infect Immun* 1976;13:1301–1306.

18. Forget A, Turcotte R, Benoit JC, Borduas AG: Enhancing effect of passive immunization with mycobacterial antibodies on humoral and cellular immunity in BCG-infected mice. *Ann Immunol (Inst Pasteur)* 1978;129G:255–265.

19. Gros P, Skamene E, Forget A: Genetic control of natural resistance to Mycobacterium bovis (BCG) in mice. *J Immunol* 1981;127:2417–2431.

20. Hahn H, Kaufmann SHE: The role of cell-mediated immunity in bacterial infections. *Rev Infect Dis* 1981;3:1221–1250.

21. Hawrylko E, Mackaness GB: The kinetics of lymphoïd proliferation in tuberculosis mouse spleen cells. *Cell Immunol* 1972;5:148–170.

22. Howard JG, Hale C, Liew FY: Immunological regulation of experimental cutaneous leishmaniasis. IV. Prophylactic effect of sublethal irradiation as a result of abrogation of suppressor T cell generation in mice genetically susceptible to Leishmania tropica *J Exp Med* 1981;153:557–568.

23. Hurtrel B, Hurtrel M, Lagrange PH: Genetic control of tuberculin time course in mice, correlation with natural and acquired resistance against BCG, in Skamene E (ed): *Genetic control of Host Resistance to Infection and Malignancy.* New York, Alan R. Liss, 1985, pp 305–312.

24. Kiderlen AF, Kaufmann SHE, Lohmann-Matthes ML: Protection of mice against the intracellular bacterium Listeria monocytogenes by recombinant immune interferon. *Eur J Immunol* 1984;14:964–967.

25. Kingston AE, Stagg AJ, Colston MJ: Investigation of antigen cross-reactivity of Mycobacterium leprae reactive murine T cell lines and clones. *Immunology* 1986;58:217–223.

26. Lagranse PH, Hurtrel B: Local immune response to Mycobacterium lepraemurium in C3H and C57Bl/6 mice. *Clin Exp Immunol* 1979;38:461–474.

27. Lagrange PH, Hurtrel B, Ravisse P, Grosset J: A single subcutaneous inoculation of 10^7 armadillo derived irradiated Mycobacterium leprae evokes different immunological behavior in C57Bl/6 and C3H mice. *Ann Inst Pasteur Microbiol* 1982;1338:167–168.

28. Lagranse PH, Hurtrel B, Brandely M, Thickstun PM: Immunological mechanisms controlling mycobacterial infections. *Bull Eur Physiopathol Respir* 1983;19:163–172.

29. Lagrange PH, Hurtrel B: Anergy and other immunological perturbances in mycobacterial infections: an overview, in Friedman H, Bendinelli M (eds): *Tuberculosis: Interactions with the Immune System.* New York, Plenum, 1988. pp 171–205.

30. Lagrance PH, Hurtrel B: Induction of protective immunity to mycobacterial infection. *Lepr Rev* 1986;57:231–244.

31. Lefford MJ: The effect of inoculum size on the immune response to BCG infection in mice. *Immunology* 1970;21:369–381.

32. Lefford MJ: Transfer of adoptive immunity to tuberculosis in mice. *Infect Immun* 1975; 11:1175–1181.

33. Lefford MJ: Induction and expression of immunity after BCG immunization. *Infect Immun* 1977;18:646–653.

34. Naher H, Sperling U, Hahn H: H_2k restricted granuloma formation by $Ly2^+$ T cells in antibacterial protection to facultative intracellular bacteria *J Immunol* 1985;135:569–572.

35. Nathan CF, Murray HW, Wiebe ME, Rubin BY: Identification of interferon gamma as the lymphokine that activates human macrophage oxidative metabolism and antimicrobial activity. *J Exp Med* 1983;158:670–689.

36. Orme IM, Collins FM: Infection with Mycobacterium kansaii and efficacy of vaccination against tuberculosis. *Immunology* 1983;50:581–586.

37. Orme IM, Ratcliffe MJH, Collins FM: Acquired immunity to heavy infection with Mycobacterium bovis, bacillus Calmette-Guérin, and its relationship to the development of non specific unresponsiveness in vitro. *Cell Immunol* 1984;88:285–296.

38. Orme IM, Collins FM: Cross protection against non-tuberculous mycobacterial infections by Mycobacterium tuberculosis memory immune T-lymphocytes. *J Exp Med* 1986;163:203–208.

39. Palmer CE, Long MW: Effect of infection with atypical mycobacteria on BCG vaccination and tuberculosis. *Am Rev Respir Dis* 1966;94:533–568.

40. Patel PJ, Lefford MJ: Specific and non specific resistance in mice immunized with irradiated Mycobacterium leprae. *Infect Immun* 1978; 20:692–697.

41. Pelletier M, Forget A, Bourassa D, et al: Immunopathology of BCG infection in genetically resistant and susceptible mouse strains. *J Immunol* 1982; 129:2179–2185.

42. Reggiardo Z, Middlebrook G: Failure of passive serum transfer of immunity against aerogenic tuberculosis in rabbits. *Proc Soc Exp Biol Med* 1974;145:173–175.

43. Rook GAW, Bahr GM, Stanford JL: The effects of two distinct forms of cell-mediated response to mycobacteria on the protective efficacy of BCG. *Tubercle* 1981;62:63–68.

44. Rook GAW: Suppressor cells of mouse and man—What is the evidence that they contribute to the aetiology of the mycobacterioses? *Lepr Rev* 1982;53:306–312.

45. Rook GAW: An integrated view of the immunology of the mycobacterioses in guinea-pigs, mice and men, in Ratledge C, Stanford JL (eds): *Biology of Mycobacteria.* London, Academic Press, 1983, vol 2, pp 279–319.

46. Sheppard CC, Walker L, van Landingham RM: Heat stability of Mycobacterium leprae immunogenicity. *Infect Immun* 1978;22:87–93.

47. Skamene E, Gros P: Role of macrophages in resistance against infectious diseases. *Clin Immunol Allergy* 1983;3:539–560.

48. Skamene E: Genetic control of resistance to mycobacterial infection. *Curr Top Microbiol Immunol* 1986;124:49–66.

49. Smith DW: Animal models for study immunity in infection disease. *J Infect Dis* 1973;128:800–801.

50. Stach JL, Delgado G, Tchibozo V, et al: Natural resistance to mycobacteria: antimicrobial activity and reactive oxygen intermediate releasing function of murine macrophages. *Ann Inst Pasteur Immunol* 1984;135D:25–37.

51. Stach JL, Gros P, Forget A, Skamene E: Phenotypic expression of genetically controlled natural resistance to Mycobacterium bovis. *J Immunol* 1984;132:888–892.

52. Stanford JL, Shield MJ, Rook GAW: How environmental mycobacteria may predetermine the protective efficacy of BCG. Tubercle 1981;62:55–62.

53. Suga M, Dannenberg AM, Higuchi S: Macrophage functional heterogeneity in vivo. *Am J Pathol* 1980;99:305–324.

54. Sultzer BM, Nilsson B: PPD tuberculin—a B cell mitogen. *Nature* 1972; 240n:198–200.

55. WHO: Immunological research in tuberculosis: memorandum from a WHO meeting. *Bull WHO* 1982;60:723–727.

CHAPTER 8

Geographic Variation in Vaccine Efficacy: The Polio Experience

T. Jacob John

Most of the currently available vaccines against a variety of infectious diseases were developed and tested in European or North American countries. With the popularization of routine infant and child immunization under the expanded and universal immunization programs of the World Health Organization, vaccines against tuberculosis, poliomyelitis, measles, diphtheria, whooping cough, and tetanus are increasingly being used in all nations of the world. Many national immunization programs have used or are currently using other vaccines as well, such as those against mumps, rubella, hepatitis B, meningococcal diseases. *Hemophilus influenzae* disease, Japanese encephalitis, and typhoid fever.

In most instances, when a vaccine is chosen by a prospective user country, its immunogenicity and protective efficacy as well as its safety are field-tested before adopting a national policy for its use. Such a rational approach has not been feasible in the case of most developing countries that adopted the expanded and universal immunization programs. The policy planners of the various countries assumed that the efficacy of a vaccine would be equal or near equal everywhere, as the hosts belong to one species. Although this approach is reasonable, there is evidence that the assumption is wrong in the case of some vaccines. In other words, there are geographic variations in vaccine efficacy in some cases.

Factors Determining Vaccine Efficacy

Vaccine efficacy is usually defined as the percent reduction of incidence of the target disease, the denominator being the disease incidence prior to immunization. To be precise, this measure may be called the *protective efficacy,* in contrast to the *immunogenic efficacy,* defined as the proportion of vaccinees who respond with the appropriate immune response to the vaccine. Here the denominator is the number of nonimmune vaccinees who were given the vaccine. Although immunogenic and protective efficacies are related, they need not be the same for a given vaccine. Where the measured immune response is indicative of protection from disease, the two efficacies may be nearly identical. Measles vaccine is an example; measles antibody response is synonymous with protection. In instances where the measured immune response is not indicative of protection, the two efficacies are at variance. BCG vaccine is an example; tuberculin sensitivity induced by BCG is the marker for immunogenic efficacy, but it is not synonymous with protection.

Geographic variation in the protective efficacy of BCG has been well recognized. The protective efficacy ranges from about 80% to 0% in different continents. In an elaborate study in southern India, no protective efficacy was recorded. Although part of the problem of defining protective efficacy of BCG has been the lack of definition of the target disease (i.e., primary tuberculosis versus secondary, or "adult-type," tuberculosis), there is no doubt that geographic variation does exist. It appears that the prevalence of mycobacteria other than *Mycobacterium tuberculosis* is a factor that affects BCG efficacy. Another factor may be the intensity of exposure to infection with *M. tuberculosis* itself.

The lessons to be learned from BCG are that the protective and immunogenic efficacies can be dichotomous and that epidemiologic factors may affect protective efficacy. The efficacy of measles vaccine depends on the prevalence of maternal antibody at the age at which the vaccine is given. There are geographic variations in the prevalence rates of maternal antibody; therefore there are differences in the immunogenic (and protective) efficacy when the vaccine is given at, for instance, 9 months of age. In North America the immunogenic efficacy would be relatively low, and in India the efficacy is high, reflecting the high and low prevalences of maternal antibody. This situation is another example of an "epidemiologic" factor affecting efficacy.

Geographic Variation and Efficacy of Oral Poliovirus Vaccine

In the United States, where the oral poliovirus vaccine (OPV) was developed, its protective efficacy after three doses is nearly 100%. In the U.S. experience with OPV from 1961 to 1986, no individual who had received three doses of the trivalent OPV subsequently developed paralytic poliomyelitis. The first report of OPV failure was in 1972 in India (6). Four preschool children, who had been given three doses of trivalent OPV in our pediatric clinic developed paralytic poliomyelitis of proved etiology 3 to 12 months after the last dose (6). The long time interval meant that it was not vaccine-virus-induced paralysis. These circumstances led to the hypothesis that OPV efficacy was markedly lower in India than in the western countries.

Since then hundreds of thousands of children in India have developed paralytic poliomyelitis despite swallowing three or four doses of OPV (1,5,16). In one outbreak in Madras, 10% of 516 children with poliomyelitis had earlier been given three doses of OPV (16). It may be estimated that about 10,000 to 30,000 children develop paralytic poliomyelitis *annually* in India, after having received at least three doses of OPV (12). In the face of such overwhelming evidence for OPV failure after three doses, neither the World Health Organization (WHO) nor the Ministry of Health of the Government of India recognized its occurrence. Their view was that such vaccine failure was only apparent, not real. Because OPV is temperature-sensitive and because developing countries tend to be in the tropical belt, it was assumed that OPV was reaching children after heat inactivation. Thus they adhered to the three-dose OPV schedule and concentrated on strengthening the cold chain.

The lower vaccine efficacy is not confined to India alone. In an outbreak of poliomyelitis in Taiwan, about 83 children developed paralysis despite taking three or more doses of OPV (13). OPV failure has also been recognized in some parts of Israel, and a sequential inactivated poliovirus vaccine (IPV)-OPV regimen was used to overcome vaccine failure (3).

It is unlikely that the geographic difference in vaccine efficacy is simply between tropical countries and temperate climate countries. That some tropical countries (e.g., Sri Lanka) have been able to control poliomyelitis using a three-dose OPV schedule indicates that the pattern of geographic variation is more complex. Moreover, those who argue against the existence of geographic variation in OPV efficacy have used examples such as Sri Lanka to support their view. In summary, experts are divided: There are those who accept and others who reject the phenomenon of geographic variation.

Geographic Variation and the Immunogenic Efficacy of OPV

Two considerations prompted the initiation of investigations into the immunogenic efficacy of OPV in Vellore. The first was the detection of vaccine failure, as stated earlier. The second was that by 1967 a few reports had already appeared showing that the antibody response of children to OPV was apparently lower in hot climates than in cold climates (2,14,15). In these studies the potency of OPV had not been checked; therefore it was not clear as to the real issue involved. Was antibody response poor because of poor potency of OPV (which is understandable in hot climates), or was it because of poor response despite giving fully potent vaccine? To illustrate this dilemma, there was a report on poor immunogenic efficacy of OPV in New Delhi in 1970 (4). On personal inquiry, the virologist (Dr. S. Balaya) confided that the vaccine given had very low potency. This information unfortunately did not find a place in the publication (4). Under these circumstances, it was essential to conduct systematic studies on the immunogenic efficacy of OPV (7–9).

As a result of these and other studies in India, it may be concluded that the immunogenic efficacy of OPV in India is about 75 to 80% after three doses, in contrast to more than 98% in the United States. Because immunogenic efficacy and protective efficacy have been found to be low, the conclusion is inevitable that the low immunogenic efficacy contributes to the low protective efficacy. As far as we know, serum antibody is the only marker of protection against the disease of paralytic poliomyelitis. Therefore, as in the case of measles vaccine, the immunogenic and protective efficacies of OPV are expressions of the same thing.

It is indeed curious that the national policy of three-dose immunization with OPV in India has not been field-tested. With the available data on the prevalence of poliomyelitis in Bombay, the protective efficacy of three dose of OPV is estimated to be between 63 and 80% (11). This estimate is strikingly close to the immunogenic efficacy of 75 to 80%.

However, the controversy does not end here. Because there is incontrovertible evidence to show that there is geographic variation in the immunogenic efficacy of OPV, and that in India the efficacy is lower than elsewhere, the national policy of a three-dose schedule does not appear to be on a sound basis. There is no logical answer to the argument of the protagonists of the three-dose schedule that the protective efficacy might be equal to that in the United States whereas the immunogenic efficacy is low.

Reasons for Low Immunogenic Efficacy of OPV in India

Unlike all other vaccines, which are injected, OPV is given by mouth. The vaccine viruses infect the alimentary tract, and an immune response results. Infection by vaccine virus is referred to as "take." We investigated the reasons for low immunogenic efficacy and found that the "take" rate was low; in other words, the difference between Western and Indian infants lies in their susceptibility to infection by vaccine viruses and not in their immune competence (8).

In the course of these investigations, it was found that each dose of OPV was an independent stimulus, unlike the sequential doses of other vaccines. No other live vaccine is given in several doses; only toxoids and killed vaccines are given in multiple doses. In their case the first dose "sensitizes" and the subsequent doses "consolidate" or "enhance" the immune response, building on the immunologic memory. The mechanism of action of the sequential doses of OPV is entirely different. The first dose of OPV may or may not infect the recipient with a given type of vaccine virus, e.g., type 1 poliovirus. The second dose of OPV is effective (as far as type 1 is concerned) only in the infant who had not responded to the first dose. Here the infant may or may not get infected with type 1, as in the case of the first dose. The third dose, then, is necessary only for the infant who did not get infected following the first two doses. Thus sequential doses merely increase the probability of vaccine virus take and the consequent immune response. For this simple reason, doses over and above the usual three improve the immunogenic efficacy.

It is for this reason that the recommendation to give five doses of OPV during infancy was formulated (10). The only way to test if five doses improve the protective efficacy is to conduct field tests to compare the efficacies of three- and five-dose schedules of OPV.

References

1. *Annual Reports of the Indian Council of Medical Research Enterovirus Laboratory.* Bombay, 1972–1981.
2. Drozdov SG, Cockburn WC: The state of poliomyelitis in the world, in *Proceedings of the 1st International Conference on Vaccines Against Viral and Rickettsial Diseases of Man.* Washington, DC, Pan American Health Organization, 1969, pp 198–208.
3. Gerichter CB, Lasch EE, Sever I, et al: Paralytic poliomyelitis in the Gaza strip and West Bank during recent years. *Dev Biol Stand* 1978;41:173–177.
4. Ghosh S, Kumari S, Balaya S, et al: Antibody response to oral polio vaccine in infancy. *Indian Pediatr* 1970;7:78–81.
5. Gujral VV, Narayanan I, Dutta AK: *The Underprivileged Child.* New Delhi, Lady Hardinge Medical College. 1981, pp 56–58.
6. John TJ: Problems with oral poliovaccine in India. *Indian Pediatr* 1972;9:252–256.
7. John TJ, Jayabal P: Oral polio vaccination of children in the tropics. I. The poor seroconversion rates and the absence of viral interference. Am J Epidemiol 1972;96:263–269.
8. John TJ: Oral polio vaccination of children in the tropics. II. Antibody response in relation to vaccine virus infection. Am J Epidemiol 1975;102:414–421.
9. John TJ, Christopher S: Oral polio vaccination of children in the tropics. III. Intercurrent enterovirus infections, vaccine virus take and antibody response. *Am J Epidemiol* 1975;102:422–428.
10. John TJ: Antibody response of infants in tropics to five doses of oral polio vaccine. *Br Med J* 1976;1:811–812.
11. John TJ: Poliovaccines and their use in developing/tropical countries, in *Proceedings of the Second International Seminar on Vaccinations in Africa.* Paris, Association for the Promotion of Preventive Medicine, 1981, pp 81–93.
12. John TJ: Poliomyelitis in India: prospects and problems of control. *Rev Infect Dis* 1984;6:S438–S441.
13. John TJ: Poliomyelitis in Taiwan: lessons for developing countries. *Lancet* 1985;1:872.
14. Lee LH, Wenner HA, Rosen L: Prevention of poliomyelitis in Singapore by live vaccine. *Br Med J* 1964;1:1077–1080.

15. Poliomyelitis Commission, Western Region Ministry of Health, Nigeria: Poliomyelitis vaccination in Ibadon, Nigeria during 1964 with oral vaccine. *Bull WHO* 1966;34:865–876.
16. Sundaravalli N, Narmada R, Nedunchezian S, Mukundan P: Spurt in poliomyelitis in Madras. *Indian Pediatr* 1981;18:539–544.

Part IV
Hepatitis

CHAPTER 9

Molecular Immunology of Viral Antigens in Hepatitis B Vaccination

Soumitra Roy and Girish N. Vyas

Jaundice has been historically noted in association with both sporadic and epidemic disease, but it was not until 1839 that it was recognized to be a result of hepatocellular necrosis. The series of experimental inoculations into humans with infectious sera and subsequent transmission in animal models (marmosets and chimpanzees) established the existence of two distinct hepatotropic viruses. Hepatitis A virus (HAV) and hepatitis B virus (HBV) produce clinically indistinguishable acute hepatocellular necrosis. Viral hepatitides negative for the serologic markers of HAV or HBV infection have been termed non-A non-B (NANB). Among the heterogeneous etiologies, a proportion of such NANB disease may be ascribed to antigenically distinct agents that share nuclei acid homology with HBV (7). Because of the lack of a reliable laboratory test, the NANB virus continues to be enigmatic and a major cause of posttransfusion hepatitis (4,25). The epidemic NANB hepatitis prevalent in India and Russia is due to a distinct etiologic agent. We assess here the current concepts and information about biology (genes and antigens) of the HBV and current vaccines, and the future prospects in worldwide control of HBV infection.

Prevalence

Hepatitis B virus is one of the most important causative agents of liver disease worldwide. The prevalence of HBV carriers varies greatly, being low in Europe and North America and having a 6 to 12% incidence in China and South-east Asia and one as high as 15% in some areas of West Africa. In the United States carriers number about a million; occupationally acquired HBV infection causes 1 to 2% of practicing physicians to be carriers of the virus. In India we estimate that at least 15 million to 20 million people are carriers of HBV in comparison with 200 million people in the world.

An important aspect of HBV infection is the development of the chronic carrier state. Acute hepatitis B completely resolves in 90% of cases; 1 to

2% of patients develop fatal fulminant hepatitis, and up to 10% become chronically infected with varying rates of progressive liver disease ranging from asymptomatic through chronic persistent to chronic active disease. In contrast, HBV infection leads to a chronic carrier state in at least 85% of infected neonates (1). However, neonates respond favorably to hepatitis B vaccination by producing high levels of antibodies to the hepatitis B surface antigen (HBsAg). Sex also appears to play a part; for reasons not understood, men are more likely than women to become chronic carriers.

Perhaps the most significant potential consequence of the chronic carrier state is the development of hepatocellular carcinoma. In Taiwan, where hepatocellular carcinoma (HCC) is the most common neoplasm, the persistence of HBsAg in the serum is associated with a 200-fold increased risk of acquiring HCC (2).

The epidemiologic surveys by the Centers for Disease Control (CDC) indicate that about 200,000 HBV infections occur each year in the United States. About 50,000 people get acute hepatitis with jaundice, and 12,000 to 20,000 become carriers of HBV. Chronic hepatitis B is estimated to cause 4000 deaths due to cirrhosis and 800 deaths due to primary hepatocellular carcinoma per year. At least 750,000 Americans are among the 200 million persons in the world estimated to be chronic carriers of HBsAg and serve as an epidemiologic reservoir of HBV infection. For control of this problem in the United States and worldwide, the vaccines containing HBsAg provide a real hope. The major constraints at the present time are availability and high cost.

Modes of Spread

Seroepidemiologic studies indicate that the major modes of spread are venereal, oral, and perinatal. HBsAg has been detected in saliva, semen, vaginal secretions, tears, and breast milk. Intimate contact with carriers probably accounts for most of the cases encountered in a nonhospital setting. Homosexual men, sexually promiscuous heterosexuals, and heterosexual contacts of patients or carriers, especially those in whom the hepatitis B e antigen (HBeAg) can be detected, constitute high risk groups. Vertical transmissions of the virus from an infected mother can taken place in utero or during delivery. An infant born to an HBeAg-positive mother has about a 90% chance of becoming infected.

Despite screening donated blood for HBsAg, hepatitis B accounts for at least 10% of posttransfusion hepatitis in the United States. This group probably represents donors with undetectably low levels of HBsAg. Hepatitis B is also an occupational hazard for hospital personnel, especially certain high risk categories such as dialysis technicians, emergency room nurses, laboratory personnel, intensive-care nurses, and housekeeping workers. About 13% of hospital personnel who accidentally stick themselves with needles contaminated with HBsAg-positive blood become infected. Intravenous drug abusers who share needles are also at high risk for contracting the virus.

Structure of HBV Antigens and Genes

Although HBV has been propagated only relatively recently in tissue culture, considerable knowledge has been gained from molecular cloning of HBV DNA and from animal models for HBV (chimpanzee) and for HBV-like agents found in woodchucks, squirrels, and pekin ducks. These hepatotropic DNA agents share certain biologic properties, and are termed hepadna viruses (26,31). Based on current information we provide a model for HBV in Figure 9.1.

HEPATITIS B VIRUS

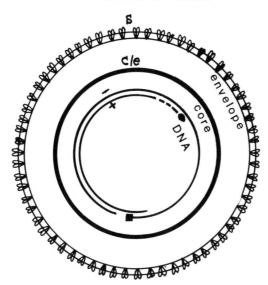

Fig. 9.1. Hepatitis B virus, with the envelope carrying the hepatitis B surface antigen(s) and the core carrying the hepatitis B core antigen (HBcAg) (C). The envelope consists of a lipid bilayer embedding 50 to 60 dimeric subunits of the virally encoded 25,000-dalton protein and its glycoprotein derivative of 29,000 daltons. Upon removal of the envelope with nonionic detergents, the HBcAg becomes detectable. When the cores are denatured by sodium dodecyl sulfate, the HBcAg activity is lost and hepatitis B e antigen activity is detected in the core peptides. The circular double-stranded DNA is contained within the hepatitis B virus core.

Hepatitis B Antigens

HBsAg-positive plasma contains abundant spherical and filamentous particles with a 22 nm diameter and which are devoid of nucleic acids; the plasma also contains a few of the 42-nm Dane particles (complete virion) consisting of the 27-nm core fully enveloped by the coat protein bearing the HBsAg. The core contains double-stranded DNA with a covalently attached protein and a DNA polymerase. If it were not for the abundant HBsAg detectable by immunoassays, it would be impossible to identify infected individuals. The infectivity of contaminated plasma is higher than the lowest amount of HBsAg that can be detected by radioimmunoassays. When the HBsAg is purified and concentrated, the Dane particles can be treated with nonionic detergents to expose the core antigen. Purified cores, when treated with sodium dodecyl sulfate, generate the HBeAg with concomitant loss of the HBcAg. Thus HBcAg is considered a conformational determinant of the assembled cores, and the HBeAg is considered a marker of the structural protein or the polypeptide encoded by the viral genome (14). However, HBsAg-positive donors are currently the best resource for the preparation of a safe and effective vaccine against hepatitis B virus infection.

Viral DNA Cloning and Sequencing

Following a DNA polymerase reaction, the fully double-stranded DNA was extracted from the concentrates of Dane particles derived from infected human plasma. The DNA extract was used for cloning HBV DNA in *Escherichia coli* using a single EcoR1 site. Analyses of the cloned HBV DNA using several restriction enzymes revealed a remarkable genetic heterogeneity, including differences in the restriction pattern within the same subtypes (27). The genetic heterogeneity was evident in the restriction patterns of the cloned DNA molecules and in the nucleotide sequence performed for the *ayw* serotype by Galibert et al (9) and the *adw* serotype [presumed from restriction analysis of Siddiqui et al (27)] performed by Valenzuela et al (35). In addition, Pasek et al (21) cloned and sequenced parts of the HBV genome. For a review of the extensive information on the antigens and genetic heterogeneity of HBV, refer to Tiollais et al (33). A simplified genetic map of HBV is shown in Figure 9.2. The circular structure of HBV molecules is maintained by the base-pairing of the 5′ extremities of the two strands over a length of about 300 nucleotides. The genetic map in Figure 9.2 is oriented in a clockwise mode using the single EcoR1 site as a point of reference. The minus-strand has the 5′ end located at a fixed position around nucleotide 1,800 with a nick or gap between 1800 and 1818 nucleotides (21). The biologic role of the protein covalently attached at the 5′ end is unknown, but it is possible that the nick may be involved in the integration of HBV DNA in the host genome. The 5′ end of the plus-strand is located at nucleotide 1560 and the 3′ end at a variable

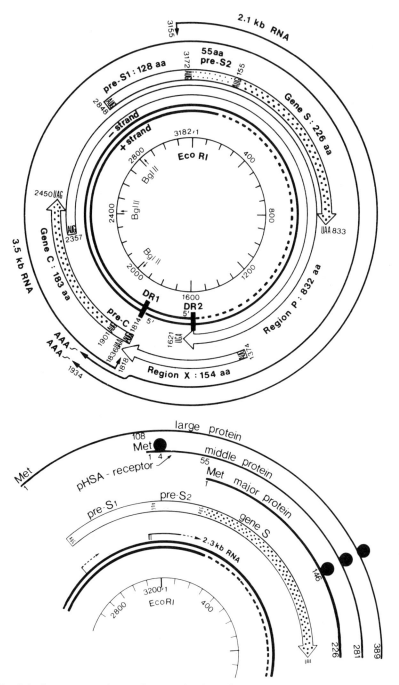

Fig. 9.2. Structure and genetic organization of the HBV genome **(top)** and region pre-S + gene S of HBV **(bottom)**. (From Michel, M-L, Tiollais P: Structure and expression of hepatitis B virus genome. *Hepatology* 7:615, 1987, © by Am. Assoc. for the Study of Liver Diseases.)

position in different molecules. The endogenous DNA polymerase functions to elongate the plus-strand to its full length.

The complete sequencing of the nucleotides revealed that HBV DNA has approximately 3200 base-pairs. The minus-strand of virion DNA contains four open reading frames based on the three reading phases of the nucleotide sequence and the position of the AUG start codon and the TAA, TGA, or TAG stop codons. A comparison of the nucleotide sequence of the reading frames with the available information on the gene products HBsAg and HBcAg (molecular weight, amino acid composition and sequence) allowed identification and localization of these viral genes in the viral genome (Fig. 9.2). These coding regions include the region that codes for HBsAg (gene S) and a contiguous region upstream termed presurface region (preS gene), the region that codes HBcAg (gene C), and two hypothetical regions coding for DNA polymerase (gene P) and the X protein (gene X). Through extensive overlapping of gene P with genes preS/S, X, and C (Fig. 9.2), the small HBV genome can code for all four genes that account for the known functions of the virus. The pre-S is further divided into pre-S1 and pre-S2, the latter encoding 55 amino acids upstream of the HBsAg sequence. The pre-S2 region contains a binding site for polymerized human serum albumin (pHSA) and an antigenic determinant that has been shown to be important in evoking an immune response to HBsAg in the genetically nonresponder strains of mice (18,19). How important it would be in overcoming nonresponsiveness to HBsAg vaccine in a small group of 5 to 10% of vaccinees has yet to be determined (see below).

Based on the amino-terminal and the carboxy-terminal amino acid sequence in the structural polypeptide of the envelope protein (22), the gene S was precisely located in the nucleotide sequence of HBV (34). Although the amino acid sequence of the core protein was not determined, the C gene was localized by expression of the product encoded by the open reading frame of the gene defined by the nucleotide sequence determination (14,21).

Replication of Hepatitis B Virus Genome

Based on work with the Pekin duck hepatitis virus (15,31), replication of HBV is strikingly different from all other animal DNA viruses and involves the reverse transcription of an RNA intermediate by a virus-encoded reverse transcriptase. The life cycle is typical for RNA-containing retroviruses, many of which are tumor viruses (36).

Upon internalization into host cells, the infecting viral genome is made double-stranded, and it serves as a template for the transcription of its minus-strand into a full-length viral RNA molecule, which serves as either messenger RNA or as template ("pregenome") for the reverse transcription into minus-strand DNA. These minus-strand DNA species are then made partially double-stranded to give mature virion DNA. Viral DNA replication is therefore asymmetric, with separate pathways for plus- and minus-strand synthesis, and it involves the reverse transcription of an RNA intermediate. In addition to the Pekin duck hepatitis virus (15,31), this mechanism has now been demonstrated to be operative in ground squirrel hepatitis virus (44) and human HBV (5,8) as well, indicating that this mode of replication is central to the life cycle of all hepadna viruses.

Approaches to Vaccine Development

HBsAg Vaccine from Plasma

The human experiments of protection afforded by immunization with a 1:10 dilution of HBsAg-positive serum after heat inactivation of 100°C for 1 minute laid the foundation for use of HBsAg as a vaccine against HBV infection (12). The 22-nm particles of envelope protein isolated from the plasma of HBsAg carriers was demonstrated by Purcell and Gerin (24) to be a useful vaccine. Since then, a large body of data on clinical trials, safety, and efficacy of the HBsAg vaccines has been published in the proceedings of several symposia (16,32,38,41). We have succinctly summarized the current knowledge and prospective improvements of HBsAg vaccine in the control of HBV infection (42). In the United States a safe and effective vaccine produced by Merck, Sharp & Dohme (MSD) has been licensed for clinical use; similarly, in Europe vaccines made by Pasteur Production of France as well as MSD have been licensed. Both of these vaccines utilize highly purified HBsAg following inactivation with formalin and using alum as adjuvant. The Central Laboratory of The Netherlands Red Cross (CLRC) employed a different process of purification of HBsAg with heat inactivation at 101°C for 90 seconds and the final product heated at 65°C for 11 hours. This vaccine is also safe and effective in the clinical trials reported at the 1982 WHO/IABS Symposium in Athens. Thus MSD, Pasteur, and CLRC vaccines containing 20, 5, and 3 μg of HBsAg per dose, respectively, are highly immunogenic, safe, and effective in providing sustained protection against acute and chronic hepatitis B. Although long-term follow-up is not adequate, most of the vaccinees have serologic evidence of anti-HBs for as long as 3 years and several of the early vaccinees have shown protection and anti-HBs for a period exceeding 5 years. Most importantly, HBsAg of one subtype is protective against HBV infection of other subtypes. It must be emphasized that the inactivation procedures have been shown to destroy infectivity of all of the viruses tested, which is strong ensurance that no transmissible agent possibly present in human plasma could escape these inactivation procedures.

Intramuscular injections of the vaccine are given in three doses over a period of 6 months. Following the first dose, more than 50% of the vaccinees produce serologic evidence of anti-HBs response. The second dose of the vaccine is given 1 to 2 months later, and a third dose is administered 6 months after the primary immunization. After completion of the course of three doses, more than 95% of the individuals are successfully immunized. The nonresponders and the hyporesponders may be given a second set of injections. At the present time there is no recommendation for the management of nonresponders.

The perinatal transmission of HBV infection can be interrupted in 75% of the babies by hepatitis B immune globulin (HBIG) given at birth, within the first 48 hours, and at intervals of 1 to 3 months for the first year of life (29). Because the infants and babies responded well to the hepatitis B vaccine trials conducted in Africa, Asia, and Europe, several clinical trials are now in progress to evaluate the efficacy of HBIG and hepatitis B vaccine in passive–active immunization of the high risk neonates born to HBeAg-positive carrier mothers. These studies promise a means of controlling propagation of the chronic carrier state. We await the outcome of these clinical trials with high expectation. In China every day more than 1800 newborns become chronic carriers, thus perpetuating the epidemiologic reservoir of HBV infection. The risk of infection is virtually 100% in babies born to HBeAg-positive Chinese women (6% of the population). Hence a vaccine is urgently needed for mass-scale use in China and must be produced from indigenous plasma resource. For control of this problem in the world, the vaccines containing HBsAg provide a real hope. The major constraints at the present time are availability and high cost.

The HBsAg carrier's plasma is a unique resource for large-scale production of the vaccine. The immunochemical characteristics of HBsAg (39) are summarized in Table 9.1. Because the plasma of the most infectious donors (those HBeAg-positive) often contains the highest amount of HBsAg, it is the best resource for isolating HBsAg. Because the turn-

Table 9.1. Properties of HBsAg derived from human plasma.

HBsAg in HBeAg+ plasma	500.0 mg/L
HBsAg in anti-HBe+ plasma	50.0 mg/L
Half-life in humans	3.3 days
Excretion of catabolized HBsAg in urine	25.0 mg/day
Common antigenic determinants most immunogenic in humans	a (subtypes a_1, a_2^1, a_2^3, a_3), Re
Other antigenic determinants poorly immunogenic in humans	d/y, w/r, g, j, k, q, t, x
Composition of viral envelope purified from plasma	
Lipids	22.5%
Carbohydrates	7.5%
Proteins	70.0%
Polypeptide P1	25,000 daltons
Polypeptide P2 (glycosylated P1)	29,000 daltons
Polypeptide P6 (albumin + HBsAg)	68,000 daltons

HBeAg+ = hepatitis Be antigen-positive; anti-HBe+ = positive for antibody to HBeAg.

Table 9.2. ACIP recommended treatment.

	HBIG		Vaccine	
Exposure	Dose	Recommended timing	Dose	Recommended timing
Perinatal	0.5 ml i.m.	Within 12 hours of birth	0.5 ml (10 μg) i.m.	Within 7 days[a]; repeat at 1 and 6 months
Percutaneous	0.06 ml/kg i.m. or 5 ml	Single dose within 24 hours	1.0 ml (20 μg) i.m.[b]	Within 7 days[a]; repeat at 1 and 6 months
		or[c]		
	0.06 ml/kg i.m. or 5 ml for adults	Within 24 hours; repeat at 1 month	—	—
Sexual	0.06 ml/kg i.m. or 5 ml for adults	Within 14 days of sexual contact	[d]	—

[a]The first dose can be given the same time as the HBIG dose but at a different site.
[b]For persons under 10 years of age, use 0.5 ml (10 μg).
[c]For those who choose not to receive hepatitis vaccine.
[d]Vaccine is recommended for homosexually active men and for regular sexual contacts of chronic HBV carriers.

over rates are rapid (3.3. days), carriers can be repeatedly plasmapheresed to obtain a sustained supply of carrier plasma. The safety and efficacy of current hepatitis B vaccines renders the attempts to make vaccines of component polypeptides of HBsAg less practical and unattractive.

The CDC Committee on Immunization practices has recommended the schedule of prophylaxis outlined in Table 9.2.

Expression of Cloned HBV DNA Gene Products

The cost of screening donors, selecting suitable donors for plasmapheresis, and processing the plasma for isolating HBsAg increases the cost of the HB vaccines. Therefore a second generation of technology to produce HBsAg in vitro by recombinant DNA techniques or cell culture techniques is being actively pursued, as shown in Table 9.3. If these alternative means of producing HBsAg prove to be more economical than production from plasma, a wider use of the vaccine is possible. Particularly in Africa and Asia, control of the perinatal transmission and chronic carrier state is most warranted for worldwide strategies to control HBV infection, chronic liver disease, and primary hepatocellular carcinoma, and it may be provided by a recombinant vaccinia virus with an HBsAg gene inserted in the thymidine kinase region (28).

A yeast-recombinant hepatitis B vaccine has been licensed by the U.S. Food and Drug Administration and is now available (30). To assess the

Table 9.3. Current approaches to hepatitis B vaccine production.

Source	Form	Status
Plasma	Subviral HBsAg particles	} Licensed
	Micelle	} Investigational
	Polypeptide	
Recombinant, plasmid		
Yeast	} Subviral HBsAg particles	} Licensed
E. coli		
Mammalian cells	} Subviral HBsAg particles	} Investigational
Recombinant, viral		
Vaccinia	Viccinia virus	}
SV-40 in mammalian cells	Subviral HBsAg particles	} Investigational
Hepatoma cell line	Subviral HBsAg particles	Investigational
Anti-idiotypic antibodies	Internal image of HBsAg	Investigational
Synthetic	Peptides	Investigational

efficacy of the yeast-recombinant vaccine, we administered the vaccine in combination with hepatitis B immune globulin to high-risk newborns. If infants whose mothers were positive for both HBsAg and the e antigen receive no immunoprophylaxis, 70 to 90% become infected with the virus, and almost all become chronic carriers. Among infants in this study who received hepatitis B immune globulin at birth and three 5-μg doses of yeast-recombinant hepatitis B vaccine, only 4.8% became chronic carriers—a more than 90% level of protection and a rate comparable with that seen with immune globulin and plasma-derived hepatitis B vaccine. These data suggest that in this high-risk setting the yeast-recombinant vaccine is as effective as the plasma-derived vaccine in preventing HBV infection and the chronic carrier state.

Peptide Analogues of HBsAg

When the amino acid sequence of the envelope protein was derived from the nucleotide sequence of the S gene, a sequence of 226 residues was obtained. The molecular weight of this protein was 25,400 daltons and close to the original molecular weight determination of 25,000 daltons observed by Vyas et al (37). From predictive structural analyses of the sequence shown in Figure 9.3, several investigators have synthesized peptide analogues of the sequence and determined the immunogenicity of various synthetic oligopeptide analogues of HBsAg (HBsPA) by injection into experimental animals (3,6,13,20,23). It is anticipated that one or more of these HBsPAs producing immune response to the common *a* determinant of HBsAg, similar to the physiologic response after infection or immunization, may provide a synthetic vaccine useful in active immunization. Most recently we have demonstrated that a dimeric form of HBsPA(139–147), after coupling it with tetanus toxoid, produced an excellent immune

```
  1  Met Glu Asn Ile Thr Ser Gly Phe Leu Gly Pro Leu Leu Val Leu Gln Ala Gly Phe Phe

 21  Leu Leu Thr Arg Ile Leu Thr Ile Pro Gln Ser Leu Asp Ser Trp Trp Thr Ser Leu Asn

 41  Phe Leu Gly Gly Ser Pro Val Cys Leu Gly Gln Asn Ser Gln Ser Pro Thr Ser Asn His

 61  Ser Pro Thr Ser Cys Pro Pro Ile Cys Pro Gly Tyr Arg Trp Met Cys Leu Arg Arg Phe

 81  Ile Ile Phe Leu Phe Ile Leu Leu Leu Cys Leu Ile Phe Leu Leu Val Leu Leu Asp Tyr

101  Gln Gly Met Leu Pro Val Cys Pro Leu Ile Pro Gly Ser Thr Thr Thr Ser Thr Gly Pro

121  Cys Lys Thr Cys Thr Thr Pro Ala Gln Gly Asn Ser Met Phe Pro Ser Cys Cys Cys Thr

141  Lys Pro Thr Asp Gly Asn Cys Thr Cys Ile Pro Ile Pro Ser Ser Trp Ala Phe Ala Lys

161  Tyr Leu Trp Glu Trp Ala Ser Val Arg Phe Ser Trp Leu Ser Leu Leu Val Pro Phe Val

181  Gln Trp Phe Val Gly Leu Ser Pro Thr Val Trp Leu Ser Ala Ile Trp Met Met Trp Tyr

201  Trp Gly Pro Ser Leu Tyr Ser Ile Val Ser Pro Phe Ile Pro Leu Leu Pro Ile Phe Phe

221  Cys Leu Trp Val Tyr Ile
```

Fig. 9.3. Amino acid sequence derived from the nucleotide sequence of the gene S. The sequence corresponds to the HBsAg/adw subtype. One of the two hydrophilic regions between residues 110 and 156 has been shown to contain the major antigenic specificities. The *a* determinant common to all serotypes is determined by a minimum sequence of 139 to 147 residues, whereas the *d/y* specificities are determined by a minimum sequence of 125 to 137 residues.

response specific for HBsAg/*a* (40). The hydrophilic part of the amino acid sequence 110–158 in Figure 9.3 contains the HBsPA(122–137), and the HBsPA(139–147) was shown to contain the *y* determinant and the *a* determinant of HBsAg by Melnick et al (17) and Bhatnagar et al (3), respectively. We propose to evaluate the possibility of immunointervention in the chronic carrier state using the dimeric form of HBsPA(139–147) coupled with tetanus toxoid, with or without an additional cyclic form of HBsPA(122–137), which also contains a part of the *a* determinant as shown by Melnick et al (17). In addition to the possible clinical usefulness of the HBsPAs in vaccination or immunomanipulation of chronic carriers of HBsAg (40), HBsPAs may have utility in newer diagnostics in concert with monoclonal antibodies to HBsAg or HBcAg. The potential application of these new approaches are of great interest to medical virologists (43).

The sequence of 19 amino acids (14 to 32 from the N-terminus) of the pre-S$_2$ region, representing an area of high local hydrophilicity, is shared by all HBV strains. A synthetic peptide analogue of this pre-S$_2$ region has proved to be immunogenic in chimpanzees and protective from challenge with 10^6 chimp infectious dose of HBV (10). Chemical synthesis of the oligopeptides mimetic of the epitopes of HBsAg, particularly the common *a* determinant(s), offer yet an alternative third generation of vaccines. We find this alternative even more attractive than other forms of HBsAg because of its potential application in the termination of the chronic carrier state. Not only can a dimeric form of HBsAg peptide analogues (HBsPAs) coupled with tetanus toxoid serve as a vaccine, it can conceivably terminate the chronic carrier state by overcoming the specific immunologic tolerance to HBsAg. A safe and effective outcome of this pursuit may bring to fruition the purpose of our sustained and progressive investigations of the molecular immunology of HBsAg.

Prospects

Imaginative experiments with the use of anti-idiotypic antibodies as a vaccine have been carried out by Kennedy et al (11). Similarly, Itoh et al (10) have demonstrated the prophylactic value of a 19-amino-acid synthetic peptide analogue of pre-S_2. However, the practical utility of such innovative approaches for mass vaccination remain to be substantiated. For the present time only the plasma-derived and yeast-produced vaccines can be considered useful for mass vaccination.

Among the 80% of the 200 million carriers of HBsAg who reside in Southeast Asia, perinatal transmission accounts for a major proportion of carriers. Interruption of the chronic carrier state can be achieved effectively by a combination of passive–active immunization with HBIG and HBsAg vaccination. The opportunity to prevent perpetuation of HBV infection in highly endemic areas warrants a vigorous worldwide prophylaxis program focused on Asia and Africa. However, the major impediment is the cost of current vaccines. Although various methods, including recombinant DNA methodology, may provide an alternate vaccine at a minimum cost, it is possible for the Southeast Asian countries to locally manufacture HBIG and HBsAg derived from the plasma of immunized persons and carriers, respectively. Mass immunization programs are currently contemplated in China, and we can look forward to the day when the chronic carrier state, HBV-related chronic liver disease, and hepatocellular carcinoma will be eliminated by universal vaccination strategies in Asia and Africa. The progress in immunology/virology of HBV has invaluably contributed to developing the means to detect and control chronic HBV infection. We must now dedicate our efforts to wide-scale implementation of these strategies in order to make HBV infection a disease of the past, hopefully before we usher in the next century.

Acknowledgment. Supported in part by grant PO1 HL36589 and contract NO1-HB-7020 from the National Heart, Lung and Blood Institute.

References

1. Beasley RP, Trepo C, Stevens CE, Szmuness W: The e antigen and vertical transmission of hepatitis B surface antigen. *Am J Epidemiol* 1977;105:94–98.
2. Beasley RP, Hwang LY, Lin CC, Chien CS: Hepatocellular carcinoma and hepatitis B virus. *Lancet* 1981; 2:1129–1132.
3. Bhatnagar PK, Papas E, Blum HE, et al: A synthetic analogue of hepatitis B surface antigen sequence 139-147 produces immune response specific for the common a determinant. *Proc Nat Acad Sci* USA 1982;79:4400–4404.
4. Blum HE, Vyas GN: Non-A, non-B: a contemporary assessment. *Haematologia (Budap)* 1982;15:153–173.
5. Blum HE, Haase AT, Harris JD, et al: Asymmetric replication of hepatitis B virus DNA in human liver: demonstration of cytoplasmic minus-strand DNA by Southern blot analyses and in situ hybridization. *Hepatology* 1983; 3:840 (Abstract).
6. Dreesman GR, Sanchez Y, Ionescu-Matiu I, et al: Antibody to hepatitis B surface antigen after a single inoculation of uncoupled synthetic HBsAg peptides. *Nature* 1982;295:158–160.
7. Figus A, Blum HE, Vyas GN, et al: Hepatitis B virus nucleotide sequences in patients with non-A, non-B or type B chronic liver disease. *Hepatology* 1984;4:364–368.
8. Fowler MJF, Monjardino J, Tsiquaye KN, et al: The mechanism of replication of hepatitis B virus: evidence of asymmetric repliation of the two DNA strands. *J Med Virol* 1984;13:83–91.
9. Galibert F, Mandart E, Fitoussi F, et al: Nucleotide sequence of hepatitis B virus genome (subtype ayw) cloned in E. coli. *Nature* 1979;281:646–650.
10. Itoh Y, Takai E, Ohnuma H, et al: A synthetic peptide vaccine involving the pre-S2 region of hepatitis B virus: protective efficacy in chimpanzees. *Proc Nat Acad Sci* USA 1986;83:9174–9178.
11. Kennedy RC, Eichberg JW, Lanford RE, Dreesman GR: Anti-idiotypic antibody vaccine for type B viral hepatitis in chimpanzees. *Science* 1986;232:220–223.
12. Krugman S, Giles JP, Hammond J: Viral hepatitis type B (MS-2 strain): studies on active immunization. *JAMA* 1971;217:41–45.
13. Lerner RS, Green N, Alexander H, et al: Chemically synthesized peptides predicted from the nucleotide sequence of the hepatitis B virus genome elicit antibodies reactive with the native envelope protein of Dane particles. *Proc Nat Acad Sci* USA 1981;78:3403–3407.
14. Mackay P, Lees J, Murray K: The conversion of hepatitis B core antigen synthesized in E. coli into e antigen. *J Med Virol* 1981;8:237–243.
15. Mason WS, Aldrich C, Summers J, Taylor JH: Asymmetric replication of duck hepatitis B virus DNA: free minus-strand DNA. *Proc Nat Acad Sci* USA 1982;79:3997–4001.
16. Maupas P, Guesry P: *Hepatitis B Vaccine.* Amsterdam, North Holland-Elsevier, 1981.
17. Melnick JL, Dreesman GR, Hollinger FB: Hepatitis B vaccine: expectations and realities for the prevention of infection and of primary hepatocellular carcinoma. *Infect Dis* 1982;12:1–8.

18. Milich DR, Thornton B, Neurath RA, et al: Enhanced immunogenicity of the pre-S region of hepatitis B surface antigen. *Science* 1985;228:1195–1199.
19. Milich MA, McLachlan F, Chisari S, Thornton G: Immune response to the pre-S(1)-specific T cell response can bypass nonresponsiveness to the pre-S(2) and S regions of HBsAg. *J Immunol* 1986;137:315–322.
20. Neurath AR, Kent SBH, Strick N: Specificity of antibodies elicited by a synthetic peptide having a sequence in common with a fragment of a virus protein, the hepatitis B surface antigen. *Proc Nat Acad Sci* USA 1982;79:7871–7875.
21. Pasek M, Goto T, Gilbert W, et al: Hepatitis B virus genes and their expression in E. coli. *Nature* 1979;282:575–579.
22. Peterson DL, Chien DY, Vyas GN, et al: Characterization of polypeptides of HBsAg for the proposed "U.C. Vaccine" for hepatitis B, in Vyas GN, Cohen SN, Schmid R (eds): *Viral Hepatitis* Philadelphia, Franklin Institute Press, 1978, pp 542–625.
23. Prince AM, Ikram H, Hopp TP: Hepatitis B virus vaccine: identification of HBsAg/a and HBsAg/d but not HBsAg/y subtype antigenic determinants on a synthetic immunogenic peptide. *Pro Nat Acad Sci* USA 1982;79:579–582.
24. Purcell RH, Gerin JL: Hepatitis B subunit vaccine: a preliminary report of safety and efficacy tests in chimpanzees. *Am J Med Sci* 1975;270:395–399.
25. Robinson WS: The enigma of non-A, non-B hepatitis. *J Infect Dis* 1982;145:387–395.
26. Robinson WS, Marion P, Fettelson M, Siddiqui A: The hepadna virus group: hepatitis B and related viruses, in Szmuness W, Alter HJ, Maynard JE (eds): *Viral Hepatitis*. Philadelphia, Franklin Institute Press, 1982, pp 57–68.
27. Siddiqui A, Sattler F, Robinson WS: Restriction endonuclease cleavage map and location of unique features of the DNA of hepatitis B virus, subtype adw$_2$. *Proc Nat Acad Sci* USA 1979;76:4664–4668.
28. Smith GL, Mackett M, Moss B: Infectious vaccinia virus recombinants that express hepatitis B surface antigen. *Nature* 1983;302:490–495.
29. Stevens CE, Beasley RP, Lin GC, et al: Perinatal hepatitis B virus infection: use of hepatitis B immune globulin, in Szmuness W, Alter HJ, Maynard JE (eds): *Viral Hepatitis*. Philadelphia, Franklin Institute Press, 1982, pp 527–535.
30. Stevens CE, Taylor PE, Tong MJ, et al: Yeast-recombinant hepatitis B vaccine. *JAMA* 1987;257:2612–2616.
31. Summers J, Mason W: Replication of the genome of a hepatitis B-like virus by reverse transcription of an RNA intermediate. *Cell* 1982;29:403–415.
32. Szmuness W, Alter HJ, Maynard JE (eds): *Viral Hepatitis*. Philadelphia, Franklin Institute Press, 1982.
33. Tiollais P, Charnay P, Vyas GN: Biology of hepatitis B virus. *Science* 1981;213:406–411.
34. Valenzuela P, Gray P, Quiroga M, et al: Nucleotide sequence of the gene coding for the major protein of hepatitis B virus surface antigen. *Nature* 1979;280:815–819.
35. Valenzuela P, Quiroga M, Zaldivar J, et al: Necleotide sequence of hepatitis B virus DNA, in Fields B, Jaenisch R, Fox CF (eds): *Animal Virus Genetics*. New York, Academic Press, 1980, pp 57–69.
36. Varmus HE: Form and function of retroviral proviruses. *Science* 1982;216:812–820.

37. Vyas GN, Williams EW, Klaus GB, Bond HE: Hepatitis associated Australia antigen-protein, peptides and amino acid composition of purified antigen with its use in determining sensitivity of the hemagglutination test. *J Immunol* 1972;108:1114–1118.
38. Vyas GN, Cohen SN, Schmid R (eds): *Viral Hepatitis*. Philadelphia, Franklin Institute Press, 1978.
39. Vyas GN: Molecular immunology of HBsAg, in Maupas P, Guesry P (eds): *Hepatitis B Vaccine*. Amsterdam, Elsevier, 1981, pp 227–237.
40. Vyas GN, Bhatnagar PK, Blum HE, et al: Appraisal and prospects of a dimeric synthetic peptide coupled with tetanus toxoid for a bifunctional vaccine against hepatitis B virus infection. *Dev Biol Stand* 1983;54:93–102.
41. Vyas GN, Dienstag JL, Hoofnagle JH: *Viral Hepatitis and Liver Disease*. New York, Grune & Stratton, 1984.
42. Vyas GN: A brief overview of the new vaccines against hepatitis B virus infection: immunogenic gene products and peptide analogues of antigenic epitopes. *Dev Biol Stand* 1986;63:141–146.
43. Wands JR, Bruno RR, Carlson RI, et al: Monoclonal IgM radioimmunoassay for hepatitis B surface antigen high binding activity in serum that is unreactive with conventional antibodies. *Proc Nat Acad Sci* USA 1982;79:1277–1281.
44. Weiser B, Ganem D, Seeger C, Varmus HE: Closed circular viral DNA and asymmetrical heterogeneous forms in liver from animals infected with ground squirrel hepatitis virus. *J Virol* 1983;48:1–9.

CHAPTER 10

Biologic Significance of Pre-S Antigen and Anti-Pre-S Antibodies in Hepatitis B Virus Infection

A. Budkowska, P. Dubreuil, and J. Pillot

The outer membrane of the hepatitis B virus (HBV) consists of host lipids and the protein components' major (p 25, gp 29), middle (gp 33, gp 36), and large (p 39, gp 41) envelope polypeptide (8). These proteins are encoded for by a large open reading frame of HBV DNA, which has three in-phase translation start codons (17). The major protein, encoded by the S gene, is the predominant structural constituent of noninfectious, 22-nm spherical or tubular particles and virions and expresses conformationally dependent hepatitis B surface antigen (HBsAg). The DNA sequence corresponding to the S gene and the region preceding the S gene (pre-S region) encode for the polypeptides larger than the S polypeptide. The "middle protein" represents S protein with an additional 55-amino-acid, conformation-independent fragment at the N-terminus encoded by the pre-S_2 domain (10). The "large protein" is encoded for by the S, pre-S_2, and pre-S_1 domain; the latter is located either 108 or 119 codons upstream of the pre-S_2 region. The large envelope polypeptide is preferentially expressed on HBV particles and filaments relative to 22-nm spheres (8).

Antigenic determinants located on the pre-S_2 encoded sequence of the middle protein have important biologic implications:

1. The protein expressing pre-S_2 determinants is more immunogenic than S protein located on the same HBsAg particle. The presence of pre-S_2 sequences enhances the immune response to S protein and may circumvent the immunologic nonresponsiveness to S protein in nonresponder mouse strains (12).

2. The protein expressing pre-S_2 determinants may be involved in the virus attachment to hepatocytes (10,13). The pre-S_2 encoded sequence is responsible for the species-specific receptor activity for polymerized human serum albumin (pHSA) of HBV, and hepatocytes have organ-specific receptors for pHSA. HBV attachment and penetration would be mediated by pHSA polymers that occur in the circulation as the result of molecular aging. Studies have shown that pre-S_2-specific antibodies along with anti-pre-S_1 inhibit the attachment of HBV to hepatocytes (15).

3. Expression of pre-S-encoded determinants related to pHSA receptor activity on HBsAg varies considerably among patients in different phases of HBV infection (e.g., acute phase, recovery) (1,2,16).

4. The antibodies to the pre-S_2 region may be relevant to recovery from HBV infection (1,2) and effectively protect the chimpanzee against infectious HBV (9).

The initial studies of Alberti et al. (2) reported detection of antibodies reacting with antigenic determinants on Dane particles distinct from HBsAg and HBcAg in sera of patients with acute-type B hepatitis and suggested the role for this antiviral response in the recovery from HBV infection. HBV-envelope antibody was probably directed to the peptide expressing pHSA receptor (pHSAR) activity, and anti-pHSAR antibody could be relevant in reducing infectivity and neutralizing HBV.

These concepts led us to investigate a possible relation between the expression of the antigenic determinants encoded by the pre-S region of HBV genome and pHSAR activity. In the first stage, we explored the use of antibodies from patients recovering from acute-type hepatitis for a solid-phase radioimmunoassay (RIA) to detect pre-S encoded antigenic specificity (3,4).

The presence of pre-S coded sequences on HBsAg particles was determined by RIA involving presumed human anti-pre-S on the solid phase. The antigen bound by the pre-S determinants was detected by radiolabeled anti-HBs. Immunoglobulin G (IgG) fraction of anti-pre-S_2 activity was obtained from the serum of a patient recovering from acute HBV infection, which had been also found to be positive for HBsAg, anti-HBe, and anti-HBc. Anti-pre-S-coated solid phase effectively bound purified Dane particles and recombinant HBsAg particles (11) (pre-S and S gene product synthesized in CHO cells transfected with S and pre-S and S fragment of HBV genome) (kindly provided by M. L. Michel, Institut Pasteur, Paris) but not pHSAR-negative HBsAg from human serum. The reaction was quantitatively inhibited by pHSA, and 100% of the sera positive for pre-S determinants were also bound on pHSA-coated beads. The expression of the pre-S encoded epitopes on viral or subviral particles paralleled the expression of pHSA binding sites. However, high concentrations of pHSA needed for complete inhibition of the reaction suggested steric hindrance and the existence of different sites of fixation for pHSA and anti-pre-S. Because more pre-S encoded determinants were localized on purified virions (Dane particles) and in the DNA polymerase-positive fraction than on pHSAR-positive noninfectious forms, the expression of pre-S determinants was possibly related to virus replication activity. These observations favored the presence of anti-pre-S_1 in addition to anti-pre-S_2 specificity on the solid phase. Consequently, our polyclonal RIA could determined both pre-S_1 and pre-S_2 epitopes expressed on the portion of the HBV envelope responsible for pHSAR activity.

Pre-S determinants were expressed during the acute phase of disease, as was HBeAg, and disappeared rapidly with recovery. In contrast, continuous expression of pre-S antigenic sites on HBsAg was observed in chronically infected patients seropositive for either HBeAg or anti-HBe.

The anti-pre-S antibody (determined by inhibition of the reaction for pre-S determinants) was detected in sera of patients with acute hepatitis followed by recovery. Anti-pre-S usually appeared during HBsAg antigenemia (Table 10.1), and anti-pre-S evolution correlated well with HBsAg clearance, followed by seroconversion to anti-HBs. No anti-pre-S antibody was observed in patients with acute infection progressing to chronic disease or in patients with chronic hepatitis who had a continuing presence of pre-S. The kinetics of elaboration of antibody directed to pre-S-encoded determinants in the course of uncomplicated, resolved hepatitis B, always followed by HBsAg elimination, strongly suggested the role of this antibody in immunologic neutralization and clearance of infectious virus.

These results prompted us to prepare monoclonal antibodies against pre-S determinants for immunochemical characterization and the development of immunoassays for determinants of the pre-S_2 sequence and corresponding antibodies. Monoclonal antibodies specific for the pre-S region antigenic determinants were prepared using HBV particles or recombinant HBsAg particles as immunogens (5,6). Initial screening of the hybridoma culture supernatants was performed by a solid-phase RIA (Ausab, Abbott Laboratories) to eliminate clones producing anti-HBs antibodies. Anti-pre-S antibodies were screened by hemagglutination inhibition (3) and by polyclonal inhibition RIA (4). Hybridomas producing anti-pre-S antibodies were cloned by limiting dilution, and mouse ascitic fluids were obtained. Three monoclonal anti-pre-S_2 antibodies yielded stable clones and were selected for further study (6). Isotype analysis revealed that one of the cell lines produced antibodies of the IgM class (F 52), whereas the two others were of IgG_1 subclass. The antibodies reacted with three epitopes localized in HBV "middle" protein, sensitive to V 8 protease from *Staphylococcus aureus* but not with HBsAg particles without pre-S_2 sequences whether from human sera or from mice cells carrying chromosomally integrated copies of HBV S gene (kindly provided

Table 10.1. Prevalence of pre-S determinants and anti-pre-S antibody in 146 sera of hepatitis B virus-infected patients.

Serum	No.	HBsAg(+)	anti-HBs(+)	HBeAg(+)	anti-HBe(+)	HBeAg(−) anti-HBe(−)
Pre-S(+)	86	86	0	57	23	6
Anti-pre-S(+)	31	28	2	6	13	12

A total of 146 sera from HBV-infected patients with different viral markers were tested for pre-S determinants by polyclonal RIA. Sera found negative for pre-S were assayed for anti-pre-S by inhibition RIA. Other HBV markers were determined by routine RIAs from Abbott Laboratories.

by M.L. Michel). The first antibody characterized, McAb-F 124, was used to develop RIAs for determining pre-S_2a epitope, recognized by this antibody in sera of hepatitis B patients (5). In the assays, McAb-F 124 was used for coating the solid phase, with ^{125}I-anti-HBs as the radioactive reagent in the detection system. Anti-pre-S_2a was efficiently detected by inhibiting the reaction for the pre-S_2a epitope. We have shown that the pre-S_2a epitope of Gp 33 Gp 36, exposed on the surface of the viral or subviral particles, is immunogenic in animals; and the antibodies to this pre-S_2a epitope are elicited during the early phase of infection in patients recovering from acute B hepatitis.

Figure 10.1A shows the typical time course of the development of antibodies specific to pre-S_2a epitope in patients recovering from uncomplicated acute-type hepatitis B. The presence of a pre-S_2a epitope determined by the monoclonal antibody-based assay in the early phase of acute HBV infection correlated well with the other markers of active viral replication (HBeAg and HBV DNA). The development of anti-pre-S_2a was observed in 13 of 13 patients with acute resolved hepatitis; and in 70% of cases anti-pre-S_2 appeared during the third month.

The pre-S_2a epitope was detected in sera of all 13 tested patients with acute hepatitis B in the early phase of infection and was no longer detectable in any serum samples from these patients after 6 months from the onset of symptoms (Table 10.2). In contrast, 50 of 50 serum samples from chronic hepatitis B patients were positive for the pre-S_2a epitope (Fig. 10.1B). Interestingly, 60% of these patients were HBeAg-positive and 40% were anti-HBe-positive. These results confirmed our previous observations (4) concerning the expression of pre-S-encoded determinants in chronic hepatitis B patients and were in accordance with those of Alberti et al (1,2) and Pontisso et al (16), who have demonstrated pHSAR activity in chronic HBsAg carriers. When the presence of serum HBV DNA was determined by a molecular hybridization technique and compared with pre-S_2a activity in serum, it was shown that pre-S_2 was expressed on HBsAg particles in 28 of 28 of HBV DNA-positive sera; however, only 15 of 50 serum samples from chronic HBsAg carriers positive for pre-S_2a were positive for HBV DNA (Table 10.2). This finding was conclusive for the presence of pre-S_2a epitope on infectious virions as well as on nonreplicating forms of HBV.

Anti-pre-S_2 could be detected by inhibition of RIA in healthy individuals after active immunization with HBV vaccine licensed in France (Hevac B, Pasteur Institute) before anti-HBs response (Fig. 10.2). However, we noted that high anti-HBs concentrations (400 mUI), when present in tested sera, interfered in the assay system (5).

Other monoclonal antibodies F 376 and F 52 specific for pre-S_2 sequence recognized the epitopes different from pre-S_2a. No competition was observed between F 376 and F 124 in experiments involving labeled monoclonal antibodies (6). In addition, F 124 was unreactive with the synthetic

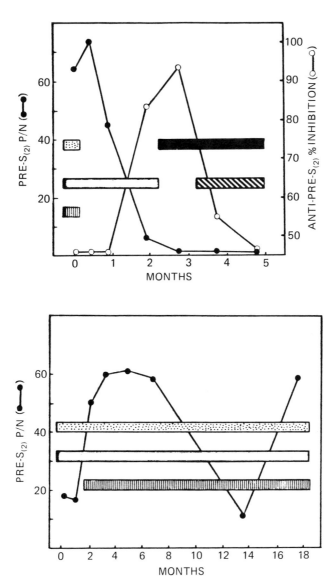

Fig. 10.1. A: Typical acute episode of type B hepatitis followed by recovery, with appearance of the anti-pre-S$_2$ antibody. **B:** Serologic profile of acute-type hepatitis with chronic evolution. There is a continuous presence of pre-S$_2$ epitope on HBsAg particles. ● pre-S$_2$ epitope; ○ anti-pre-S$_2$; ▨ HBsAg; ▦ HBeAg; ◪ anti-HBs; ▥ anti-HBe; ■ HBV DNA (5).

Table 10.2. Prevalence of pre-S_2 and anti-pre-S_2 in various forms of HBV infections and in persons with immunity to HBV.

Patients	Pre-S_2 after 6 months	HBV DNA($+$)	Anti-pre-S_2($+$)
Acute hepatitis with recovery			
13 Cases	0/13	10/13	13/13
70 Samples	0/70	13/70[a]	
Chronic evolution		3/6	0/6
6 Cases	6/6		
29 Samples	29/29	12/29	
Chronic hepatitis			
35 Cases	35/35	12/35	0/35
50 Samples	50/50	15/50	
Naturally acquired immunity			
to HBV (163 cases)	0/163	nt	57/163
Recipients of HBV vaccine			
(112 cases)	0/112	nt	49/112

Sera from patients with acute or chronic HBV infection and a recent history of hepatitis (anti-HBc- and anti-HBs-positive) and from healthy recipients of HBV vaccine (HEVAC B, Pasteur Vaccins) were tested for the presence of pre-S_2 or anti-pre-S_2 by monoclonal RIAs or immunoassays. HBV DNA was determined by the molecular hybridization technique.
(nt) not tested.
[a]DNA was detected in early samples from this group of patients.

peptides corresponding to pre-S_2 132-145 amino acid sequence recognized by F 376 (14). The epitope specificity of Mo 52 could not be directly determined by competition experiments. However, the reactivity of this monoclonal antibody with different antigenic subtypes of HBsAg suggested recognition of an epitope different from those recognized by F 124 and F 376 (A. Budkowska, unpublished observations).

Monoclonal anti-pre-S_2 antibodies with different specificities may provide useful tools in delineating antigenic structure of the pre-S_2 region and in evaluating the immune response to the individual pre-S_2 epitopes. Therefore pre-S_2a- and pre-S_2b-specific F 124 and F 376 monoclonal antibodies were employed to develop highly sensitive immunoenzymatic assays for corresponding pre-S_2 determinants and anti-pre-S_2 antibodies in sera (6,7). The assays with F 52 (anti-pre-S_2c) are in progress.

The test for pre-S_2a or pre-S_2b epitopes was based on the solid-phase sandwich principle in which the two epitope-specific antibodies were used as immunosorbents and as enzyme-labeled probes (6). The assay for epitope-specific anti-pre-S_2 response was based on the competition of anti-pre-S_2 in serum with HRPO-labeled monoclonal antibody probes for pre-S_2 epitopes on recombinant, pre-S_2-positive HBsAg particles. The systems utilizing different monoclonal antibodies permitted detection of pre-S_2a or pre-S_2b epitopes and corresponding antibodies. Both pre-S_2a and pre-S_2b epitopes were detected simultaneously in pre-S_2 sequences in sera with ay or ad HBsAg subtypes. Moreover, anti-pre-S_2 antibodies

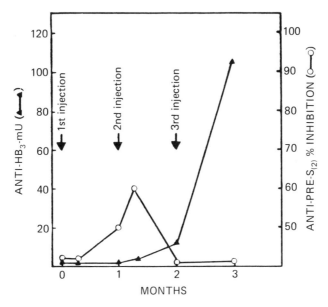

Fig. 10.2. Immune response to HBV envelope proteins in a healthy recipient of HBV vaccine. ▲ anti-HBs; ○ anti-pre-S$_2$ (5).

directed against different pre-S$_2$a and pre-S$_2$b epitopes were elicited in patients recovering from HBV infection and in recipients of HBV vaccine. Antibody to both pre-S$_2$ epitopes emerged early in the course of HBV infection, suggesting strong immunogenicity of the pre-S$_2$ sequence in vivo. Anti-pre-S$_2$ was detected in patients recovering from HBV infection, in patients with naturally acquired immunity to HBV, and in recipients of HBV vaccine (Table 10.2).

In accordance with previous results obtained utilizing polyclonal assays (4) and Mo-RIA with F 124 (5), anti-pre-S$_2$ was never found in patients with active viral replication and chronic active liver disease. All 51 patients with high amino transferase levels and/or HBV-DNA were continuously positive for pre-S$_2$ epitopes. These data indicate that expression of pre-S$_2$ determinants on HBAg particles is related to virus replication activity, in agreement with previous reports of our group (4.5) and those of others (1,2).

Expression of pre-S$_2$ have important clinical and prognostic implications. The pHSA receptor levels determined by others (16) were proved to be predictive of the subsequent serologic outcome, being significantly higher in patients with active liver disease than in those with mild inflammatory activity and favorable outcome. In our studies, we found low or no pre-S$_2$ activity in most asymptomatic HBsAg carriers, in contrast to high pre-S$_2$ levels in patients actively replicating HBV or in those with severe liver lesions (A. Budkowska, unpublished observations). Pre-S$_2$ elimination and elaboration of anti-pre-S$_2$ may be considered the first sign of recovery and is accompanied by clinical and biochemical resolution of liver disease. In contrast, the continuous presence of pre-S$_2$ epitopes and the absence of an anti-pre-S$_2$ response can act as a prognostic marker predicting the development of a chronic course of HBV infection and a worse prognosis.

Testing for pre-S$_2$ levels in acute and chronic HBV infection may be useful for assessing the state of virus replication and the degree of the liver injury. Furthermore, pre-S$_2$ and anti-pre-S$_2$ levels may serve as prognostic markers to predict favorable outcome of type B hepatitis.

The correlation of the appearance of anti-pre-S$_2$ response with recovery and the lack of anti-pre-S$_2$ in persistently individuals suggest that the immune response against pre-S$_2$ may play an important role in the host mechanisms of virus neutralization and clearance.

References

1. Alberti A, Pontisso P, Chemello L, et al: Virus receptor for polymerized human serum albumin and anti-receptor antibodies in hepatitis B virus infection, in Vyas GN, Dienstag J, Hoofnagle J (eds). *Viral Hepatitis and Liver Disease.* Orlando, Florida, Grune & Stratton, 1984.
2. Alberti A, Pontisso PP, Shavion E, et al: Antibody precipitating Dane particles in acute hepatitis type B: relation to receptor sites that bind polymerized human serum albumin on virus particles. *Hepatology* 1984;4:220–226.
3. Budkowska A, Briantais MJ, Dubreuil P, et al: Detection of antibodies directed against the pre-S gene product of hepatitis B virus: relationship between anti-pre-S response and recovery. *Ann Inst Pasteur Immunol* 1985;136D:57–65.
4. Budkowska A, Dubreuil P, Capel F, et al: Hepatitis B virus pre-S gene encoded antigenic specificity and anti-pre-S antibody: relationship between anti-pre-S response and recovery. *Hepatology* 1986;6:360–369.
5. Budkowska A, Riottot MM, Dubreuil P, et al: Monoclonal antibody recognizing pre-S 2 epitope of hepatitis B virus: characterization of pre-S$_2$ epitope and anti-pre-S$_2$ antibody. *J Med Virol* 1986;20:111–125.
6. Budkowska A, Dubreuil P, Briantais MJ, et al: A monoclonal antibody enzyme immunoassay for detection of epitopes encoded by the pre-S$_2$ region of the hepatitis B virus genome. *J Immunol Methods* 1987;97:77–85.
7. Budkowska A, Dubreuil P, Pillot J: Detection of antibodies to pre-S$_2$ encoded epitopes of hepatitis B virus by monoclonal antibody enzyme immunoassay. *J Immunol Methods* 1987;102:85–92.
8. Heerman K, Goldman V, Schwartz W, et al: Large surface proteins of hepatitis B virus containing the pre-S sequence. *J Virol* 1984;52:396–402.
9. Itoh Y, Takai E, Ohnuma H, et al: A synthetic peptide vaccine involving the product of the pre-S$_2$ region of hepatitis B virus DNA: protective efficacy in chimpanzees. *Proc Natl Acad Sci USA* 1986;83:9174–9178.

10. Machida A, Kishimoto S, Ohnuma H, et al: A polypeptide containing 55 aminoacid residues coded by the pre-S region of hepatitis B virus deoxyribonucleic acid bears the receptor for polymerized human as well as chimpanzee albumins. *Gastroenterology* 1984;86:910–918.
11. Michel ML, Pontisso P, Sobczak E, et al: Synthesis in animal cells of hepatitis B surface antigen particles carrying a receptor for polymerized human serum albumin. *Proc Natl Acad Sci USA* 1984;81:7708–7712.
12. Milich DR, Thompson GB, Neurath AR, et al: Enhanced immunogenicity of the pre-S region of hepatitis B surface antigen. *Science* 1985;228:1195–1199.
13. Neurath AR, Kent SBH, Strick N, et al: Hepatitis B virus contains pre-S gene encoded domains. *Nature* 1985;315:154–156.
14. Neurath AR, Adamowicz P, Kent SBH, et al: Characterization of monoclonal antibodies specific for the pre-S$_2$ region of the hepatitis B virus envelope protein. *Mol Immunol* 1986;23:991–997.
15. Neurath AR, Kent SBH, Strick N, et al: Identification and chemical synthesis of a host cell receptor binding site on hepatitis B virus. *Cell* 1986;46:429–436.
16. Pontisso P, Alberti A, Bortolotti F, et al: Virus associated receptors for polymerized human serum albumin in acute and chronic hepatitis B virus infection. *Gastroenterology* 1983;84:220–226.
17. Tiollais P, Charnay P, Vyas GN: Biology of hepatitis B virus. *Science* 1981;213:403–411.

CHAPTER 11

Enterically Transmitted Hepatitis Viruses: Prospects for Control

Robert H. Purcell

Five distinct hepatitis viruses are currently recognized. Three of them—hepatitis B virus, hepatitis D (δ) virus, and blood-borne non-A non-B hepatitis virus—are transmitted predominantly by exposure to contaminated blood, blood products, and bodily secretions. Two of the viruses, hepatitis A virus (HAV) and enterically transmitted non-A non-B hepatitis virus (ENANB), are spread predominantly by oral exposure to contaminated feces. Although these two viruses share a similar epidemiology and cause a similar clinical disease, they are different in several respects. The purpose of this review is to summarize current knowledge about these two enterically transmitted hepatitis viruses and to evaluate prospects for their control.

Hepatitis A Virus

Hepatitis A virus generally causes less severe disease than some of the other hepatitis viruses (e.g., hepatitis B virus) and never progresses to chronicity; it is, however, the cause of significant human morbidity and loss of productivity (23). Approximately 25% of clinical hepatitis in many developed countries and up to 80% of such disease in selected developing countries is caused by HAV. In the United States 38% of cases reported to the Centers for Disease Control is hepatitis A, and until recently it was the most frequently diagnosed hepatitis; the rising incidence of hepatitis B and the continuing fall in the incidence of hepatitis A in the United States, however, has resulted in a change in the relative importance of these two viruses (8).

However, HAV continues to be highly endemic in many underdeveloped countries and, like many other enterically transmitted viruses, infects almost 100% of the population by age 5 to 10 years. Such infections are often unrecognized because hepatitis A is milder in infants and children than in older individuals. As sanitary conditions improve in such countries and the average age at which exposure to HAV occurs increases, a paradoxical increase in clinical hepatitis A occurs. Because just such improvements in sanitation are occurring in many developing countries throughout the world, there will be a need for prophylaxis against hepatitis A for the foreseeable future. Passive immunoprophylaxis (administration of immune serum globulin prior to or shortly after exposure) is effective but of only temporary benefit, and active immunization (vaccination) would be a more practical approach to the control of hepatitis A where HAV is endemic.

Hepatitis A virus is a picornavirus that was first visualized in 1973 (14) and isolated in cell culture in 1979 (31). Although classified as enterovirus 72 (24), it is sufficiently unique that it will probably be reclassified into a separate genus (46). The virus is difficult to isolate in cell culture as it grows slowly and to relatively low titer; most strains are not cytopathogenic in vitro.

Because HAV can be propagated in cell culture and because it does not cause chronic or malignant disease or serious sequelae, the "classical" approaches to vaccine development are feasible for this virus. These approaches include the development of inactivated whole-virus vaccine, analogous to the Salk vaccine for polio, and live attenuated oral (or parenteral) vaccine, analogous to the Sabin polio vaccines, as well as the application of new approaches based on recombinant DNA technology. The feasibility of killed whole-virus vaccines was demonstrated when HAV, extracted from the liver of experimentally infected marmosets, was inactivated with formalin and used as a vaccine to protect marmosets against challenge with virulent HAV (30). This crude "vaccine," though unacceptable for human use, did provide impetus for the development of suitable vaccines.

Serial passage of HAV in cell culture has resulted in more rapid growth, and several strains that attain relatively high titers in vitro have been developed. Such strains yield sufficient viral antigen to make a killed hepatitis A vaccine marginally feasible economically, and experimental vaccines have been developed from several of these strains (4, 15, 34, 40). Most are produced in diploid cell lines that are approved for vaccine development. Preliminary results indicate that such killed vaccines are immunogenic and capable of stimulating protective antibody in marmosets, owl monkeys, or chimpanzees.

Phase 1 clinical trials in man have begun with some of these vaccines. For example, eight adult volunteers were vaccinated intramuscularly with an experimental killed vaccine prepared from HAV grown in MRC-5 cells (42). The unpurified vaccine was inactivated with formalin but not adjuvanted. All eight volunteers developed serum neutralizing antibody to HAV following four doses of vaccine, and six of the eight also became weakly positive for anti-HAV by radioimmunoassay (HAVAB, Abbott Laboratories).

Live attenuated HAV vaccines have also been developed. The virulence of HAV appears to be attenuated rapidly by serial passage in cell culture or in lower primates (6, 19, 32, 33, 36). As few as ten serial passages of one strain of HAV (HM-175) in primary African green monkey kidney cell culture resulted in partial attenuation of virulence for chimpanzees but not marmosets (36). Additional serial passage in cell culture resulted in attenuation for both species (19). Similar results were obtained with another strain of HAV (CR-326) in a study conducted at Merck Sharp and Dohme (33). Quantitative data for diminished replication of cell culture-attenuated HAV (strain HM-175) has been obtained. When compared with replication of the wild-type virus in these animals, a 10^3- to 10^4-fold decrease in viral replication was observed. The attenuated strain produced little or no disease in the experimentally infected animals but did produce high levels of neutralizing antibody (19). Diminished replication of an attenuated CR-326 strain of HAV has also been reported (33).

The CR-326 strain of HAV has been evaluated at two levels of attenuation in adult volunteers (33). Following 15 cell culture passages in semicontinuous fetal rhesus kidney cells and eight passages in MRC-5 cells, (variant F) the virus stimulated RIA antibody (HAVAB) to HAV in 23 of 37 volunteers inoculated subcutaneously, but 9 of the 37 also had mild elevations of liver enzymes, indicating slight residual virulence of the variant for man. Following eight additional passages in MRC-5 cells (variant F') the virus stimulated RIA antibodies (HAVAB) in 6 of 11 volunteers, none of whom had evidence of hepatitis. These results paralleled those obtained in marmosets and chimpanzees. Thus these animals appear to be useful models for evaluating the suitability of candidate HAV vaccines for man.

The molecular basis of such attenuation of candidate live HAV vaccines is being evaluated as well. To determine the molecular basis of the attenuation of the HM-175 strain of HAV, cDNA was prepared and cloned from the genomes of the virulent and attenuated viruses (2, 9, 10). Sequencing of the cDNA representing the two genomes revealed that they differed from each other by only 24 nucleotide changes. Full-length cDNA clones were constructed, and one of these clones, derived from the attenuated virus, was shown to be infectious in cell culture (11). When the progeny virus derived from this infectious clone was inoculated into marmosets, it was shown to have the attenuated phenotype (11). Infectious cDNA molecules that are molecular hybrids between the genomes of the virulent and attenuated viruses have been constructed and are being characterized in vitro and in vivo to determine the exact molecular basis for attenuation. Should these or other viruses yield useful candidate human vaccines, it will represent the first time that a human live vaccine has been so characterized before use in man. Should the candidate vaccine prove too attenuated or too virulent, it may be possible to fine-tune the degree of attenuation by molecular alteration of its genome. Thus the combination of "classical" vaccine development methods and modern recombinant DNA technologies may prove useful in the construction of this and other future vaccines.

Direct application of recombinant DNA technology to the development of nonvirion HAV vaccines is less likely to succeed in the immediate future. Like other picornaviruses, the single large open reading frame of the HAV genome appears to be translated into a single large polyprotein that is posttranslationally cleaved by viral proteases into a number of structural and nonstructural proteins (2, 10, 27, 45). The four structural proteins (VP1 to VP4) are probably the most important for vaccine development. Although epitopes that stimulate neutralizing antibody have been identified on VP1, VP2, and VP3 of other picornaviruses (16, 38), all currently available neutralizing monoclonal antibodies to HAV appear to be directed to a single region of overlapping neutralization sites of VP3 (17, 44). Furthermore, the epitopes that stimulate neutralizing antibody to most picornaviruses are conformational and not linear epitopes. That is, they exist in the intact virion but not in the individual denatured proteins. Therefore a recombinant HAV vaccine that is expressed in eukaryotic cells would probably have to be capable of autocleavage and self-assembly to be an efficient immunogen. The success achieved with expressed antigens of foot-and-mouth disease virus has not been matched with other picornaviruses to date (22).

Similarly, hepatitis A vaccines prepared from synthetic peptides will be difficult to develop for the same reasons (5). Nevertheless, synthetic peptides representing different regions of viral capsid proteins have proved to be useful, along with monoclonal antibodies, in mapping the important epitopes of HAV. If methods can be found to "fix" synthetic peptides or expressed viral capsid proteins into the three-dimensional configuration that they assumed in the intact virion (16, 38), useful synthetic vaccines prepared in this way may be possible.

Enterically Transmitted Non-A Non-B Hepatitis

The ENANB virus was first recognized in 1980 (20, 48) and first visualized in 1983 (1). The disease caused by this virus was originally thought to be hepatitis A, but the high prevalence of clinical disease in young adults, the high mortality associated with disease in pregnant women, and the high clinical attack rate during water-borne epidemics occurring in regions where virtually 100% of the population was immune to HAV pointed to another etiology. The detection of antibody to HAV in acute-phase sera from virtually all such patients and failure to demonstrate IgM antibody or a serologic response to HAV in paired sera from the patients confirmed that the disease was not hepatitis A but a newly recognized form of viral hepatitis.

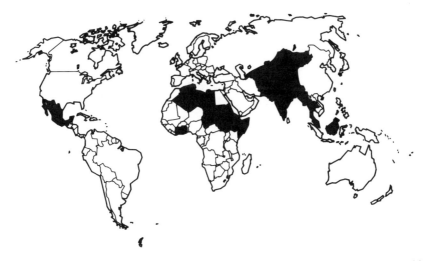

Fig. 11.1. Geographic distribution of ENANB hepatitis (1, 3, 13, 20, 25, 26, 28, 29, 39, 41, 43, 48). [From Purcell and Ticehurst (37), with permission.]

The geographic distribution of ENANB appears to be more restricted than that of HAV (Fig. 11.1); the absence of practical serologic tests for the diagnosis of ENANB hepatitis has limited seroepidemiologic studies, however, and the recognized range of this virus is probably underestimated. Disease caused by ENANB virus is significant and accounts for more than 50% of clinical hepatitis in some countries in which the virus is highly endemic (21, 28). ENANB is a major cause of "sporadic" and endemic hepatitis as well as the more spectacular water-borne epidemics. ENANB hepatitis, like hepatitis A, appears never to progress to chronicity. Although the mortality in the general population may be slightly higher than that of hepatitis A, ENANB hepatitis is particularly virulent in pregnant women, especially those in the third trimester of pregnancy, in whom the mortality rate may exceed 20% (21, 47, 48). Attempts to transmit the disease to primates have been partially successful, especially with chimpanzees and certain species of monkey (1, 7, 18, 29). Based on limited serologic comparisons utilizing the technique of immune electron microscopy, there is probably only one serotype of ENANB virus, a situation analogous to that with HAV (7, 12, 37, 49).

The etiologic agent of ENANB hepatitis is believed to be a 27-nm virus that is transmitted by the fecal-oral route. The virus does not appear to be serologically related to the picornaviruses, and limited cross-hybridization studies with cDNA derived from picornavirus genomes has failed to reveal a taxonomic relation with the picornaviruses (J.R. Ticehurst et al, unpublished data). The virus may represent the first recognized member of a new group of viruses. Methods for the control of ENANB hepatitis are less certain than for the more extensively studied HAV. It is clear that improved conditions of sanitation diminish the incidence of water-borne outbreaks, and such improvements are probably responsible for the virtual absence of ENANB from developed countries. There is some evidence for successful passive immunoprophylaxis against ENANB hepatitis by immune serum globulin prepared in countries where the virus is highly endemic. It is unlikely that immune serum globulin prepared in Western countries protects against ENANB hepatitis.

Active immunoprophylaxis against ENANB hepatitis probably must await the isolation and characterization of the etiologic agent in cell culture and the establishment of a reproducible animal model in which to test vaccines.

References

1. Balayan MS, Andzhaparidze AG, Savinskaya SS, et al: Evidence for a virus in a non-A, non-B hepatitis transmitted via the fecal-oral route. *Intervirology* 1983;20:23–31.
2. Baroudy BM, Ticehurst JR, Miele TA, et al: Sequence analysis of hepatitis A virus cDNA coding for capsid proteins and RNA polymerase. *Proc Natl Acad Sci USA* 1985;82:2143–2147.
3. Belabbes E-H, Bouguermouh A, Benatallah A, Illoul G: Epidemic non-A, non-B viral hepatitis in Algeria: strong evidence for its spreading by water. *J Med Virol* 1985;16:257–263.
4. Binn LN, Bancroft WH, Lemon SM, et al: Preparation of a prototype inactivated hepatitis A virus vaccine from infected cell cultures. *J Infect Dis* 1986;153:749–756.
5. Bittle JL, Houghten RA, Alexander H, et al: Protection against foot-and-mouth disease by immunization with a chemically synthesized peptide predicted from the viral nucleotide sequence. *Nature* 1982;298:30–33.
6. Bradley DW, Schable CA, McCaustland KA, et al: Hepatitis A virus: growth characteristics of in vivo and in vitro propagated wild and attenuated virus. *J Med Virol* 1984;14:373–386.
7. Bradley DW, Krawczynski K, Cook EH Jr, et al: Enterically transmitted non-A, non-B hepatitis: serial passage of disease in cynomolgus macaques and tamarins and recovery of disease-associated 27- to 34-nm viruslike particles. *Proc Natl Acad Sci USA* 1987;84:6277–6281.
8. Centers for Disease Control: ACIP: recommendations for protection against viral hepatitis. *MMWR* 1985;34:313–335.
9. Cohen JI, Rosenblum B, Ticehurst JR, et al: Complete nucleotide sequence of an attenuated hepatitis A virus: comparison with wild-type virus. *Proc Natl Acad Sci USA* 1987;84:2497–2501.
10. Cohen JI, Ticehurst JR, Purcell RH, et al: Complete nucleotide sequence of wild-type hepatitis A virus: comparison with different strains of hepatitis A virus and other picornaviruses. *J Virol* 1987;61:50–59.
11. Cohen JI, Rosenblum B, Feinstone S, et al: Use of infectious hepatitis A virus cDNA to study viral attenuation, in Chanock RM, Lerner RA, Brown, F,

Ginsberg H (eds): *Modern Approaches to New Vaccines Including Prevention of AIDS (Proceedings 1987 Conference).* Cold Spring Harbor, New York, Cold Spring Harbor Laboratory, 1987 (in press).

12. De Cock KM, Bradley DW, Sandford NL, et al: Epidemic non-A, non-B hepatitis in patients from Pakistan. *Ann Intern Med* 19 ;106:227–230.
13. Favorov MO, Khukhlovich PA, Zairov GK et al: Clinical and epidemiological features and diagnosis of viral non-A, non-B hepatitis with fecal-oral transmission mechanism. *Vopr Virusol* 1986;1:65–69.
14. Feinstone SM, Kapikian AZ, Purcell RH: Hepatitis A: detection by immune electron microscopy of a virus-like antigen associated with acute illness. *Science* 1973;182:1026–1028.
15. Flehmig B, Haage A, Pfisterer M: Immunogenicity of a hepatitis A virus vaccine. *J Med Virol* 1987;22:7–16.
16. Hogle JM, Chow M, Filman DJ: Three-dimensional structure of poliovirus at 2.9 Å resolution. *Science* 1985;229:1358–1365.
17. Hughes JV, Stanton LW, Tomassini JE, et al: Neutralizing monoclonal antibodies to hepatitis A virus: partial localization of a neutralizing antigenic site. *J Virol* 1984;52:465–473.
18. Kane MA, Bradley DW, Shrestha SM, et al: Epidemic non-A, non-B hepatitis in Nepal: recovery of a possible etiologic agent and transmission studies in marmosets. *JAMA* 1984;252:3140–3145.
19. Karron RA, Daemer R, Ticehurst J, et al: Studies of prototype live HAV vaccine in primate models. *J Infect Dis* 1987;000:000–000 (in press).
20. Khuroo MS: Study of an epidemic of non-A, non-B hepatitis: possibility of another human hepatitis virus distinct from post-transfusion non-A, non-B type. *Am J Med* 1980;68:818–924.
21. Khuroo MS, Duermeyer W, Zargar SA, et al: Acute sporadic non-A, non-B hepatitis in India. *Am J Epidemiol* 1983;118:360–364.
22. Kleid DG, Yansura D, Small B, et al: Cloned viral protein vaccine for foot-and-mouth disease: responses in cattle and swine. *Science* 1981;214:1125–1129.
23. Lemon S: Type A viral hepatitis: new developments in an old disease. *N Engl J Med* 1985;313:1059–1067.
24. Melnick JL: Classification of hepatitis A virus as enterovirus type 72 and of hepatitis B virus as hepadnavirus type 1. *Intervirology* 1982;18:105–106.
25. Morrow RH, Sai FT, Edgcomb JH, Smetana HF: Epidemiology of viral hepatitis in Accra, Ghana. *Trans R Soc Trop Med Hyg* 1969;63:755–767.
26. Myint H, Soe MM, Khin T, et al: A clinical and epidemiological study of an epidemic of non-A non-B hepatitis in Rangoon. *Am J Trop Med Hyg* 1985;34:1183–1189.
27. Najarian R, Caput D, Gee W, et al: Primary structure and gene organization of human hepatitis A virus. *Proc Natl Acad Sci USA* 1985;82:2627–2631.
28. Pattanayak S: *Magnitude of the Problem of Hepatitis and Directions for Research, Prevention and Control.* New Delhi, World Health Organization, WHO Regional Office for South-East Asia, 1986, pp 1–13.
29. Pillot J, Sharma MD, Lazizi Y, et al: Immunological characterization of a viral agent involved in epidemic and sporadic non-A, non-B hepatitis. *Ann Inst Pasteur Virol* 1987;138:145–158.
30. Provost PJ, Hilleman MR: An inactivated hepatitis A virus vaccine prepared from infected marmoset liver. *Proc Soc Exp Biol Med* 1978;159:201–203.

31. Provost PJ, Hilleman MR: Propagation of human hepatitis A virus in cell culture in vitro. *Proc Soc Exp Biol Med* 1979;160:213–221.
32. Provost PJ, Giesa PA, Banker FS, et al: Studies in chimpanzees of live, attenuated hepatitis A vaccine candidates. *Proc Soc Exp Biol Med* 1983;172:357–363.
33. Provost PJ, Bishop RP, Gerety RJ, et al: New findings in live, attenuated hepatitis A vaccine development. *J Med Virol* 1986;20:165–175.
34. Provost PJ, Hughes JV, Miller WJ, et al: An inactivated hepatitis A viral vaccine of cell culture origin. *J Med Virol* 1986;19:23–31.
35. Public Health Service, Centers for Disease Control: Enterically transmitted non-A, non-B hepatitis—East Africa. *MMWR* 1987;36:241–244.
36. Purcell RH, Feinstone SM, Ticehurst JR, et al: Hepatitis A virus, in Vyas GN, Dienstag JL, Hoofnagle JH (eds): *Viral Hepatitis and Liver Disease*. Orlando, Florida, Grune & Stratton, 1984, pp. 9–22.
37. Purcell RH, Ticehurst JR: Enterically transmitted non-A, non-B hepatitis: epidemiology and clinical characteristics, in Zuckerman A (ed): *Viral Hepatitis*. New York, Alan R. Liss, 1987, (in press).
38. Rossman MG, Arnold E, Erickson JW, et al: Structure of a human common cold virus and functional relationship to other picornaviruses. *Nature* 1985;317:145–53.
39. Shakhgildyan IV, Khukhlovich PA, Kuzin SN, et al: Epidemiological characteristics of viral non-A, non-B hepatitis with fecal-oral mode of transmission. *Vopr Virusol* 1986;31:175–179.
40. Shimojo H: Development of inactivated hepatitis A vaccine in Japan. Presented at the Seventh Joint Conference on Hepatitis: The United States-Japan Cooperative Medical Science Program; Session III, 1986.
41. Shrestha SM, Kane MA: Preliminary report of an outbreak of non-A/non-B viral hepatitis in Katmandu Valley. *J Inst Med (Nepal)* 1983;5:1–10.
42. Sjogren MH, Eckels KH, Binn LM, et al: Safety and immunogenicity of an inactivated hepatitis A vaccine, in *Viral Hepatitis and Liver Disease*. New York, Alan R. Liss, 1987.
43. Sreenivasan MA, Arankalle VA, Sehgal A, Pavri KM: Non-A, non-B epidemic hepatitis: visualization of virus-like particles in the stool by immune electron microscopy. *J Gen Virol* 1984;65:1005–1007.
44. Stapleton JT, Lemon SM: Neutralization escape mutants define a dominant immunogenic neutralization site on hepatitis A virus. *J Virol* 1987;61:491–498.
45. Ticehurst JR, Racaniello VR, Baroudy BM, et al: Molecular cloning and characterization of hepatitis A virus cDNA. *Proc Natl Acad Sci USA* 1983;80:5885–5889.
46. Ticehurst, JR: Hepatitis A virus: clones, cultures, and vaccines. *Semin Liver Dis* 1986;6:46–55.
47. Viswanathan R, Sidhu AS: Infectious hepatitis: clinical findings. *Indian J Med Res* 1957;45(suppl):49–58.
48. Wong DC, Purcell RH, Sreenivasan MA, et al: Epidemic and endemic hepatitis in India: evidence for a non-A, non-B hepatitis virus aetiology. *Lancet* 1980;2:882–885.
49. Zairov GK, Stakhanova VM, Listovskaya EK, et al: Electron microscopic investigations in non-A, non-B viral hepatitis with fecal-oral transmission mode. *Vopr Virusol* 1985;31:172–175.

CHAPTER 12

Immunological Characterization of a Viral Agent Involved in Epidemic and Sporadic Non-A Non-B Hepatitis

J. Pillot, M.D. Sharma, Y. Lazizi, A. Budkowska, C. Dauguet, and J.L. Sarthou

In humans, acute and/or chronic hepatitis are due to virus A, B, δ, and others not yet known. A virus possibly involved in epidemic and sporadic cases of non-A non-B (NANB) hepatitis has been isolated. This virus is infectious for African green and *Saimiri* monkeys, and the antibodies produced in these nonhuman primates have been used to develop an ELISA detecting a NANB-associated antigen in stool. This virus grows in the hepatocyte PLC/PRF 5 cell line but not in HEP G2. Two types of particle were observed in association with the NANB antigen: large particles of 60 to 80 nm diameter (presumed complete virions), and smaller, lighter homogeneous particles of 25 to 30 nm. Using the inhibition of ELISA, antibodies have been detected in sera from African patients with epidemic or endemic NANB hepatitis as well as from European patients with endemic NANB hepatitis.

Introduction

Viral hepatitis may be caused by hepatitis A virus (HAV), hepatitis B virus (HBV), hepatitis δ virus (HDV), and an unknown number of not yet identified NANB hepatitis viruses (10). In developed countries where HBV prophylaxis is efficient, NANB viruses are the most important cause of hepatitis. This disease is believed to involve at least three viruses: Two are transmitted by contaminated blood and blood products, and the third, which causes sporadic NANB hepatitis, accounts for about 25 to 30% of reported hepatitis cases in Western countries (4); its mode of transmission is unknown. In developing countries, NANB virus appears also to be associated with large-scale outbreaks, with a water-borne, fecal-oral mode of transmission. Seven well documented epidemics have been reported since 1980: in India (6), Russia (2), Japan (14), Nepal (5), Algeria (3), Tchad (7), and the Ivory Coast (11). The mortality rate was high in pregnant women.

Our knowledge of the viruses causing NANB hepatitis is frustratingly sketchy, being limited to clinical and epidemiological data. Inoculation of chimpanzees with NANB posttransfusion hepatitis virus has been successful, and filtrable agents have been identified (10); infectious blood products are available. For NANB epidemic hepatitis few experimental infection studies have been performed: The infection has been reported to be transmissible to marmosets *(Saguinus mystax)* (5), macaque monkeys *(Macacus cynomolgus),* a human volunteer (2), and African green monkeys *(Cercopithecus aethiops)* and *Erythrocebus patas)* (11), although a definite infectious agent has not yet been identified. For posttransfusion hepatitis, despite a plethora of reports of promising putative NANB antigen–antibody systems, ultrastructural changes, and virus-like particles, no system or virus has been identified that fulfills accepted serological criteria for a specific causal association. For epidemic hepatitis, 27-nm spherical virus-like particles have been visualized only in some stools, and the immunological characterization was performed only by electron microscopic agglutination (1, 2, 5, 13). Viral replication has never been obtained in cell cultures for NANB viruses. It is believed that the agent(s) of the water-borne form of NANB is unrelated to the agent transmitted by blood because the latter has a high propensity to induce chronic hepatitis (not seen with the water-borne form). Whether the virus(es) producing epidemics could account for nonblood-product-transmitted sporadic hepatitis is not known.

Ivory Coast Epidemic

The outbreak of NANB hepatitis in the Ivory Coast provided the opportunity for new approaches. We report here:

1. Infection of monkeys with NANB agent and the utilization of their antibodies and antibodies from human convalescents allowed development of an ELISA test for specific detection of antigen associated with virus and antiviral antibodies
2. Production of a viral antigen in a continuous hepatocyte cell line with a cytopathic effect
3. Visualization of viral particles in purified preparations from stool specimens
4. Characterization of the same causative agent in epidemic as in sporadic NANB viral hepatitis

Three *Cercopithecus aethiops* and one *Erythrocebus patas* were orally inoculated with 2 ml of a mixture of 10% aqueous extract of stools from Ivorian patients with presumed epidemic NANB hepatitis, collected within the first few days after the onset of symptoms. The mixture was previously centrifuged and filtered through a 0.22-μm Millipore filter to eliminate bacteria and parasites. The serum samples and stool specimens were collected from these monkeys before administration of the stool extract and on the 27th day after inoculation. The liver function of these monkeys was not studied; the animals did not have evident clinical manifestations (11). One *Saimiri* monkey was inoculated under the same conditions with 0.5 ml of an extract from an Ivorian patient different from those used for African green monkeys. A second *Saimiri*, which had been inoculated 3 months before with a fecal preparation of hepatitis A virus and seroconverted to immunoglobulin M (IgM) anti-HAV antibodies on day 29, was inoculated intravenously with 0.5 ml of the same filtered stool specimen. The two *Saimiri* were bled twice weekly for biochemical and serological investigations, and stools were examined daily.

Both *Saimiri* died on day 28 without particular symptoms. Neither monkey exhibited convincing biochemical evidence of liver dysfunction, and postmortem histological examination showed no parenchymal lesion, only liver congestion.

Sera from these monkeys were fractionated on Sephacryl S 300 in 0.1 *M* Tris NaCl buffer pH 8, and the fraction containing IgM was collected. IgM was used as catcher antibodies in an immunoenzyme assay. IgM was chosen for its high capture capacity and to prevent false positive reactions due to rheumatoid factor or rheumatoid factor-like reactant. Indeed these factors have been shown to simulate NANB antigen in sera (12). Monkey IgG could also be used, but the reaction appeared less specific for some preparations. To reveal the fixation of stool or cell culture antigens on solid-phase antibody, IgG different from that in the convalescent sera was used. IgG was fractionated by DEAE-trisacryl chromatography and conjugated to β-galactosidase or peroxidase. Three labeled IgG samples were used for antigen detection: IgG from a *Cercopithecus* monkey experimentally inoculated; IgG from an Ivorian patient convalescing from epidemic NANB hepatitis; and IgG from an Algerian patient recovered from Medea epidemics in 1981–1982 (3). The three IgGs gave equivalent results. Monkey preinoculation IgM coated on the solid phase was used as a control of specificity for the postinoculation IgM.

In the first phase of the study, stools of inoculated monkeys were examined for the presence of a presumed NANB antigen in an ELISA test. Antigenic activity was detected in the stools of both *Saimiri* monkeys (Table 12.1). Ultrasonic treatment increased the amount of detected antigen and sensitivity of the method. Antigen appeared as early as day 4 after inoculation and persisted until day 13. Stools from one *Saimiri* monkey inoculated with a stool extract from a patient suffering from a presumed cytomegalovirus liver infection (with specific IgM) served as a control and gave consistently negative results.

In the second phase of the study, stools from patients suffering from presumed acute NANB hepatitis, either the epidemic form (Ivorian patients) or the sporadic form (French patients), were studied. The diagnosis of NANB hepatitis was established according to the classical immuno-

Table 12.1. NANB antigen detection in feces of monkeys infected with stool extracts from patients with epidemic NANB hepatitis.

Days after inoculation	P/N without ultrasonic treatment		P/N with ultrasonic treatment	
	Monkey 1: intravenous inoculation	Monkey 2: oral inoculation	Monkey 1	Monkey 2
0	1.15	1.14	1.18	1.07
4	*2.47*	1.60	*5.10*	0.79
7	*6.48*	*3.28*	*8.78*	*5.52*
9	*3.64*	*2.24*	*4.72*	*5.82*
11	1.12	1.40	1.95	*5.20*
13	1.30	1.46	*2.40*	0.80
15	1.51	1.68	0.93	0.90
19	1.53	1.51	1.20	0.80
21	1.44	1.42	1.12	0.91
24	1.31	1.53	1.00	1.53
27	0.93	1.57	0.91	0.91

NUNC immunoplates were coated with 100 μl of the IgM fraction, 50 μg/ml, sampled 27 days after inoculation, in phosphate-buffered saline (PBS) (1 hour at 37°C + overnight at 4°C). The plates were overcoated with 200 μl of PBS containing 1% bovine serum albumin (BSA) and 0.1% Tween 20 (3 hours at 37°C). A 100-μl aliquot of the stool extract (1:10 w/w) filtered on 0.22 μm porosity was incubated on an IgM-coated plate at 37°C for 2 hours and overnight at 4°C. After washings, 100 μl of immune monkey IgG labeled with peroxidase (5 g/ml) in 1% BSA plus 0.1% Tween 20 in Tris buffer pH 7.6 (TNB) containing 0.1% Na_3N was used for antigen detection. After a 1-hour incubation at 37°C and washings, the reaction was developed using 100 μl orthophenylene diamine in 0.005 M citric buffer pH 5.6 containing 1 μl H_2O_2 ml. The results are expressed by the P/N ratio: optical density (O.D.) obtained for the tested samples/mean of five negative controls. The background for negative specimens was in the range of 0.04 to 0.06 O.D. at 492 nm. Samples with a P/N ratio of more than 2:1 were considered positive. The stools (1:10 w/w) were examined with or without ultrasonic treatment (Brenson apparatus), for 3 minutes at 4°C, 4 output intensity for a volume of 2 to 3 ml, followed by centrifugation and filtration through a 0.22-μm porosity filter. Positive results are italicized.

logical criteria of the three types of hepatitis virus (A, B, δ) infection: the presence of IgM and IgG anti-HAV for virus A infection; HBsAg, anti-HBc, and IgM anti-HBc, anti-HBs, for HBV infection; and eventually δ antigen and anti-δ for HDV infection. For Ivorian patients, yellow fever markers were researched. All patients were also investigated for cytomegalovirus infection by immunocapture of IgM antibodies and for Epstein-Barr virus infection by IgM anti-VCA detection with immunofluorescence staining. NANB hepatitis-associated antigen was detected in epidemic as well as sporadic cases (Table 12.2). No antigen was detected in 90 control extracts from European patients with other gastrointestinal disorders. In most cases these results were confirmed several times with different solid-phase IgM or IgG and different labeled IgG. Some positive samples were studied with preimmune IgM-coated plates, and no antigen

Table 12.2. NANB antigen detection in stools of patients with epidemic NANB and sporadic hepatitis.

	Stool from Ivory Coast: epidemic cases	Stool from France: sporadic cases	Controls from France
NANB antigen positive	17	7	0
NANB antigen negative	44	10	90

Ninety stools from patients with gastrointestinal troubles were used as controls. All specimens from patients with presumed NANB hepatitis and 17 of 90 controls were submitted to ultrasonic treatment. Specific IgM from monkeys and peroxidase- or β-galactosidase-tagged human IgG were used as antigen catcher and tracer, respectively.

was detected. We concluded that the detected antigen appears to be closely associated with NANB virus infection.

Therefore in the third phase, continuous cell lines, two from hepatocytic origin (PLC/PRF 5 and HEP G2) and one of fibroblastic origin (MRC 5), were cultivated in Dulbecco's medium plus 10% of fetal calf serum. They were inoculated with antigen-positive and antigen-negative stool extracts after filtration through a 0.22-μm porosity filter. When the culture was confluent, a subculture was performed after EDTA-trypsin addition. In some cases, toxic substances present in the stools caused the death of hepatocytic cells early in the first week. The other NANB antigen-positive stools, which were not toxic to the cell line, induced cell syncitial formation and PLC/PRF 5 cell death after two or three passages. After washing and ultrasonic treatment of the cells, NANB virus-associated antigen was released in the liquid phase of these infected PLC/PRF 5 cells. HEP G2 and MRC 5 cells treated under the same conditions did not display a cytopathic effect or NANB virus-associated antigen production. NANB antigen appeared firmly bound to the cells. No definite viral particles could be detected in the cell extract; and ultrasonic treatment affected the viral morphology. Liver extracts from inoculated monkeys did not display cellular toxicity. PLC/PRF 5 cells inoculated with a 20% solution of liver extract from infected *Saimiri* monkeys showed antigen positivity, as in the case of stool specimens (Table 12.3). These results suggested viral replication and seemed to confirm the hepatotropism of the NANB agent. Whatever the process, a large amount of NANB antigen was obtained. Viral material could not be immunologically detected directly in liver extract because of unspecific binding.

In the fourth phase of the study, virus-associated antigen was purified from stool specimens of different origins. Three profiles of antigen distribution were observed (Fig. 12.1). In fresh specimens the antigenic activity banded at a density of 1.29 to 1.32 g/cm^3 in CsCl. Some stool specimens stored at 4°C for weeks or months and fresh specimens submitted to ultrasonic treatment showed antigenic activity only at a density of 1.10 to 1.15 g/cm^3, whereas others showed activity at both densities. Antigenic

Table 12.3. Detection of NANB antigen in hepatoma cell line inoculated with a human stool extract from epidemic NANB hepatitis patient and a liver extract from an infected *Saimiri* monkey.

Hepatoma cell line	Cells inoculated with stool extract of Ivorian patient	Cells inoculated with liver extract of infected *Saimiri* monkey	Cells inoculated with liver extract of control *Saimiri* monkey	Noninoculated cells
PLC/PRF 5	8.19[a]	4.39	2	1.01
HEP G 2	1.62	1.14	2	1.07

Liver extract *(Saimiri* monkey 1) was prepared in PBS (10% w/w) with an ultraturax homogenizer. After centrifugation, the pellet was resuspended in 5 volumes of TNB and submitted to ultrasonic treatment in the presence of phenylmethylsulfonide chloride (PMSF) 0.005 mg/ml. A liver extract of one *Saimiri* monkey inoculated with a stool from a patient suffering from presumed cytomegalovirus hepatitis (with specific IgM) served as control. After inoculation, the medium was replaced three times per week. The precedent supernatant was present at a concentration of 10%. Inoculated cell cultures, after three passages, were centrifuged. The pellet was taken in TNB, and after adding PMSF the cells were submitted to ultrasonic treatment. Numbers indicate the P/N ratio. Solid-phase monkey IgG and labeled monkey IgG were used for ELISA.
[a]After dilution at 1:100, the P/N ratio was 6.29.

material of low density appearing under these conditions is believed to be the result of a distorting effect of stool enzymes or physical treatment. Fractions from the CsCl gradient positive for viral antigen were submitted to a rate run in 10 to 60% sucrose, and reactive fractions were analyzed by electron microscopy. The NANB antigen was found in 40.0 to 41.5% sucrose but predominantly in 26 to 28% sucrose. Of nine NANB antigen-positive specimens studied, seven showed virus-like particles: One sample was from an inoculated *Saimiri* monkey (the second monkey was not examined), two (of three tested) were from patients with acute NANB epidemic hepatitis (Ivory Coast), and four (of five tested) were from patients with NANB sporadic hepatitis (France). Two types of spherical viral particle were seen: large particles, about 60 to 80 nm in diameter with a double membrane structure, and smaller, homogeneous particles 25 to 30 nm in diameter without a defined internal structure. Some tubular structures, about 15 nm in diameter (or width) and up to 1 μm in length, were also found more or less attached to the particles (Fig. 12.2). These filaments resemble bacterial pili frequently seen in stool preparations (normal and infected individuals). The most characteristic particles were observed in fresh specimens not submitted to ultrasonic treatment. Large spherical particles were seen mostly in fractions of CsCl density 1.29 to 1.32 g/cm^3 and in 40.0 to 41.5% sucrose; only small particles were observed in the 1.10 to 1.15 g/cm^3 CsCl fractions and 26 to 28% sucrose. Whether small particles of low density represent a released form from the original virus is not clear. Six stool specimens negative for NANB antigen were submitted to ultracentrifugation in CsCl gradient and were studied, with con-

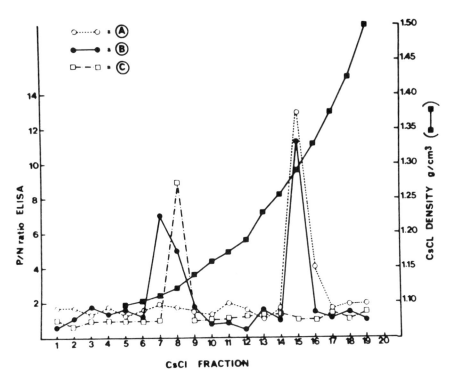

Fig. 12.1. Antigen distribution in a CsCl gradient after isopycnic ultracentrifugation: Three milliliters of stool (10% in RPMI) after 0.22-μm filtration were submitted to isopycnic centrifugation in CsCl (1.1 to 1.6 g/ml) density gradient for 48 hours at 78,000 × g. Fractions (0.6 ml) were collected and diluted 1:2 in PBS and tested for the presence of NANB antigen with specific monkey IgM and specific labeled human IgG. A = fresh stools; B and C = conserved or ultrasonically treated stools.

trols, by electron microscopy. No virus-like structure was found in any of these preparations. In conclusion, the large particle represents a possible candidate for the NANB virus.

Lastly, detection of NANB antigen in stools from French patients suggested that the virus of epidemic NANB hepatitis may be involved in sporadic NANB hepatitis as well. Antigen associated with NANB virus was effectively detected in 7 of 17 stool specimens from patients with acute sporadic NANB hepatitis (Table 12.2). Serological investigations confirmed these results: Of eight patients whose sera were available, seven displayed anti-NANB serum antibody. Because epidemics of NANB hepatitis are usually observed in developing countries, it was of interest to look for a role of the same virus in sporadic cases in these countries. Therefore 28 sera from Moroccan patients suffering from sporadic acute

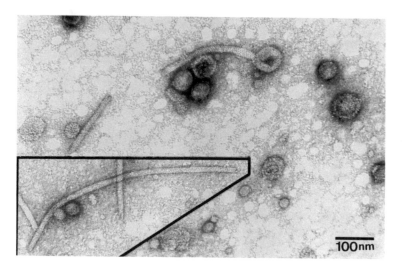

Fig. 12.2. Characterization of viral particles by electron microscopy in the stools of NANB-infected monkeys and patients with NANB hepatitis. After isopycnic ultracentrifugation as described in Figure 12.1, positive fractions were pooled and dialyzed against TNB and then submitted to rate zonal ultracentrifugation in sucrose gradient (10 to 60%,) at 78,000 × g for 5 hours. Positive fractions were pooled, dialyzed, concentrated, and examined by electron microscopy. One drop of viral preparation was spread onto a 200-mesh carbon-coated grid; and after absorption on filter paper, 1 drop of 2% uranyl acetate in distilled water was added. After 2 minutes the grid was absorbed by filter paper and studied using a JEM 100 C × II ultramicroscope.

NANB hepatitis were examined for the presence of antibodies to the virus of epidemic NANB hepatitis: 68% of the patients displayed antibodies to NANB virus, whereas the antibodies were detected in only 10% of sera from blood donors in the same country ($p < 0.0001$). Hemodialysis or laboratory work with human blood products did not seem to increase the risk of infection by this NANB virus (Table 12.4). Furthermore, for antibody-positive sera, one from Morocco and three from France, were studied by ELISA against two NANB antigenic preparations from the Ivory Coast and two others from France. Cross-reactivity was observed, indicating antigenic similarity. Thus the role of the virus of epidemic NANB hepatitis in the sporadic form of NANB appears to be confirmed.

Conclusions

A new viral agent involved in NANB hepatitis was characterized in this study. This viral agent appears to be responsible for the epidemic form as well as the sporadic form of NANB hepatitis. It is different from HAV.

Table 12.4. Detection of antibody specific to NANB antigen in sera of patients with sporadic NANB hepatitis.

| Country | Source of sera | | | |
	Blood donors	Laboratory staff	Hemodialysis patients	NANB sporadic hepatitis
Morocco	8/78	0/12	1/12	19/28
France	0/26	—	—	7/8[a]

Antibody presence was demonstrated by inhibition of the reaction described in Table 12.1. For inhibition, 50 μl of a reference stool extract from an infected monkey was incubated for 1 hour at 37° and overnight at 4°C with 50 μl undiluted serum. All the other steps of the reaction were performed as for antigen detection in human stools (Table 12.2.). The P/N ratio of this reference antigen was in the range of 6 to 8. Twenty-four undiluted normal human sera were tested for inhibition against reference antigen compared to PBS; none of them inhibited. In further experiments, 50 μl of a mixture of these sera served to establish the cutoff point of the reaction. Tested samples were considered positive if inhibition exceeded 50%.
[a]Five patients suffered from acute NANB hepatitis, and in all of them NANB antigen was detected in the stools; only one of the five was sero-negative for anti-NANB antibody. Two other patients had a history of acute liver disease that had occurred 9 and 3 months before and antigen was no longer detected in the stools. The last patient had presented in the past a NANB hepatitis with detectable antibody that had disappeared in 3 years.

1. Epidemic NANB virus infection was observed in one monkey and in patients with anti-HAV immunity.
2. Usual markers of acute HAV infection (IgM anti-HAV) were not detected.
3. This virus induced a specific immune response, and the antibodies did not react with a preparation of HAV (Abbott Laboratories); sera from patients convalescing from hepatitis A were negative in the test for detecting antibodies to epidemic NANB virus.

Epidemic NANB virus is also distinct from HBV and from HDV.

1. The disease evolved in some patients having postinfection protective anti-HBs antibodies as well as in a vaccinated European patient with strong anti-HBs immunity.
2. HBsAg and HBeAg generally were not detected in sera; if HBV markers preexisted in the NANB hepatitis (in some Ivorian patients), anti-HBc IgM and anti-HDV did not appear.
3. IgM from monkeys infected with epidemic NANB virus and used for viral antigen detection did not react with either HBV (8) or HBsAg purified preparations (9), or with sera from patients actively replicating HBV; anti-HBs antibodies did not interfere in the test for antibodies to NANB virus detection.

The NANB virus described here seems to be widely distributed. In addition to the cases described in our study, the virus-associated antigen was also found in a Senegalese patient, and antibodies were found in serum from a patient convalescing from the Algerian epidemics (3). The epidemic NANB agent is classically considered to be a water-borne virus. Nevertheless, its role in sporadic hepatitis, particularly in a country with good hygienic conditions, suggests some other route of transmission.

References

1. Andjaparidze AG, Balayan MS, Savinov AP, et al: Non A non B hepatitis transmitted by the fecal oral mode experimentally produced in monkeys. *Vopr Virusol* 1986;31:73–81.
2. Balayan MS, Andjaparidze AG, Savinska SS, et al: Evidence for a virus in non A non B hepatitis transmitted via fecal oral route. *Intervirology* 1983;20:23–31.
3. Belabbes EH, Bouguermough A, Bentallah A, Illoul G: Epidemic non A non B viral hepatitis in Algeria: strong evidence for its spreading by water. *J Med Virol* 1985;16:275–281.
4. Francis DP, Hadler SC, Prendergast TJ: Occurrence of hepatitis A, B and non A non B in the United States: CDC Sentinel County hepatitis study. *Am J Med* 1984;76:69–74.
5. Kane MA, Bradley DW, Shrestha SM, et al: Epidemic non A non B hepatitis in Nepal: recovery of a possible etiologic agent and transmission in marmosets. *JAMA* 1984;252:3140–3145.
6. Khuroo MS: Study of an epidemic of non A non B hepatitis: possibility of another human hepatitis virus distinct from post-transfusion non A non B type. *Am J Med* 1980;68:818–824.
7. Molinie C, Roue E, Saliou P, et al: Hépatite aigue non A non B epidémique: étude clinique de 38 cas observés au Tchad. *Gastroenterol Clin Biol* 1986;10:475–479.
8. Petit MA, Capel F, Pillot J: Demonstration of a firm association between hepatitis B surface antigen proteins bearing polymerized human albumin binding sites and core-specific determinants in serum hepatitis B viral particles. *Mol Immunol* 1985;22:1279–1287.
9. Pillot J, Petit MA: Immunochemical structure of the hepatitis B surface antigen vaccine. I. Treatment of immobilized HBsAg by dissociation agents with or without enzymatic digestion and identification of polypeptides by protein blotting. *Mol Immunol* 1984;21:53–60.
10. Prince AM: Non A non B hepatitis viruses. *Ann Rev Microbiol* 1983;37:217–232.
11. Sarthou JL, Budkowska A, Sharma MD, et al: Characterization of an antigen-antibody system associated with epidemic non A non B hepatitis in West Africa and experimental transmission of an infectious agent to primates. *Ann Inst Pasteur Virol* 1986;137E:225–232.
12. Shiraishi H, Alter HJ, Feinstone SM, Purcell RH: Rheumatoid factor-like reactants in sera proven to transmit non A non B hepatitis: a potential source of false positive reactions in non A non B assays. *Hepatology* 1985;5:181–187.

13. Sreenivasan MA, Arankale VA, Seghal A, Pavri KM: Non A non B epidemic hepatitis: vizualization of virus-like particles in the stools by immune electron microscopy. *J Gen Virol* 1984;65:1005–1007.
14. Yamauchi M, Nakajima H, Kimura K, et al: An epidemic of non A non B hepatitis in Japan. *Am J Gastroenterol* 1983;78:652–655.

Part V
Gastrointestinal Infections

CHAPTER 13

Attenuated Oral Typhoid Vaccine Ty 21a

R. Germanier* and S.J. Cryz, Jr.

Parenterally administered killed, whole-cell vaccines against typhoid fever were introduced almost a century ago and have been used in essentially the same form since. These vaccines are, however, not wholly satisfactory due to the frequency of adverse reactions and variable efficacy. Inactivated whole-cell vaccines for oral application, on the other hand, are well tolerated, although their efficacy could not be demonstrated in controlled field trials.

There is no reliable serological test available that can predict vaccine efficacy in man. Therefore, experimental work on immunity against *Salmonella typhi* infection has been carried out in a murine "typhoid" model, using *Salmonella typhimurium*. From such studies it appears that optimal protection can be achieved only by vaccination with live attenuated bacteria. This situation is particularly evident when the oral route of application is used. Multiple doses of orally-administered killed vaccine can delay the proliferation of the challenge bacteria in macrophages. However, the invading organisms eventually multiply sufficiently to kill the host. In contrast, live-attenuated *S. typhimurium* vaccine strains can provide complete protection against a lethal challenge.

The first live oral-attenuated typhoid vaccine evaluated in humans was a streptomycin-dependent strain of *Salmonella typhi*. This vaccine proved to be safe in volunteer studies but failed in its final lyophilized form to induce clinical protection (4).

Based upon results obtained in the murine model system, a mutant strain of *S. typhi* termed Ty 21a, which lacks the enzyme UDP-4-galactose-epimerase, has been developed as a candidate vaccine strain. The critical attribute of this attenuated strain is that lipopolysaccharide, the expression of which is essential for both virulence and protective capacity, is synthesized only under conditions that induce autolysis of the bacteria. This mechanism thereby renders the *S. typhi* Ty 21a strain avirulent despite the presence of smooth LPS essential to evoke a protective immune response.

---

*Deceased

Clinical Trials

The *S. typhi* Ty 21a strain was tested for stability, safety and efficacy as a live oral vaccine in 155 American volunteers (3). These individuals ingested five to eight doses of up to 1×10^{11} freshly prepared viable bacteria without significant adverse reactions. Despite this high dosage, the vaccine strain was seldom excreted in the feces on the day after vaccination and never for longer than three days post-vaccination. Of the 958 *S. typhi* isolates from coprocultures, none showed any sign of reversion to a wild type phenotype.

The vaccine afforded 87% protection against wild type *S. typhi* challenge in volunteer trials (3). These excellent results were obtained in the face of a challenge dose that caused typhoid fever in 53% of the unimmunized control group. In prior studies of a similar design, parenteral vaccines afforded no protection against such a high challenge dose. In addition to protection against overt clinical disease, vaccination also reduced the time in which the challenge organisms were excreted in the feces.

In 1978, a placebo-controlled field trial was initiated in Alexandria, Egypt, to evaluate the safety and efficacy of the Ty 21a vaccine strain in an area endemic for typhoid fever (6). A total of 32,388 school-aged children participated. Each child received three doses of vaccine or placebo. Each dose of vaccine contained an average of 3.5×10^9 lyophilized Ty 21a bacteria reconstituted immediately before administration in a sucrose-buffer solution. To neutralize gastric acidity, 1 g of bicarbonate as a chewable tablet was given immediately before vaccination. These children as well as a group of 25,628 unvaccinated age-matched children were followed over a 3-year period. Twenty-two blood-culture confirmed cases of typhoid fever were detected in the placebo group, 38 cases in the nonvaccinated group, and only one case in the vaccine group. Protection afforded by the vaccine was estimated to be 96%.

The dose regimen used in the Egyptian trial was selected more or less arbitrarily with the hope that three closely spaced immunizations would induce a good immune response. It was conceivable that one or two doses would afford the same level of protection. In addition, a vaccine formulation more practical for large-scale immunization was subsequently developed. This entailed placing the lyophilized vaccine strain in enteric-coated capsules, resistant to low pH, thereby circumventing the need to neutralize gastric acidity.

To evaluate fewer vaccine doses and the new formulation, a large controlled field trial was initiated in 1982 in Santiago, Chile where the attack rate for typhoid fever was approximately three times that seen in Alexandria, Egypt (1). In a randomized double-blind study, 92,000 schoolchildren were orally vaccinated with one or two doses of enteric-coated vaccine or placebo.

Table 13.1. Efficacy of one or two doses of Ty 21a vaccine in enteric-coated capsules after 4 years of surveillance.

Parameter	Placebo	One dose	Two doses
Year 1			
Incidence/10^5	210	176	109
Efficacy (%)		16	48
Year 2			
Incidence/10^5	141	85	40
Efficacy (%)		39	72
Year 3			
Incidence/10^5	69	70	54
Efficacy (%)		0	21
Year 4			
Incidence/10^5	78	100	79
Efficacy (%)		0	0

The highly optimistic expectations dictating the design of this trial were only partially met. After 4 years of surveillance one dose of vaccine gave 14% protection and two doses 38% (Table 13.1).

Because at least three parameters were changed in this trial compared to the Egyptian trial (i.e., the vaccine formulation, the number of doses, and the field trial area) the data obtained raised more questions than they answered. Therefore, in 1983 a second field trial was initiated in Santiago, Chile. A group of 141,000 school-aged children were given three doses of vaccine in enteric-coated capsules with an interval of either 2 days or 21 days between vaccinations. The efficacy was 67% and 49%, respectively after 3 years of surveillance (5). The fact that an immunization schedule with a greater time interval between doses was less effective indicates that multiple doses serve to effect a vaccine "take" confirming the relation between the number of doses and vaccine efficacy (Table 13.2).

These results were confirmed in a third field trial in Santiago, Chile (1). A group of 248,000 children received two, three or four doses of enteric-coated vaccine. For ethical reasons no placebo group was included in this study. The incidence of typhoid fever after 1 year in the four-dose group was 53% lower than in the three-dose group and 62% lower than in the two-dose group (Table 13.3).

Discussion

Several placebo-controlled field trials have clearly established the safety and efficacy of the Ty 21a vaccine strain against endemic typhoid fever. While the currently available enteric-coated formulation is convenient to

Table 13.2. Efficacy of three doses of Ty 21a vaccine in enteric-coated capsules after 3 years of surveillance.

Parameter	Placebo	Short interval	Long interval
Year 1			
Incidence/10^5	122	32	37
Efficacy (%)		74	70
Year 2			
Incidence/10^5	83	36	46
Efficacy (%)		56	44
Year 3			
Incidence/10^5	104	36	69
Efficacy (%)		65	33
Average of 3 years			
Incidence/10^5/year	103	33	47
Efficacy (%)		67	49

administer and provides about 70% protection, this is considerably less than the 96% efficacy rate seen in Alexandria, Egypt. At present, it is not clear why protection varied between these 2 trial sites, but there are several distinct possibilities. First, the attack rate for typhoid fever was about three times greater in Santiago than in Alexandria. However, analysis of vaccine efficacy on a month by month basis showed that protection did not wane during the "typhoid season" in Santiago when disease incidence was highest. It must also be taken into consideration that a higher attack rate reflects the fact that a certain proportion of the population will be exposed to numbers of *S. typhi* sufficient to overcome vaccine-induced immunity. We know from prior volunteer challenge studies that immunity can be readily overcome by increasing the challenge dose (3). A more likely explanation is that the difference in efficacy was due to vaccine formulation. In Alexandria, the vaccine was administered in a liquid form after neutralization of gastric acidity whereas in Santiago the vaccine was in the form of enteric-coated capsules. It may be that administering the vaccine as a liquid "bolus" is more effective than enteric-coated capsules at delivering viable organisms to the ileum where they induce a protective immune response.

To address the above question, two additional field trials are now being

Table 13.3. Comparison of the efficacy of two, three, or four doses of Ty 21a vaccine in enteric-coated capsules.

Parameter	Two doses	Three doses	Four doses
Vaccinees (No.)	94,387	95,543	58,614
Typhoid cases (No.)	121	111	32
Incidence per 10^5	128	116	54

conducted in Plaju, Indonesia and in Santiago, Chile comparing Ty 21a given as either a liquid or as an enteric-coated vaccine formulation. Preliminary results after 1 year of surveillance based on overall incidence indicate that the liquid formulation is considerably more effective in the Santiago field trial area and marginally better in the Plaju trial area. Detailed analysis will have to wait for approximately 1 year when the code is broken.

References

1. Black RE, Levine MM, Clements ML, et al: Efficacy of one or two doses of Ty 21a Salmonella typhi vaccine in enteric-coated capsules during the first year of a field trial in Santiago, Chile, in *Twenty-third Interscience Conference on Antimicrobial Agents and Chemotherapy, Las Vegas* 1983, pp 1–12.
2. Germanier R, Fürer E: Isolation and characterization of Gal E mutant Ty 21a of Salmonella typhi: a candidate strain for a live, oral typhoid vaccine. *J Infect Dis* 1975;131:553–558.
3. Gilman RH, Hornick RB, Woodward WE, et al: Evaluation of a UDP-glucose-4-epimeraseless mutant of Salmonella typhi as a live oral vaccine. *J Infect Dis* 1977;136:717–723.
4. Levine MM, DuPont HL, Hornick RB, et al: Attenuated, streptomycin-dependent Salmonella typhi oral vaccine. *J Infect Dis* 1976;133:424–429.
5. Levine MM, Ferreccio C, Black RE, et al: Large-scale field trial of Ty 21a live oral typhoid vaccine in enteric-coated capsule formulation. Lancet 1987-*i*:1049–1052.
6. Wahdan MH, Sérié C, Cerisier Y, et al: A controlled field trial of live Salmonella typhi strain Ty 21a oral vaccine against typhoid: three-year results. *J Infect Dis* 1982;145:292–295.

CHAPTER 14

Oral B Subunit–Whole Cell Vaccine Against Cholera: From Basic Research to Successful Field Trial

Jan Holmgren, John Clemens, David Sack, and Ann-Mari Svennerholm

Cholera is an important cause of morbidity and mortality in many developing countries. Cholera is also the prototype for a large group of diarrheal diseases—the "enterotoxic enteropathies"—which may be responsible for roughly half of all diarrheal disease episodes in the world. These diseases are caused by various bacteria that produce one or more toxins, which by upsetting normal fluid transport processes in the gut give rise to watery diarrhea.

Because parenteral vaccination against cholera has yielded only modest and short-term protection, attention has turned to the development of oral vaccines that stimulate intestinal immunity more efficiently. Furthermore, because natural protection from cholera appears to result from local intestinal antibacterial and antitoxic antibodies, which act synergistically, special efforts have been made to develop vaccines that could effectively stimulate these antibodies.

Basic research at the University of Göteborg, Sweden, and subsequent studies in collaboration with the International Centre for Diarrhoeal Disease Research in Bangladesh (ICDDR,B), Institut Mérieux in France, The National Bacteriological Laboratory in Sweden, and the University of Maryland in the United States have resulted in the development and systematic testing of a promising oral vaccine against cholera. This vaccine, which combines the B subunit component of the cholera toxin molecule with formalin- or heat-inactivated classical and E1 Tor cholera vibrios, has given promising results in the first clinical trials and now also in a large field trial in Bangladesh.

Development of the Vaccine

The development of this vaccine took its departure from basic studies of the mechanisms of disease and immunity in cholera (6, 7). Of prime importance was clarification of the molecular properties of cholera toxin and its mode of action (6), which made the pathogenesis of cholera better

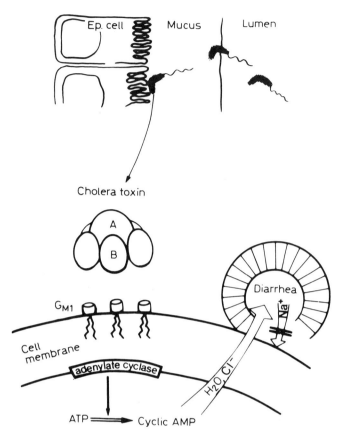

Fig. 14.1. Pathogenic mechanisms of cholera. Immune factors may interfere at all levels before toxin binding and action on target cells.

understood than that of any other infectious disease (Fig. 14.1). Cholera toxin consist of a binding region composed of five B subunits into which is inserted the toxic-active component, the A subunit. The toxin binds to specific receptors on the brush border membrane of the intestine, which consist of the GM2 monosialoganglioside. This step leads to translocation of the A subunit across the membrane into the cell where, by mechanisms now well understood (enzymatic ADP-ribosylation of a regulatory protein of adenylate cyclase), it activates adenylate cyclase, which results in an increase of intracellular cyclic AMP. The cyclic AMP then causes diarrhea and fluid loss by both inhibiting uptake of sodium chloride and stimulating active electrolyte secretion by intestinal epithelial cells (6).

Another important line of research for the development of improved cholera vaccines has focused on the gut mucosal immune system and its optimal stimulation by immunogens. These studies have clearly indicated the important role of locally produced immunoglobulin A (IgA) antibodies and IgA immunologic memory for protection against cholera as well as several other enteric infections. They have also shown that the oral route of administration of immunogens is usually superior to parenteral immunization for stimulation of protective immunity, intestinal IgA antibody formation, and gut mucosal immunologic memory (7) (see Chap. 18).

A third important observation for the design of this vaccine has concerned the synergistic cooperation between antitoxic and antibacterial immune mechanisms in cholera (13). In cholera-infected individuals cholera toxin is elaborated by cholera vibrios (*Vibrio cholerae* 01) adhering to and colonizing the small intestine. Studies of cholera immunity in humans and animals have identified two main protective antibodies, one being directed against the *V. cholerae* cell wall lipopolysaccharide (LPS) and the other against cholera toxin (7). These two antibodies can independently protect against experimental cholera by inhibiting bacterial colonization and toxin action, respectively; and when present together in the gut, the anti-LPS and anti-cholera-toxin antibodies have been found to have a strikingly synergistic protective effect (13).

Cholera toxin neutralizing antibodies are mainly directed against the B subunit, and their main action is to prevent the binding of toxin to mucosal receptors. The identification of anti-LPS antibodies being mainly responsible for the antibacterial cholera immunity does not preclude the possible contribution of additional protective antibodies. Antibodies to cell-bound *V. cholerae* hemagglutinins and outer membrane proteins may be of particular interest in this regard, as these protein antigens may have a role as bacterial adhesins in the pathogenesis of the disease.

Based on this knowledge efforts to develop an improved vaccine against cholera have followed two alternative, equally logical approaches. One, used by us, has been to prepare an oral killed vaccine that combines immunogenic amounts of a suitable "toxoid" antigen and the main protective somatic *V. cholerae* antigens (5). The other has been to develop, using genetic methods, a live attenuated vaccine strain that may produce and release these antigens when the vaccine strain multiplies in the gut of vaccinated individuals. So far, candidate live vaccine strains prepared according to the second approach have caused modest, yet unacceptable diarrhea when fed to human volunteers, which has precluded their further testing as potential vaccines (9). In contrast, extensive testing of a combined B subunit–whole cell vaccine (5) administered by the oral route has revealed no significant side effects and, as described below, has shown strong mucosal immunogenicity and protective efficacy in both adults and children.

The composition of the B subunit–whole cell cholera vaccine (B + WCV) (Table 14.1) is such that it provides sufficient amounts of the most suitable antigens to evoke protective mucosal antitoxic and antibacterial cholera immunity.

Table 14.1. Composition of one dose of oral B subunit–whole cell cholera vaccine.

Component	Amount
B subunit	Oral cholera toxin B subunit: 1 mg
WCV	Oral inactivated whole-cell cholera vaccine: 1×10^{11} cells
Contents	Heat-killed *V. cholerae* Inaba: classical biotype (strain Cairo 48) 2.5×10^{10} cells
	Heat-killed *V. cholerae* Ogawa: classical biotype (strain Cairo 50) 2.5×10^{10} cells
	Formalin-killed *V. cholerae* Inaba: El Tor biotype (strain Phil 6973) 2.5×10^{10} cells
	Formalin-killed *V. cholerae* Ogawa: classical biotype (strain Cairo 50) 2.5×10^{10} cells

The purified B subunit component (a pentamer of individual B subunits) of cholera toxin seems to be an almost ideal "toxoid" immunogen for an oral cholera vaccine. The exclusion of the enzymatically active portion of cholera toxin, the A subunit, eliminates the risk of reversion to toxicity but does not result in a significant loss of protective antigenic determinants (5). Antibodies to the B subunits are more active in neutralizing cholera toxin activity than antibodies to the A subunit and are also formed in much greater quantity. The B subunit may be particularly good as an oral immunogen because it is able to bind to the intestinal epithelium, and this property is known to be important for stimulating mucosal immunity in animals (12). The B subunit has been shown to be more immunogenic than chemically inactivated toxoids; however, it is less immunogenic than cholera toxin because, like any other nontoxic preparation, it lacks the adenylate cyclase activating activity of the intact toxin that makes cholera toxin a strong mucosal adjuvant (10).

The heat-killed organisms in the whole-cell vaccine provide appropriate amounts of LPS antigens of the Inaba and Ogawa serotypes, and the formalin-killed E1 Tor and classical organisms provide heat-labile antigens that may add to the antibacterial immunogenic effect.

Because of the acid sensitivity of particularly the B subunit component (the B pentamer dissociates with a loss in immunogenicity at pH ≤4) the vaccine has been administered after and/or together with a sufficient volume of a 2.8% sodium bicarbonate/1.1% citric acid buffer solution to ensure adequate neutralization of stomach acidity for preservation of the vaccine when it passes through the stomach.

Vaccine Production

The procedures for preparation and characterization of the B + WCV are well defined. The B subunit component is prepared from the culture filtrate of fermentor-grown *V. cholerae* 569B bacteria by affinity chromatography on a specific toxin-binding Spherosil-GM1 ganglioside column followed by gel filtration on a Sephadex column in acid buffer (17). By this procedure about 50 g of highly purified B subunit can be obtained from each 1000-L culture. The heat-killed organisms in the whole cell vaccine are prepared by standard procedures for cholera vaccines using established Inaba and Ogawa cholera vaccine strains (Table 14.1). The formalin-killed organisms that are also included in the whole-cell vaccine represent an E1 Tor and a classical *V. cholerae* strain that have been selected because they are particularly good, stable producers of the main cell-associated hemagglutinins associated with E1 Tor and classical *V. cholerae* organisms, respectively. These organisms are grown and killed using procedures found to optimize the expression and preservation of these heat-labile hemagglutinins as well as other surface protein antigens (Svennerholm et al., unpublished).

Safety Studies

The safety of the B-WCV in humans was first evaluated in several small-scale clinical trials in Sweden and Bangladesh (1, 8, 14–16). These studies showed that oral administration of the vaccine in a bicarbonate solution did not result in any detectable side effects. In an initial study in Sweden, ten adult volunteers received one or two doses of B+WCV containing 0.5 mg of B subunit and 10^{10} killed vibrios. In no instance were any local or systemic side effects observed within 1 month after immunization. In subsequent studies, 30 adult Swedes and 16 Bangladeshis were given two doses of vaccine with 0.5 to 2.5 mg of B subunit and 5×10^{10} vibrios in each. Surveillance for side reactions was performed for 10 days after each immunization. In no instance did the oral immunizations give rise to any detectable side effects.

These initial studies have been followed by administration of the complete vaccine or the B subunit component alone to more than 1000 volunteers of various ages in Bangladesh, Sweden, and the United States without any side effects that can be ascribed to the vaccines. A pretrial in 1200 children and adults in Matlab, Bangladesh in the autumn of 1984 had as its primary objective to make sure that the vaccine and placebo preparations that were later to be used in a large efficacy field trial (see below) were free of side effects, which was convincingly shown to be the case (1–3).

Mucosal Immunogenicity Studies in Humans

The combined B-WCV has been tested for its ability to stimulate mucosal antitoxic and antibacterial antibody responses in the intestine of human volunteers in both Bangladesh and Sweden (8, 14–16). In an important study in Bangladesh (16) groups of healthy adults were given two oral doses of vaccine 1 month apart and were compared with a group of patients with severe cholera during their convalescence from disease and subsequently after 1 month following a single oral immunization with the B + WCV. In this study the whole-cell vaccine dose was 5×10^{10} killed organisms and various B subunit doses: 2.5 mg and 0.5 mg were used in separate groups of volunteers. Mucosal antibody responses were monitored by performing intestinal lavage at various times after immunization and examining the fluid specimens for IgA antitoxin and antibacterial (anti-LPS) antibodies by ELISA methods. Total IgA was also determined in these fluids by ELISA to permit the expression of all titers in relation to total IgA.

A single peroral administration with the 2.5 mg B subunit dose stimulated a local intestinal IgA antitoxin response in all of the volunteers and in most cases a local IgA anti-LPS antibody response as well (Table 14.2). Two immunizations and then also the 0.5 mg dose of B subunit functioned well, evoking a significant gut mucosal response in both antitoxin and anti-LPS antibodies in almost all cases. These initial and memory responses were found to be closely comparable to those in convalescents from severe cholera disease with regard to the frequency, magnitude, and kinetics of intestinal IgA antitoxin and anti-LPS (16).

From other studies there is evidence that the mucosal immune response to at least the B subunit component of the vaccine is associated with long-lasting immunologic memory. In one study evidence of such memory was found 15 months after the initial immunization and was characterized by a more rapid response, which could also be triggered by a lower antigen dose than that needed to elicit a primary immune response (15). In other, still ongoing studies, evidence of memory has been found for several years

Table 14.2. Gut mucosal IgA antitoxic and antibacterial immunity after immunization with B&WCV and after cholera disease.

			Intestinal IgA responses[b]			
			Antitoxin		Anti-LPS	
Immunization[a]	B subunit dose (mg)	WCV dose (No. of organisms)	%[c]	Rise[d]	%	Rise
First	0.5	5×10^{10}	65	5.0	57	2.6
	2.5	5×10^{10}	100	8.0		
Second	0.5	5×10^{10}	76	6.9	92	9.0
Cholera disease	—	—	89	5.0	89	10

[a]The interval between the first and second immunizations was 28 days.
[b]Specific IgA antibody titer increases in post- versus preimmunization intestinal lavage specimens after adjustment to the same total IgA level.
[c]Freqency responders with ≥ two fold titer increase response.
[d]Geometric mean fold increase in response for whole group.
Combined data from refs. 10 and 14.

after the initial oral immunizations with B+WCV (Jertborn et al., unpublished data).

Immunologic memory in the gut mucosal IgA system may be important for the duration of protection after cholera disease or vaccination. Thus Lycke and Holmgren (11) have shown a practically life-long cholera antitoxic memory in the gut of mice that had been perorally immunized with cholera toxin, and they found that this memory could be triggered by a single low-dose antigen exposure to mount a rapid, strong anamnestic response resulting in a three-fold increase in specific IgA antitoxin-producing cells in lamina propria within a 16-hour period. This kind of rapid, low-dose triggered IgA response is clearly a possible explanation for the long-term protective immunity against cholera seen in convalescents from disease and is hoped to be achieved also by orally administered cholera vaccines.

Challenge Studies in Volunteers

Two studies of vaccine efficacy have recently been performed at the Center for Vaccine Development in Baltimore (1). In the first trial 19 adult volunteers received three oral doses of combined B+WCV vaccine at 2-week intervals. Each dose of vaccine consisted of 5 mg of B subunit and 5×10^{10} heat-killed classical Inaba, 5×10^{10} classical Ogawa, and 1×10^{11} formalin-treated E1 Tor Inaba organisms, i.e., the first three components of the whole-cell vaccine (Table 14.1). In the second trial volunteers were tested with the same whole-cell vaccine preparation in the absence of any B subunit. Five weeks after the last immunization, the vaccinees and unvaccinated controls were challenged with 2×10^6 live V. cholerae organisms to determine the protective efficacy of the vaccines.

In the unvaccinated controls this challenge dose caused diarrheal illness in 100% (first trial) and 75% (second trial), respectively (Table 14.3). The

Table 14.3. Protective efficacy of oral B+WCV or WCV alone in North-American volunteers against challenge with 2×10^6 El Tor Inaba V. cholerae.

Immunization	No. of vaccinees	No. with any diarrhea	No. with ≥2 l liter diarrhea
B+WCV	11	4[a] (PE = 64%)	0[c] (PE = 100%)
Controls	7	7[a]	4[c]
WCV	9	3[b] (PE = 56%)	0[d] (PE = 100%)
Controls	8	6[b]	3[d]

[a] $p = 0.01$.
[b] $p = 0.11$.
[c] $p = 0.01$.
[d] $p = 0.08$ (Fisher's exact test for all p values).
Data from Black et al. (1).

combination vaccine with the B subunit provided 64% protection, and the whole-vibrio vaccine given alone gave 56% protection. In addition, illnesses in vaccinees were significantly milder than those in controls and developed later, and both vaccines provided complete protection against more severe disease predefined as diarrhea of ≥ 2 L (Table 14.2).

Monitoring the serological responses verified the previous clinical trials that both the whole-cell and B subunit components of the combination vaccine stimulated significant antibody responses both in serum and locally in the gut. Thus the combination vaccine stimulated fourfold or greater serum vibriocidal antibody responses in 89% of the volunteers as well as a significant rise in secretory IgA antibodies to LPS in jejunal fluid in half of the vaccinees. The B subunit component of the oral combination vaccine stimulated a serum antitoxin response in all of the vaccinees and an intestinal response in 74%. These levels of serum and local antibody responses after vaccination approached those stimulated by cholera itself in previous volunteer studies. With the vaccine, as with cholera, local antitoxic responses were short-lived, returning to baseline levels after 2 months.

The substantial level of protection demonstrated in these studies in North American volunteers against a dose of *V. cholerae* that caused cholera in nearly 90% of controls, including complete protection against more severe disease, suggest that these vaccines might provide at least as high a level of protection if given to the population of an endemic area. The mechanisms of protective immunity are unknown, but it is likely that local intestinal IgA is a critical component. As mentioned, there is evidence in animal models that antibacterial and antitoxic antibodies work synergistically in providing protection against cholera. Because of the limitations of the volunteer model (small number of individuals and the variable ability in response to the challenge dose) it was not possible to demonstrate such synergism in these studies, although the combination vaccine appeared to result in somewhat enhanced protection in comparison to the whole-vibrio vaccine alone. A field efficacy trial in an endemic area would provide the opportunity to assess the true level of protection, and a comparison of the whole vibrio vaccine alone and with B subunit would determine the importance of synergy between antibacterial and antitoxic mechanisms of immunity.

Field Trial of Oral Cholera Vaccines in Bangladesh

The ultimate test of the level of protection afforded by cholera vaccines is a field trial for efficacy in a cholera endemic area. The promising findings obtained with particularly the oral cholera vaccine consisting of a combination of killed cholera vibrios and the B subunit component of cholera toxin (B + WCV) but also with the whole-cell constituents of this vaccine alone (WCV) prompted the International Centre for Diarrhoeal Disease Research, Bangladesh (ICDDR,B), in collaboration with the government of Bangladesh and the World Health Organization, to mount a randomized, double-blind field trial of the two oral vaccines in rural Bangladesh.

Before undertaking the main trial a pretest was performed in which 1257 Bangladeshi villagers were randomized to three doses of one of the two vaccines, a heat-killed *Escherichia coli* K12 strain placebo, or distilled water. The lots of vaccines and placebo for the main trial caused no detectable side effects and elicited the expected rises in serum vibriocidal and anti-cholera-toxin antibodies (3).

The protective efficacy of the oral B + WCV and WCV vaccines was then assessed in 63,498 Bangladeshi children aged 2 to 15 years and women over 15 years of age (4). Each received three doses (Table 14.1) of B + WCV, WCV alone, or K12 placebo in a randomized, double-blind fashion. The field trial was conducted in the Matlab field study area of the ICDDR,B. Matlab lies in the Ganges delta 45 km southeast of Dhaka; 190,000 persons currently reside in the area. Cholera is endemic in Matlab, with a median yearly incidence of 1.5 hospital cases/1000 general population. Vaccination proceeded in three 6-week rounds beginning in January 1985. The vaccines and placebo were administered by 69 vaccination teams. Surveillance for diarrhea was then maintained at the three diarrheal treatment centers serving the Matlab population. Uniform information about the clinical status at the time of presentation and during the subsequent clinical course was entered onto data forms for computer entry. A stool specimen or rectal swab was obtained from each patient and was cultured to identify *V. cholerae* 01 and to determine the biotype and serotype of each isolate.

Table 14.4 presents the overall results for the initial 6-month surveillance period. The combined B + WCV vaccine conferred a high degree of protection against cholera, with no less than 85% protective efficacy ($p < 0.0001$, one-tailed). The WCV-only vaccine also afforded statistical significance, though less impressive protection (protective efficacy 58%; $p < 0.01$, one-tailed). The group vaccinated with B + WCV had 64% fewer

Table 14.4. 1985 field trial of oral cholera vaccines: overall 6-month protective efficacy.

Treatment group	Total No.	No. with cholera	Protective efficacy (%)
B + WCV	21,141	4	85 ($p < 0.0001$)[a]
WCW	21,137	11	58 ($p < 0.01$)
K12	21,220	26	—

[a]Relative to K12 placebo; PE for B + WCV versus WCV was 64% ($p < 0.05$).
All tests were one-tailed.
Data from Clemens et al. (4).

Table 14.5. 1985 Field trial: vaccine
efficacy by age, duration of follow-up,
and severity of illness.

Subgroups	Vaccine recieved (%)	
	B + WCV	WCV
Age (years)		
2–10	92	53
>10	77	62
Severity		
Severe	89	44
Nonsevere	75	87
Follow-up (weeks)		
0–11	87	60
12–25	82	54

Data from ref. 4.

cases of cholera than the group receiving the WCV-only vaccine (p = 0.04, one-tailed).

Three subgroup analyses of particular interest are shown in Table 14.5.

1. The protective efficacy for both vaccines was consistent for children aged 2 to 10 years and older persons. That both vaccines protected children and adults to a similar degree is of considerable interest. In endemic areas children are at highest risk for cholera, and parenteral cholera vaccines have conferred disappointing levels of protection in children.
2. For the combined vaccine, the efficacy was the same for severe cholera and nonsevere cholera. The WCV vaccine alone seemed to have its greatest effect against nonsevere cholera, but this finding needs further follow-up before any conclusion can be drawn.
3. Finally, although it should be emphasized that only short-term protection has been determined in this trial, it is noteworthy that no important decline in efficacy was observed for either vaccine between the first and second 3-month periods of observation (Table 14.5). This finding contrasts with the experience with parenteral vaccines, where the ~50% protective efficacy attained (in adults) shortly after vaccination decreases precipitously after 2 to 3 months to be essentially absent at 6 months after vaccination.

Thus the results from this trial show that oral vaccination with three doses of B + WCV induces high-grade short-term protection against clinically important cholera in both adults and children, and that vaccination with oral WCV alone induces moderate protection. These data provide support for the concept that intestinal immunity is most relevant for protection against cholera; they further indicate that when given with the

WCV component the B subunit elicits anticholera toxin immunity, which reduces by nearly two-thirds the attack rate of cholera seen in persons vaccinated with WCV.

The results obtained with the oral B + WCV described are clearly promising, and they urgently warrant further research to determine the most practical and cost-effective production, formulation, and immunization schedule for the vaccine. The prospects for an efficacious, practical oral cholera vaccine by the year 1990 now seem bright.

Acknowledgments. We thank the many co-workers at the University of Göteborg (Sweden), the International Centre for Diarrhoeal Disease Research (Bangladesh), the Center for Vaccine Development of the University of Maryland (Baltimore, U.S.A.), the Institut Mérieux (Lyon, Fränce), and the National Bacteriological Laboratory (Stockholm, Sweden) who participated in and made the studies described in this chapter possible. We also gratefully acknowledge the research grants for these studies from the Swedish Medical Research Council (16X-3382), the Swedish Agency for Research Cooperation with Developing Countries, and the Diarrheal Disease Control Program of the World Health Organization.

References

1. Black RE, Levine MM, Clements ML, et al: Protective efficacy in humans of killed whole-vibrio oral cholera vaccine with and without the B subunit of cholera toxin. *Infect Immun* 1987;55:1116–1120.
2. Clemens JD, Jertborn M, Sack DA, et al: Effect of neutralization of gastric acid on immune responses to an oral B subunit–killed whole-cell cholera vaccine. *J Infect Dis* 1986;154:175–178.
3. Clemens JD, Stanton BF, et al: B subunit-whole cell and whole cell-only oral vaccines against cholera: studies on reactogenicity and immunogenicity. *J Infect Dis* 1987;155:79–85.
4. Clemens JD, Sack DA, Harris JR, et al: Field trial of oral cholera vaccines in Bangladesh *Lancet* 1986;2:124–127.
5. Holmgren J, Svennerholm A-M, Lönnroth I, et al: Development of improved cholera vaccine based on subunit toxoid. *Nature* 1977;269:602–604.
6. Holmgren J: Actions of cholera toxin and the prevention and treatment of cholera. *Nature* 1981;292:413–417.
7. Holmgren J, Svennerholm A-M: Cholera and the immune response. *Progr Allergy* 1983;33:106–119.
8. Jertborn M, Svennerholm A-M, Holmgren J: Gut mucosal, salivary and serum antitoxic and antibacterial antibody responses in Swedes after oral immunization with B subunit-whole cell cholera vaccine. *Int Arch Allergy Appl Immunol* 1984;75:38–43.
9. Levine MM, Kaper JB, Black RE, Clements ML: New knowledge on pathogenesis of bacterial enteric infections as applied to vaccine development. *Microbiol Rev* 1983;47:510–550.

10. Lycke N, Holmgren J: Strong adjuvant properties of cholera toxin on gut mucosal immune responses to orally presented antigens. *Immunology* 1986;59:301–308.
11. Lycke N, Holmgren J: Long-term cholera antitoxin memory in the gut can be triggered to antibody formation associated with protection within hours of an oral challenge immunization. *Scand J Immunol* 1987;25:407–412.
12. Pierce NF: The role of antigen form and function in the primary and secondary intestinal immune responses to cholera toxin and toxoid in rats. *J Exp Med* 1978;148:195–206.
13. Svennerholm A-M, Holmgren J: Synergistic protective effect in rabbits of immunization with V. cholerae lipopolysaccharide and toxin/toxoid. *Infect Immun* 1976;13:735–740.
14. Svennerholm A-M, Sack DA, Holmgren J, Bardhan PK: Intestinal antibody responses after immunization with cholera B subunit. *Lancet* 1982;1:305–308.
15. Svennerholm A-M, Gothefors L, Sack DA, et al: Local and systemic antibody responses and immunological memory in humans after immunization with cholera B subunit by different routes. *Bull WHO* 1984;62:909–918.
16. Svennerholm A-M, Jertborn M, Gothefors L, et al: Mucosal antitoxic and antibacterial immunity after cholera disease and after immunization with a combined B subunit-whole cell vaccine. *J Infect Dis* 1984;149:884–893.
17. Tayot J-L, Holmgren J, Svennerholm L, et al: Receptor-specific large scale purification of cholera toxin on silica beads derivatized with lyso-GM$_1$-ganglioside. *Eur J Biochem* 1981;113:249–258.

CHAPTER 15

Prospects of Immunization Against Cholera by Adhesive Antigen

Brahm S. Srivastava, Ranjana Srivastava, and Alice Jacob

Epidemic and endemic cholera in humans is caused by *Vibrio cholerae* 01. Infection occurs through the oral route by ingestion of contaminated food and water. If vibrios successfully escape the acidic environment of the stomach, they arrive in the small intestine, where a series of pathogenic events occur that result in colonization of vibrios and release of cholera toxin. The net result is the onset of diarrhea and associated clinical symptoms of cholera.

Several attempts have been made to develop a cholera vaccine that will provide long-lasting immunity without undesirable effects, but as yet, no such vaccine has been established. These vaccines include a variety of immunizing agents (7). Depending on the nature of the immunizing antigen, the cholera vaccines may be grouped into the following classes: (1) killed whole-cell vaccines; (2) toxoid/B subunit vaccines; and (3) live attenuated vaccines obtained by mutagenesis or application of recombinant DNA techniques.

A new class of vaccine may be considered that will be composed of protective surface antigens of *V. cholerae* capable of generating a strong antibacterial immune response. Knowledge of the events in the pathogenesis of cholera and identification of the protective antigens is required to develop such vaccines. Our interest in cholera vaccines centers around this class of immunogen.

Pathogenesis of Cholera

Vibrios arriving in the lumen of the gut begin to multiply and increase in number. They penetrate the mucous layer and arrive in close proximity to the epithelial surface of the small intestine. The motility of the vibrios is understood to play a role. *V. cholerae* bears an apical flagellum that is responsible for the motility of the organism. Nonmotile mutants neither penetrate the mucous layer nor exhibit virulence in experimental cholera, although they are fully enterotoxigenic (1,3,4,11,14).

Motile strains of *V. cholerae* appearing in large numbers in intervillous spaces and crypts adhere to the epithelial surface of the small intestine. The time course of adherence of vibrios to rabbit intestine and beginning of fluid outpouring in the ileal loops was studied (9,14), a good correlation was found between bacterial adherence and pathogenicity. Strains capable of adhering with high efficiency were pathogenic, whereas a nonadhesive strain was found to be nonpathogenic (14). Adherence of vibrios is therefore considered to be an essential step in the pathogenesis of cholera.

The role of cholera enterotoxin is undisputable in the pathogenesis of the organism. Nevertheless, immune mechanisms targeted toward inhibition of motility and adherence of vibrios seem to be important in the prophylactic control of cholera.

Immunity to cholera involves antibodies directed toward vibrios (antibacterial) and the toxin released by them (antitoxin). Rabbits immunized with bacterial vaccine or toxin, or both, showed resistance to challenge. Antibacterial or antitoxin antibodies, or both, appeared in the circulation depending on the immunogen(s) administered to rabbits. There is convincing evidence that antibacterial and antitoxin immunities work synergistically (12,16). Antitoxin antibodies neutralize cholera toxin, which appears to be the mechanism of antitoxin immunity (10).

The data obtained in our laboratory suggest strongly that antibacterial immunity might act by inhibiting adherence of vibrios on the surface of the intestine (12). A group of rabbits were immunized with killed whole-cell bacterial vaccine to induce antibacterial immunity and were then challenged with a homologous pathogenic strain in the ileal loop model (2). Sensitivity to challenge was judged by accumulation of fluid in the ileal loop (positive), whereas those without fluid (negative) were considered resistant to challenge. When the negative and positive loops were examined for the number of vibrios adherent to the intestine, it was found that few vibrios (0.5 to 2.0%) were adherent in the case of negative loops, whereas 25 to 90% of vibrios were adherent in the positive loops. Thus there was a good correlation between low adherence and resistance to challenge. This finding suggested that antibacterial immunity provided protection mainly by interfering with the adherence of vibrios to intestine and that the adhesive antigen could be one of the key antigens in cholera immunity.

22-Kilodalton Protein

In our laboratory a nonadhesive mutant (CD11) of *V. cholerae* was isolated by mutagenic treatment of a wild-type pathogenic strain (KB207). Although the mutant was motile and chemotactive, it adhered poorly to rabbit intestinal mucosa (13), exhibited reduced virulence in experimental cholera (14), and did not colonize the gut of infant mice (unpublished data).

Because CD11 was found to be motile but nonadhesive, it was concluded that it lacked the surface antigens involved in the adherence of vibrios. Based on this conclusion, it was argued that adsorption of immune serum (raised in rabbits by immunization with whole cell KB207) with CD11 would remove antibodies to most of the surface antigens of KB207 but would not remove from the serum antibodies produced in response to the antigens absent in CD11. Therefore rabbits were immunized with whole-cell KB207 by subcutaneous injections. The immune serum thus obtained was absorbed with fresh cultures of CD11 until the absorbed serum no longer agglutinated CD11 but did agglutinate KB207. The CD11-absorbed serum, which supposedly contained antibodies to the adhesive antigen, was purified on a DEAE-cellulose column and concentrated by polyethylene glycol when required (6).

Immunoglobulin G (IgG) present in the absorbed serum significantly inhibited adherence of KB207 to rabbit intestine. About 90% inhibition of adherence occurred compared to that with the control without antibodies. In a parallel experiment keeping an identical input of antibodies, adherence was inhibited to about the same extent by antibodies present in the unabsorbed immune serum (6). Because the extent of inhibition of adherence by absorbed and unabsorbed sera was the same, the antibodies present in the absorbed serum may be considered specific to the adhesive antigens of *V. cholerae* strain KB207.

The above conclusion was further strengthened when the affinity of the absorbed and unabsorbed sera to the cell-free lysates of KB207 and CD11 was examined by Ouchterlony test. Whereas the former reacted to only KB207 lysate by the appearance of a sharp precipitin line, the latter reacted to both KB207 and CD11 lysates (6). This finding clearly indicated that the antibodies in the absorbed serum were specific to the antigen(s) present in KB207 but absent in the nonadhesive mutant CD11.

Hence the IgG present in the absorbed serum was coupled to CNBr-activated Sepharose 4B, and an affinity column was prepared to isolate the specific proteins to these antibodies. Cell-free lysate of KB207, containing 100 mg total protein, was passed through the column and the bound proteins eluted. The recovery was 1.2 mg. As expected, the recovered protein reacted to both absorbed and unabsorbed sera in the Ouchterlony test.

Characteristics of the isolated protein were determined, and some of its properties are described below (5).

1. The protein was analyzed on sodium dodecyl sulfate polyacrylamide gel electrophoresis (SDS-PAGE). The relative mobility (R_f) was compared with the R_f value of the proteins of known molecular weight. The isolated antigen migrated in a single band of relative molecular weight (M_r) 22 kilodaltons (kDa).

2. When rabbits were immunized with 100 μg of the 22-kDa antigen, complement-mediated vibriocidal antibodies appeared in the circulation.

3. The intestine of the rabbits immunized with 22-kDa protein was taken, and the intestinal discs were challenged with KB207 to measure adherence of bacteria as described before (13). Compared to the control, 75% inhibition of adherence was observed.

4. The effect of antibodies to 22-kDa protein on the adherence of KB207 to unimmunized rabbit intestine was measured. Compared to the control without antibodies, 99% inhibition of adherence was observed.

5. The immunogenicity of 22-kDa antigen as a vaccine was evaluated in the ileal loops of adult rabbit as a model of experimental cholera (2). Three groups of rabbits were used. One group was kept as unimmunized controls, and the second and third groups were immunized, respectively, with the 22-kDa antigen only and with 22-kDa antigen with B subunit of cholera toxin. Rabbits received two doses of 100 μg of single or both antigens at 21-day intervals. Rabbits were challenged with 10^5 viable counts of KB207 in 1 ml saline. Protection was measured as described earlier (12). Compared to controls, 82% protection was observed in rabbits immunized with 22-kDa antigen alone. A combined vaccine of the surface antigen and B subunit toxin offered 100% protection.

In conclusion, the 22-kDa protein was found to be antigenic, induced vibriocidal antibodies that inhibited adherence of *V. cholerae* to the epithelial surface of intestine, and protected rabbits significantly from challenge of *V. cholerae*.

Gene Encoding 22-kDa Protein

Srivastava et al. (15) have reported construction of the genomic library of *V. cholerae* strain KB207. EcoRI-digested DNA was cloned in pBR325 and expressed in *Escherichia coli* HB101 (λcI857) lysogen. Antibodies were used as probes to detect cloned genes coding for the surface antigens of KB207. Transformants were grown at 32°C on plates containing antibodies. Lysogen was induced at 42°C to release expressed antigens. The antigen–antibody reaction produced a halo around the positive clones (15).

Using the antibodies present in CD11-absorbed serum as a probe to detect the gene encoding 22-kDa antigen, it was possible to detect the clone. The recombinant plasmid was larger than pBR325, and the size of the insert was found to be 1.3 kilobases (kb) (15).

The antigen encoded by the 1.3-kb insert has been isolated. The *Escherichia coli* clone was grown in broth at 32°C. Bacteria were harvested, and the cell-free lysate was passed through the affinity column described above. It codes for a protein of approximately the same size as that of 22-kDa protein (unpublished data).

Conclusion and Prospects

The experimental approach and the data described above suggest that 22-kDa protein could be a key antigen involved in the adherence of *V. cholerae*. Manning et al. (8) reported characterization of a 22-kDa protein as an outer membrane antigen. We have demonstrated that 22-kDa protein is a protective antigen that induces antibacterial immunity. Live or killed whole cells in a vaccine may be replaced by protective surface antigens. It is exciting that combined application of 22-kDa protein and the B subunit toxin offered complete protection to rabbits. Such a vaccine is expected to be most desirable, as it would induce antibacterial as well as antitoxin immunities without fear of reversion to toxicity and adverse effects.

The characterization of the 22-kDa protein as a protective antigen and cloning of its gene offers the prospects of synthetic vaccine. It would be desirable to determine the nucleotide sequence of the 1.3-kb DNA and subsequently begin to synthesize small peptides representing immunogenic determinant domains of the antigens.

References

1. Bhattacharjee JW, Srivastava BS: Adherence of the wild type and mutant strains of Vibrio cholerae to normal and immune intestinal tissues. *Bull WHO* 1979;87:123–128.
2. De SN, Chatterji DN: An experimental study of the mechanism of action of Vibrio cholerae on the intestinal mucous membrane. *J Pathol Bacteriol* 1953;66:559–567.
3. Freter F, Jones DW: Adhesive properties of Vibrio cholerae: nature of the interaction with intact mucosal surfaces. *Infect Immun* 1976;14:246–256.
4. Guentzel MN, Berry LJ: Motility as a virulence factor for Vibrio cholerae. *Infect Immun* 1975;11:890–897.
5. Jacob A, Sahib MK, Sinha VB, Srivastava BS: Adhesive antigens of Vibrio cholerae. II. Isolation and characterization of a 22 KDal protein. Submitted for publication.
6. Jacob A, Sinha VB, Srivastava BS: Adhesive antigens of Vibrio cholerae. I. Antibodies inhibiting bacterial adherence. Submitted for publication.
7. Levine MM, Kaper JB, Black RB, Clements ML: New knowledge on pathogenesis of bacterial enteric infections as applied to vaccine development. *Microbiol Rev* 1983;47:510–550.
8. Manning PA, Bartowsky EJ, Leavesly DI, et al: Molecular cloning using immune sera of a 22 KDal minor outer membrane protein of Vibrio cholerae. *Gene* 1985;34:95–103.
9. Nelson ET, Clements JD, Finkelstein RA: Vibrio cholerae adherence and colonization in experimental cholera: electron microscopic studies. *Infect Immun* 1976;14:527–547.
10. Peterson JW, Hejtmancik KE, Markel DE, et al: Antigenic specificity of neutralizing antibody to cholera toxin. *Infect Immun* 1979;24:774–779.

11. Schrank GD, Verwey WF: Distribution of cholera organisms in experimental Vibrio cholerae infections: proposed mechanism of pathogenesis and antibacterial immunity. *Infect Immun* 1976;13:195–203.
12. Srivastava R, Sinha VB, Srivastava BS: Re-evaluation of antibacterial and antitoxin immunities in experimental cholera. *Indian J Med Res* 1979;70:369–373.
13. Srivastava R, Srivastava BS: Isolation of a non-adhesive mutant of Vibrio cholerae and chromosomal localization of the gene controlling mannose-sensitive adherence. *J Gen Microbiol* 1980;117:275–278.
14. Srivastava R, Sinha VB, Srivastava BS: Events in the pathogenesis of experimental cholera: role of bacterial adherence and multiplication. *J Med Microbiol* 1980;13:1–9.
15. Srivastava R, Khan AA, Srivastava BS: Immunological detection of cloned antigenic genes of Vibrio cholerae in Escherichia coli. *Gene* 1985;40:267–272.
16. Svennerholm AM, Holmgren J: Synergistic protective effect in rabbits of immunization with Vibrio cholerae lipopolysaccharide and toxin/toxoid. *Infect Immun* 1976;13:735–740.

CHAPTER 16

Protective Antigens of *Vibrio Cholerae*

Asoke C. Ghose and Chiranjib Dasgupta

The disease cholera in the epidemic form is mainly caused by the gram-negative bacterium *Vibrio cholerae,* belonging to the 01 serovar. The organism also exists in two serotypes (Inaba and Ogawa) and biotypes (classical and El Tor). It is known that the diarrheagenic syndrome in cholera patients is primarily caused by an enterotoxin (cholera toxin) secreted by the multiplying vibrios in the lumen of the gut (10,13,19). It is also known that the cholera toxin (CT) binds to the ganglioside (GM_1) receptor in the epithelial cell membrane, thereby triggering activation of the enzyme adenylate cyclase, which in turn leads to an increase in intracellular adenosine $3'$-$5'$-cyclic monophosphate (cyclic AMP) levels and subsequent loss of water and electrolytes across the intestinal epithelium to the lumen (17,24). Apart from *V. cholerae* 01, organisms belonging to non-01 serovars have often been identified as the causative agents of sporadic cases and localized outbreaks of cholera-like diarrhea, which is sometimes accompanied by blood and mucus (28). Some of these clinical non-01 isolates also produce CT or CT-like substances (9).

Currently available data suggest that clinical cholera gives rise to antibodies against bacterial somatic antigens as well as cholera toxin (42). Studies with human volunteers have provided useful information regarding the quality and duration of immunity in cholera (39). It was observed that clinical infection due to *V. cholerae* 01 provided significant protection against rechallenge with vibrios of either serotype or biotype. On the other hand, oral immunization of volunteers with large, multiple doses of cholera toxoid conferred little protection against oral live vibrio challenge. It was inferred that antibacterial immunity probably played a more important role than that of antitoxic immunity in cholera. However, evidence is available suggesting that both antibacterial and antitoxic forms of immunity may act synergistically, mediating effective and lasting protection against the disease (51,59). The fact that cholera is a noninvasive disease where the organisms multiply and colonize within the gut epithelium suggests that effective immunity can be mediated only by mechanisms that act in or around the mucosal surface. The secretory IgA (SIgA) class of anti-bodies, which are produced locally in the gut and present in intestinal secretion and bile, largely serve the purpose.

Although the organism *V. cholerae* possesses multitudes of antigenic components, their relevance in terms of protective immunity is yet to be fully understood. It is possible that more than one type of antigen are involved in the induction of protective immunity in cholera. So far, the lipopolysaccharide moiety and cholera toxin are the most well studied antigens showing some immunoprophylactic potential. Our present knowledge regarding other antigens of *V. cholerae* is still incomplete and further studies are definitely needed in this area, including studies of fla-gellar antigens, outer membrane proteins, hemagglutinins (cell-associated and cell-free), and various enzymes.

Somatic and Flagellar Antigens of *V. cholerae*

Lipopolysaccharide

The heat-stable somatic (O) antigen of *V. cholerae* is lipopolysaccharide (LPS) in structure and determines the serovar as well as the serotype of the organism. A positive role of *V. cholerae* LPS in protection was sug-gested by earlier workers (63), which was more serotype-specific in nature. However, the precise mechanism by which this protection was mediated was a subject of investigation. Antibody raised against Ogawa or Inaba LPS was found to prevent the attachment of homologous strains to in-testinal mucosa in vitro (6,20). Unfortunately, none of these studies used the monovalent Fab fragments for their inhibition experiments to establish the antiadhering mechanism with certainity. Furthermore, these reports suffer from the lack of information with respect to the role of anti-LPS antibodies in inhibiting adhesion of heterologous serotypes. The problem has been reinvestigated in detail (C. Dasgupta, PhD thesis, Calcutta Uni-versity, 1986). An in vitro experimental model was used in this study where ^{14}C-labeled *V. cholerae* organisms were incubated with rabbit intestinal tissue slices for 10 min at 37°C, and the amount of tissue-bound radiolabeled vibrios was measured by solubilization of the washed tissue slices and subsequent radioactive counting. LPS was prepared from the *V. cholerae* 01 A17 Ogawa El Tor strain, and the preparation was subjected to pro-teolytic degradation and subsequent boiling to ensure removal of any re-sidual protein present in it. Anti-LPS serum was raised in rabbits, and the agglutination titer of this antiserum was determined against the parent *V. cholerae* organism. It is evident (Table 16.1) that rabbit anti-LPS serum showed moderate inhibition of the intestinal adhesion of the homologous strain at subagglutinating dilutions. The same was true for the IgG fraction of the antiserum, although its Fab fragment showed only weak inhibition.

Table 16.1. Antiadhering abilities of rabbit antiserum to LPS, its IgG fraction, and monovalent Fab fragment determined against the adherence of *V. cholerae* A17 El Tor Ogawa to Rabbit intestinal slices.

Tissue preincubation mixture	Conc. of IgG or Fab (μg protein/ml)	Adhesion index[c]	% Inhibition
KRT buffer[a] (control)	—	12.0	0
Normal rabbit serum (1:20)	—	12.1	0
Anti-LPS serum[b]			
1:200	—	7.2	40
1:500	—	12.5	0
KRT buffer (control)	—	11.1	0
IgG fraction			
Normal rabbit serum	500	11.5	0
Anti-LPS serum	12.5	6.0	46
–do–	6.3	11.2	0
KRT buffer (control)	—	11.1	
Fab (IgG) fragment of			
Normal rabbit serum	750	11.4	0
Anti-LPS serum	750	8.3	25
–do–	375	11.7	0

[a]Krebs-Ringer-Tris buffer pH 7.4.
[b]Bacterial agglutination titer (1:80).
[c]Mean values of duplicate sets are shown.

Intestinal adherence of a *V. cholerae* strain of a heterologous (Inaba) serotype could not be inhibited under identical conditions by either the anti-LPS serum or its immunoglobulin G (IgG) fraction. On the other hand, passive protection experiments carried out in experimental animal models (Table 16.2) demonstrated the importance of anti-LPS antibodies in protection against homologous and, to a lesser extent, heterologous strains. This protection was possibly mediated mainly through microscopic agglutination or immobilization of bacteria by antibodies, thereby preventing their colonization. The weak degree of inhibition as well as protection (Table 16.3) observed with the Fab fragments against the homologous organism was probably mediated by binding of these fragments to the LPS moiety, which either sterically or through a conformational change hindered the interaction between *V. cholerae* and intestinal epithelium.

Flagella

Motility of *V. cholerae* is believed to be an important virulence factor for these organisms and has been shown to be associated with their polar flagellum. *V. cholerae* 01 and non-01 strains are known to possess common flagellar (H) antigen, which is characteristic of these species (52). However,

Table 16.2. Protection by the IgG fraction of rabbit anti-LPS serum in rabbit ileal loop experiments against live *V. cholerae* challenge.

Challenge strain	Preincubation mixture	Conc. of IgG (μg protein/ml)	Fluid accumulation ratio[b] (volume/ length)	% Protection
V. cholerae A17 El Tor Ogawa (homologous strain)	KRT buffer[a] (control) IgG fraction	—	1.0	0
	Normal rabbit serum	500	1.0	0
	Anti-LPS serum	12.5	0.2	80
V. cholerae 569 B classical Inaba (heterologous strain)	KRT buffer (control) IgG fraction	—	1.2	0
	Normal rabbit serum	500	1.3	0
	Anti-LPS serum	12.5	0.6	50

[a]Krebs-Ringer-Tris buffer pH 7.4.
[b]Mean values of duplicate sets are shown.

the demonstration of common H antigen by agglutination reaction using OH or H antiserum is not always possible (58). Only specialized treatment of the bacterial cell (12,49) or potent antiserum (52), or both, is required to demonstrate the common flagellar antigen. Several reports are available in the literature (1,14,65) that suggest the importance of *V. cholerae* 01 flagellar antigen in mucosal adhesion and in the induction of protective immunity. Yancy et al (65) demonstrated the immunoprophylactic potential of a crude flagellar preparation of a *V. cholerae* 01 classical Inaba strain through active immunization in rabbits, which probably acted by inhibiting the vibrio association with the intestinal mucosa. These workers concluded that the critical immunogenic determinant of the flagellar preparation was protein in nature. Earlier, Eubanks et al (14) demonstrated passive protection against cholera in suckling mice by immunization of their mothers with a crude flagellar preparation. In contrast, the naked flagella, devoid of flagellar sheath material, was only poorly protective. These observations suggested the possible immunoprophylactic potential of flagellar sheath of *V. cholerae* 01, which was serotype-independent in nature (27). These studies were extended to *V. cholerae* non-01 strains (K. Datta-Roy and A.C. Ghose, unpublished data). Antibody to crude flagellar preparations of non-01 *V. cholerae* showed passive protection in suckling mice primarily against challenge with the parent organisms but only poorly against *V.*

Table 16.3. Protection by the Fab fragment derived from the IgG fraction of rabbit anti-LPS serum in rabbit ileal loop experiment against live *V. cholerae* challenge.

Challenge strain	Preincubation mixture	Conc. of Fab (μg protein/ml)	Fluid accumulation ratio[b] (volume/length)	% Protection
V. cholerae A17 El Tor Ogawa (homologous strain)	KRT buffer[a] (control) Fab (IgG)	—	1.3	0
	Normal rabbit serum	750	1.2	8
	Anti-LPS serum	750	0.9	31
V. cholerae 569B classical Inaba (heterologous strain)	KRT buffer (control) Fab (IgG)	—	1.3	0
	Normal rabbit serum	750	1.3	0
	Anti-LPS serum	750	1.2	8

[a]Krebs-Ringer-Tris buffer pH 7.4.
[b]Mean values of duplicate sets are shown.

cholerae belonging to other serovars. It was also noted that crude flagellar preparation or antiserum to it could inhibit in vitro adhesion of homologous (immunizing) strain to rabbit intestinal slices. On the other hand, flagellins or their antisera did not possess antiadhering or any protective activity. Interestingly, flagellins prepared from both 01 and non-01 organisms were found to be biochemically and immunologically similar. These results, again, demonstrate the importance of flagellar sheath component in protection, at least, within organisms belonging to the same serovar.

Outer Membrane Proteins

Biochemical and immunological characterization of the outer membrane proteins (OMPs) of *V. cholerae* was carried out by several groups of workers (34,35,37,46,56). The presence of common OMPs was noted between various *V. cholerae* 01 strains belonging to both biotypes and serotypes (7,34,37). Gross similarities in the OMP profiles between 01 and non-01 organisms were demonstrated by some workers (46; K. Datta-Roy and A.C. Ghose, to be published), although strain-specific antigenic epitopes were also noted (55). One common OMP of *V. cholerae* 01, immunogenic in rabbits as well as in humans, had a subunit molecular weight around 48 kilodaltons (kDa) (34,35), which probably represented a mixture of three proteins of closely similar molecular weights in the range 42 to

45 kDa (45). Manning and Haynes (46) also reported that the most common and immunogenic OMPs of *V. cholerae* had a molecular weight of 25 kDa. In a human volunteer study, Sears et al (56) evaluated the serum antibody response to the OMPs of *V. cholerae* 01. Results showed a significant rise in anti-OMP titers in 50% of clinical cholera cases, and the antibody mostly belonged to the IgG class. In preliminary studies, these workers also showed that some volunteers manifested brisk intestinal SIgA response to the OMPs of *V. cholerae* (39). However, no information was available regarding their protective role. Antigenic relatedness between OMPs of *V. cholerae* 01 and non-01 organisms has been studied in detail (K. Datta-Roy and A.C. Ghose, unpublished data). Rabbit antiserum to a purified outer membrane preparation from a non-01 organism (following removal of its anti-LPS activity) was shown to agglutinate whole cells belonging to both 01 and non-01 serovars. This antiserum also showed significant inhibition of intestinal adhesion of the immunizing non-01 strain. The protective efficacy of this antiserum was established.

Hemagglutinins

The hemagglutinating property of *V. cholerae* was first noted by Lankford (38), who hypothesized that it was a reflection of the intestinal adherence property of the organism. In fact, Finkelstein and Mukherjee (18) utilized the chicken cell agglutination phenomenon to differentiate between El Tor and classical vibrios. The term *hemagglutinin* is used for any component of bacterial surface that causes agglutination of erythrocytes, and the term *adhesin* is used to define the bacterial surface structure(s) that is responsible for binding of bacteria to cell surfaces (33). Hemagglutinin (HA) and adhesin are often used as synonymous terms, although there may be exceptions. *V. cholerae* exhibits two types of hemagglutinating activity. The first is the cell-associated hemagglutinating activity displayed by whole cells of *V. cholerae* (8). The second is the cell-free or soluble hemagglutinating activity that is present in the cell-free culture supernatants of *V. cholerae* and is precipitable at 50% saturation with ammonium sulfate (21b).

Cell-Associated Hemagglutinin

The involvement of the cell-associated hemagglutinin (CAHA) in vibrio adhesion to isolated brush border membranes of rabbits was investigated by Jones and co-workers (31,32). These workers observed that the hemagglutination and adhesion phenomena appeared to involve similar mechanism(s), as both of these activities had a Ca^{+2} requirement in common. However, a notable difference between the adhesion and hemagglutination was that the former, but not the latter, activity could be partially promoted by Sr^{+2}. Jones and Freter (32) investigated the nature of cell surface receptors involved in these processes. It was observed that adhesion to

brush border membrane was strongly inhibited by L-fucose and its derivatives and, to a lesser extent, by D-mannose. The hemagglutination reaction
involving human group O cells was, however, inhibitable only by L-fucose,
not by D-mannose. The finding that inhibition of adhesion by L-fucose
was strong but incomplete suggested that the natural mucosal receptor
was larger than a single L-fucose residue, and/or a certain stereochemical
configuration was required (32). In another study, Bhattacharjee and Shrivastava (2) described one D-mannose-sensitive HA in *V. cholerae* El Tor
strains that was also responsible for their intestinal adhesion. Holmgren
et al. (25) investigated CAHAs of *V. cholerae*. They observed that classical
V. cholerae strains agglutinated human group O erythrocytes in a reaction
that was inhibitable by L-fucose, whereas the agglutination of chicken
erythrocytes by El Tor strains was inhibited maximally by D-mannose.
Faris et al (15) described a high surface hydrophobicity of hemagglutinating
V. cholerae strains that caused both mannose-sensitive and mannose-resistant hemagglutination. The nonhemagglutinating vibrio, on the other
hand, was poorly hydrophobic. Kabir and Ali (36) also reported on the
involvement of both specific and nonspecific (hydrophobic and ionic) factors in hemagglutination and adhesion processes. In another detailed study,
Hanne and Finkelstein (22) described a total of four apparently distinct
HA, three of which were cell-associated in nature. Two of them were
exclusive for El Tor strains, and the third was found in both El Tor and
classical strains. So far, a single report is available in the literature where
the workers (4) attempted to purify a CAHA of *V. cholerae* 01, although
detailed biochemical characterization of this preparation was not available.

A CAHA has been isolated from a *V. cholerae* 01 A17 Ogawa El Tor
strain by the use of the chaotropic agent potassium thiocyanate (C. Dasgupta, PhD dissertation, 1986, Calcutta University). The preparation had
weak hemagglutinating activity against rabbit and human erythrocytes,
exhibited aggregative properties under native conditions, was present as
a high-molecular-weight complex, and contained some amounts of LPS.
The CAHA preparation was able to significantly inhibit the intestinal adherence activity of the parent organism. Furthermore, anti-CAHA serum
(made free from any anti-LPS activity by absorption) and its IgG fraction
could markedly reduce the intestinal adherence of *V. cholerae* 01 strains
belonging to both homologous and heterologous serotypes. These results
correlated well with the passive protection experiments carried out with
the anti-CAHA antiserum and its IgG fraction in experimental animals
(Table 16.4). Presumably, the major mechanism by which the protection
was mediated was antiadhering in nature.

Cell-Free (Soluble) Hemagglutinin

The CFHA, or soluble HA of Hanne and Finkelstein (22), was claimed
to be released by all *V. cholerae* 01 strains in the late logarithmic-phase

Table 16.4. Protection by the IgG fraction of rabbit anti-CAHA serum in rabbit ileal loop experiment against live *cholerae* challenge.

Challenge strain	Preincubation mixture	Conc. of IgG (μg protein/ml)	Fluid accumulation ratio[b] (volume/ length)	% Protection
V. cholerae A17 El Tor Ogawa (homologous strain)	KRT buffer[a] (control) IgG fraction	—	1.0	0
	Normal rabbit serum	500	1.0	0
		416	0	100
	Anti-CAHA serum –do–	26	0.1	90
V. cholerae 569B classical Inaba (heterologous strain)	KRT buffer (control) IgG fraction	—	1.2	0
	Normal rabbit serum	500	1.3	0
		416	0.1	92
	Anti-CAHA serum –do–	26	0.2	83

[a]Krebs-Ringer-Tris buffer pH 7.4.
[b]Mean values of duplicate sets are shown.

cultures. The soluble HA was not inhibitable by any simple sugar but required Ca^{+2} for maximal HA activity. The soluble HA of a classical Inaba *V. cholerae* 01 strain was subsequently termed "cholera lectin" by these workers, who purified it to apparent homogeneity. The protein, which contained three distinct isoelectric fractions, dissociated into 32-kDa subunits in SDS-PAGE. The purified HA exhibited considerable protease activity and was shown to be a zinc metalloendopeptidase (3). It was also shown (22) that the Fab fragment of the rabbit antibody raised against the purified "cholera lectin" was antiadhering in nature. Svennerholm et al (61) also reported purification of a soluble HA from an El Tor Ogawa *V. cholerae* 01 strain. The preparation revealed one major protein band of about 45 kDa and a weaker band around 15 kDa on SDS-PAGE. Chaicumpa et al (5) reported isolation of a soluble HA from a classical Inaba strain (569B) and showed that rabbit antibody to it afforded weak but significant passive protection against live oral homologous vibrio challenge in an infant mouse model.

A cell-free or soluble HA was obtained by Dasgupta and Ghose (unpublished data) from the *V. cholerae* 01 A17 El Tor Ogawa strain. Detailed biochemical and immunological characterization of the CFHA suggested that it was indeed remarkably similar to the CAHA of the same strain (described above) and was probably released in the culture supernatant owing to the degradation of bacterial membrane component(s) during their degenerative phase. A comparison between the SDS-PAGE patterns of the OMPs, CAHA, and CFHA of the *V. cholerae* A17 strain also suggested the membrane origin of the soluble HA. This conclusion was also substantiated by the kinetic experiments that showed the appearance of CFHA activity in the culture supernatant during the late log phase of the growth curve when CAHA activity started declining.

Enzymes

Vibrio cholerae organisms are known to produce a variety of enzymes, e.g., neuraminidase, mucinase, protease, lipase, lecithinase, and chitinase (57). Protease-deficient mutants of *V. cholerae* were shown to exhibit reduced virulence in experimental animals (54). However, the relevance of these enzymes in terms of protective immunity is yet to be established.

Enterotoxin(s) of *V. cholerae*

Cholera Toxin and Its B Subunit

Cholera toxin (CT) is a potent immunogen, the resultant antibody response being primarily directed against the B subunit (24). Virtually all of the neutralizing activity of cholera antitoxin can be absorbed out with the purified B subunit. These conclusions find support in studies on the monoclonal antibodies to CT (23,40), which demonstrated that potent neutralizing antibodies were directed against the B subunit. Antitoxic activity in convalescent cholera patients' sera primarily belonged to the IgG class, whereas such activity in the intestinal fluid and milk samples was almost exclusively mediated by SIgA antibodies (39,42). Human SIgA antibodies were shown to neutralize CT action, although their neutralizing capacity was somewhat lower than that of the IgG antibodies (43). Human SIgA antibodies also exhibited protective activities in the rabbit ileal loop assay (44). A direct correlation between serum antitoxic titers and protection against experimental cholera was demonstrated in dogs (50). Apart from the parental route, oral or intraintestinal administration of CT was found to be effective in inducing protection against subsequent toxin challenge in experimental animals (16). The protection in the latter cases was mainly mediated by intestinal SIgA antibodies. However, formalin or glutaraldehyde-treated toxoids were rather ineffective in inducing protection (39). On the other hand, the nontoxic B subunit is a logical toxoid immunogen specially for oral immunization (24). It is a strong protective immunogen against experimental cholera in rabbits when given alone or in combination with somatic antigens. The B subunit has been tested for its ability to stimulate mucosal immunity in humans, and a single oral administration stimulated a marked local SIgA antibody response in 80% of the recipients (60).

Synthetic Peptides of B Subunit

Several reports are now available that studied the immunogenic potential of synthetic peptides corresponding to the B subunit of CT (11,29). Antisera to some of these synthetic peptides showed reactivity to the intact CT molecule or its B subunit. However, only two of the peptides, one consisting of residues 50 to 64 (29,30) and another of residues 50 to 75 (11), were able to induce formation of antibodies that showed only partial toxin neutralizing activities. Two peptides corresponding to the amino acid sequences from residues 57 to 69 and 47 to 60 of the B subunit have been synthesized (21a). These peptides were chosen on the basis of their sequence identity with the B subunit of *Escherichia coli* heat labile toxin (LT$_h$) and hydrophilicity analysis. Rabbit antisera to these peptide–carrier conjugates reacted to intact CT, its B subunit, and LT$_h$ in the conventional enzyme-linked immunosorbent assay (ELISA), where the plates were initially coated with the antigens. Interestingly, these anti-peptide sera showed only weak reactivity to these proteins in GM$_1$-ELISA (40), thereby suggesting the possible involvement of these peptide-containing regions of the B subunit in its GM$_1$ receptor binding activity. However, both the anti-peptide sera possessed only weak CT neutralizing activity in the rabbit ileal loop assay. It is possible that the GM$_1$ binding site of the CT molecule is characterized by determinants that are highly conformational in nature and cannot be generated by short stretches of peptides as used in these studies.

Enterotoxic Factors Unrelated to Cholera Toxin

Several reports are now available (9,41,47,48,53) that have produced evidence in favor of the existence of various enterotoxic factors (unrelated to CT) in *V. cholerae* 01 and non-01 strains. They include shiga-like toxin (48), heat-stable enterotoxin (62), and hemolysins (26,64). However, their possible role in the induction of protective immunity against cholera is yet to be ascertained.

References

1. Attridge SR, Rowley D: The role of flagellum in adherence of Vibrio cholerae. *J Infect Dis* 1983;147:864–872.
2. Bhattacharjee JW, Shrivastava BS: Mannose sensitive hemagglutinins in adherence of Vibrio cholerae El Tor to intestine. *J Gen Microbiol* 1978;107:407–410.

3. Booth BA, Boseman-Finkelstein M, Finkelstein RA: Vibrio cholerae soluble hemagglutinin/protease is a metalloenzyme. *Infect Immun* 1983;42:639–644.
4. Chaicumpa W, Atthasistha N: The study of intestinal immunity against V. cholerae: purification of V. cholerae El Tor haemagglutinin and the protective role of its antibody in experimental cholera. *Southeast Asian J Trop Med Public Health* 1979;10:73–80.
5. Chaicumpa W, Peungjesda U, Martinez B, Atthasishtha N: Soluble haemagglutinin of classical vibrios: isolation and protection against cholera by its antibodies. *Southeast Asian J Trop Med Public Health* 1982;13:637–645.
6. Chitnis DS, Sharma KD, Kamat RS: Role of somatic antigen of Vibrio cholerae in adhesion to intestinal mucosa. *J Med Microbiol* 1982;5:53–61.
7. Cryz SJ Jr, Fürer E, Germanier R: Development of an enzyme-linked immunosorbent assay for studying Vibrio cholerae cell surface antigens. *J Clin Microbiol* 1982;16:41–45.
8. Dasgupta C, Majumdar AS, Ghose AC: Hemagglutination and adherence properties of Vibrio cholerae. *IRCS Med Sci* 1983;11:797–798.
9. Datta-Roy K, Banerjee K, De SP, Ghose AC: Comparative study of expression of hemagglutinins, hemolysins and enterotoxins by clinical and environmental isolates of non-01 Vibrio cholerae in relation to their enteropathogenicity. *Appl Environ Microbiol* 1986;52:875–879.
10. De SN: Enterotoxicity of bacteria-free culture filtrate of Vibrio cholerae. *Nature* 1959;183:1533–1534.
11. Delmas A, Guyon-Gruaz A, Halimi H, et al: Solid phase synthesis of two cholera toxin B subunit antigens. *Int J Peptide Protein Res* 1985;25:421–424.
12. Dey SK, Kusari J, Ghose AC: Detergent induced enhancement of antibody mediated flagellar agglutination of Vibrio cholerae. *IRCS Med Sci* 1983;11:908.
13. Dutta NK, Panse MV, Kulkarni DR: Role of cholera toxin in experimental cholera. *J Bacteriol* 1959;78:594–595.
14. Eubanks ER, Guentzel MN, Berry LJ: Evaluation of surface components of Vibrio cholerae as protective immunogens. *Infect Immun* 1977;15:533–538.
15. Faris A, Lindahl M, Wadstrom T: High surface hydrophobicity of hemagglutinating Vibrio cholerae and other vibrios. *Curr Microbiol* 1982;7:357–362.
16. Feeley JC: Antitoxic immunity in cholera, in Barua D, Burrows W (eds): *Cholera*. Philadelphia, Saunders, 1974, pp 307–314.
17. Field M: Ion transport in rabbit ileal mucosa. II. Effects of cyclic $3'-5'$-AMP. *Am J Physiol* 1971;221:992–997.
18. Finkelstein RA, Mukherjee S: Haemagglutination: a rapid method for differentiating V. cholerae and El Tor vibrios. *Proc Soc Exp Biol Med* 1963;112:335–339.
19. Finkelstein RA, LoSpalluto JJ: Pathogenesis of experimental cholera. *J Exp Med* 1969;130:185–202.
20. Freter R, Jones GW: Adhesive properties of Vibrio cholerae: nature of interaction with intact mucosal surfaces. *Infect Immun* 1976;14:246–256.
21a. Ghose AC, Karush F: Induction of polyclonal and monoclonal antibody responses to cholera toxin by the synthetic peptide approach. *Molec Immun* 1988;25:223–230.
21b. Ghosh AK, Ganguly R, Shrivastava DL: Studies in immunochemistry of V. cholerae. VI. Haemagglutination. *Indian J Med Res* 1965;50:1–7.
22. Hanne LF, Finkelstein RA: Characterization and distribution of the hemagglutinins produced by Vibrio cholerae. *Infect Immun* 1982;36:209–214.

23. Holmes RH, Twiddy EM: Characterization of monoclonal antibodies that react with unique and cross-reacting determinants of cholera enterotoxin and its subunits. *Infect Immun* 1983;42:914–923.
24. Holmgren J: Actions of cholera toxin and the prevention and treatment of cholera. *Nature* 1981;292:413–417.
25. Holmgren J, Svennerholm AM, Lindblad M: Receptor-like glycocompounds in human milk that inhibit classical and El Tor Vibrio cholerae cell adherence (hemagglutination). *Infect Immun* 1983;39:147–154.
26. Honda T, Arita M, Takeda T, et al: Non-01 V. cholerae produces two newly identical toxins related to Vibrio parahaemolyticus haemolysin and E. coli heat-stable enterotoxin. *Lancet* 1985;2:163–164.
27. Hranitzky KW, Mulholland A, Larson AD, et al: Characterization of a flagellar sheath protein of Vibrio cholerae. *Infect Immun* 1980;27:597–603.
28. Hughes JM, Dannie MD, Hollis G, et al: Non-cholera vibrio infections in the United States. *Ann Intern Med* 1978;88:602–606.
29. Jacob CO, Sela M, Arnon R: Antibodies against synthetic peptides of the B subunit of cholera toxin: cross reaction and neutralization of the toxin. *Proc Nat Acad Sci USA* 1983;80:7611–7615.
30. Jacob CO, Arnon R, Finkelstein RA: Immunity to heat labile enterotoxins of porcine and human E. coli strains achieved with synthetic cholera toxin peptides. *Infect Immun* 1986;52:562–567.
31. Jones GW, Abrams GD, Freter R: Adhesive properties of Vibrio cholerae: adhesion to isolated rabbit brush border membranes and hemagglutinating activity. *Infect Immun* 1976;14:232–239.
32. Jones GW, Freter R: Adhesive properties of Vibrio cholerae: nature of the interaction with isolated rabbit brush border membranes and human erythrocytes. *Infect Immun* 1976;14:240–245.
33. Jones GW: The adhesive properties of Vibrio cholerae and other Vibrio species, in Beachey EH (ed): *Bacterial Adherence (Receptors and Recognition Series B)*, Vol 6. London, Chapman & Hall, 1980, pp 221–249.
34. Kabir S: Composition and immunochemical properties of outer membrane proteins of Vibrio cholerae. *J Bacteriol* 1980;144:382–389.
35. Kabir S: Immunochemical properties of the major outer membrane protein of Vibrio cholerae. *Infect Immun* 1983;39:452–455.
36. Kabir S, Ali S: Characterization of surface properties of Vibrio cholerae. *Infect Immun* 1983;39:1048–1058.
37. Kelley JT, Parker CD: Identification of Vibrio cholerae outer membrane proteins. *J Bacteriol* 1981;145:1018–1024.
38. Lankford CE: Factors of virulence of Vibrio cholerae. *Ann NY Acad Sci* 1960;88:1203–1212.
39. Levine MM, Kaper JB, Black RE, Clements ML: New knowledge on pathogenesis of bacterial enteric infections as applied to vaccine development. *Microbiol Rev* 1983;47:510–550.
40. Lindholm L, Holmgren J, Wikstrom M, et al: Monoclonal antibodies to cholera toxin with special reference to cross-reactions with Escherichia coli heat labile enterotoxin. *Infect Immun* 1983;40:570–576.
41. Madden JM, McCardell BA, Shah DB: Cytotoxins production by members of genus vibrio. *Lancet* 1981;2:1217–1218.

42. Majumdar AS, Dutta P, Dutta D, Ghose AC: Antibacterial and antitoxin responses in the serum and milk of cholera patients. *Infect Immun* 1981;32:1–8.

43. Majumdar AS, Ghose AC: Evaluation of the biological properties of different classes of human antibodies in relation of cholera. *Infect Immun* 1981;32:9–14.

44. Majumdar AS, Ghose AC: Protective properties of anticholera antibodies in human colostrum. *Infect Immun* 1982;36:962–965.

45. Manning PA, Imbesi F, Haynes DR: Cell envelope proteins in Vibrio cholerae. *FEMS Microbiol Lett* 1982;14:159–166.

46. Manning PA, Haynes DR: A common immunogenic Vibrio outer membrane protein. *FEMS Microbiol Lett* 1984;24:297–302.

47. Nishibuchi M, Seidler RJ, Rollins DM, Joseph SW: Vibrio factors cause rapid fluid accumulation in suckling mice. *Infect Immun* 1983;40:1083–1091.

48. O'Brien AD, Chen ME, Holmes RK, et al: Environmental and human isolates of Vibrio cholerae and Vibrio parahaemolyticus produce a Shigella dysenteriae 1 (Shiga)-like cytotoxin. *Lancet* 1984;1:77–78.

49. Pastoris MC, Bhattacharyya FK, Sil J: Evaluation of the phenol induced flagellar agglutination test for the identification of the cholera group of vibrios. *J Med Microbiol* 1980;13:363–367.

50. Pierce NF, Kaniecki E, Northrup RS: Protection against experimental cholera by antitoxin. *J Infect Dis* 1972;126:606–616.

51. Pierce NF, Cray WC Jr, Sacci JB Jr: Oral immunization of dogs with purified cholera toxin, crude cholera toxin or B subunit: evidence for synergistic protection by antitoxin and antibacterial mechanisms. *Infect Immun* 1982;37:687–694.

52. Sakazaki R, Tomura K, Gomez CZ, Sen R: Serological studies on the cholera group of vibrios. *Jpn J Med Sci Biol* 1970;23:13–20.

53. Sanyal SC, Neogi PKB, Alam K, et al: A new enterotoxin produced by Vibrio cholerae 01. *J Diarrh Dis Res* 1984;2:3–12.

54. Schneider DR, Parker CD: Isolation and characterization of protease-deficient mutants of V. cholerae. *J Infect Dis* 1978;138:143–151.

55. Sciortino CV, Yang Z, Finkelstein RA: Monoclonal antibodies to outer membrane antigens of Vibrio cholerae. *Infect Immun* 1985;49:122–131.

56. Sears SD, Richardson K, Young C, et al: Evaluation of the human immune response to outer membrane proteins of Vibrio cholerae. *Infect Immun* 1984;44:439–444.

57. Shrivastava DL: The enzymology of V. cholerae. *Indian J Med Res* 1964;52:801–805.

58. Smith HL Jr: Antibody responses in rabbit to injections of whole cell, flagella and flagellin preparations of cholera and non-cholera vibrios. *Appl Microbiol* 1974;27:375–378.

59. Svennerholm AM, Holmgren J: Synergistic protective effect in rabbits of immunization with Vibrio cholerae lipopolysaccharide and toxin/toxoid. *Infect Immun* 1976;13:735–740.

60. Svennerholm AM, Sack DA, Holmgren J, Bardhan PK: Intestinal antibody responses after immunization with cholera B subunit. *Lancet* 1982;1:305–307.

61. Svennerholm AM, Stromberg GJ, Holmgren J: Purification of Vibrio cholerae

soluble hemagglutinin and development of enzyme-linked immunosorbent assays for antigen and antibody quantitations. *Infect Immun* 1983;41:237–243.

62. Takao T, Shimonishi Y, Kabayashi O, et al: Amino acid sequence of heat stable enterotoxin produced by V. cholerae non-01. *FEBS Lett* 1985;193:250–254.

63. Watanabe Y: Antibacterial immunity in cholera, in Barua D, Burrows W (eds): *Cholera*. Philadelphia, Saunders, 1974, pp 283–306.

64. Yamamoto K, Ichinose Y, Nakasone N, et al: Identity of hemolysins produced by Vibrio cholerae non-01 and V. cholerae 01, biotype El Tor. *Infect Immun* 1986;51:927–931.

65. Yancy RJ, Willis DL, Berry LJ: Flagella-induced immunity against experimental cholera in adult rabbits. *Infect Immun* 1979;25:220–228.

CHAPTER 17

Rationale for the Development of a Rotavirus Vaccine for Infants and Young Children

Albert Z. Kapikian, Jorge Flores,
Yasutaka Hoshino, Karen Midthun, Kim Y. Green,
Mario Gorziglia, Robert M. Chanock, Louis Potash,
Irene Perez-Schael, Marino Gonzalez, Timo Vesikari,
Leif Gothefors, Goran Wadell, Roger I. Glass,
Myron M. Levine, Margaret B. Rennels,
Genevieve A. Losonsky, Cynthia Christy,
Raphael Dolin, Edwin L. Anderson,
Robert B. Belshe, Peter F. Wright,
Mathuram Santosham, Neal A. Halsey,
Mary Lou Clements, Stephen D. Sears,
Marc C. Steinhoff, and Robert E. Black

Vaccinology is a relatively new term, but its very existence reflects the progress in vaccine development. The field of vaccinology embraces the breadth of biology ranging from molecular biologic initiatives to conventional, well-established techniques.

This chapter presents an overview of the rationale for the development of a human rotavirus vaccine with emphasis on (1) the need for such a vaccine, (2) the approaches to development of a vaccine, (3) the clinical application of candidate vaccines, and (4) the efficacy of candidate vaccines in preventing or modifying the targeted disease.

Rotaviruses as the Major Etiologic Agents of Infantile Diarrhea

Diarrheal diseases are well recognized as a major cause of morbidity in developed countries and of both morbidity and mortality in developing countries, with the greatest toll in infants and young children (41,45). For example, a long-term family study in the United States over a period of approximately 10 years revealed that infectious gastroenteritis was the second most common disease, accounting for 16% of more than 25,000 illnesses (19). However, mortality from diarrheal disease is rare in developed countries, as rehydration therapy is readily available. In contrast, the mortality rate from diarrheal diseases in developing countries is enormous. One estimate concerning the occurrence of major infectious diseases ranked diarrhea first in frequency as a cause of mortality in Asia, Africa, and Latin America, with a projection of 3 billion to 5 billion cases and 5 million to 10 million deaths annually (98). Table 17.1 summarizes selected studies that examined the occurrence of diarrheal disease in Asia, Africa, and Latin America (excluding China) in children under 5 years of age (88). Although this estimate is somewhat lower, the figures are still staggering. Analysis of the age distribution of diarrhea-associated deaths in various countries clearly shows that infants and young children are at highest risk of mortality from diarrheal illness and that this risk is greatest in the developing countries (58) (Table 17.2).

Despite major discoveries in bacteriology and parasitology during the past century, the etiology of most acute diarrheal illnesses remained elusive for many years. Although studies during the 1940s and 1950s revealed that oral administration of bacteria-free stool filtrates from patients with acute diarrhea could induce a similar illness in volunteers, the anticipated viral etiologic agent(s) could not be identified (42). Even the advent of tissue culture technology, which enabled the discovery of hundreds of viruses, many of which were shed in the stool, failed to yield the major etiologic agents of this ubiquitous disease (108).

The discovery in 1972 of the 27-nm Norwalk virus and its etiologic association with epidemic gastroenteritis in older children and adults, followed by the discovery in 1973 of the 70-nm human rotavirus and its etiologic association with sporadic gastroenteritis of infants and young children, marked the beginning of a new phase of research in diarrheal diseases (Fig. 17.1) (7,52). Both of these fastidious viruses were discovered using electron microscopic techniques: Norwalk virus by immune electron

Table 17.1. Estimate of morbidity and mortality from acute diarrheal illnesses in children <5 years of age in Africa, Asia (excluding China), and Latin America.

Estimate	No.
Population <5 years of age	338,000,000
Median diarrheal episodes/child/year	2.2
Diarrheal illnesses/year	744,000,000
Median diarrheal mortality rate/1000 children	13.6
Diarrheal deaths/year	,600,000[a]
Diarrheal deaths/day	12,600
Diarrheal deaths/hour	525
Case fatality ratio (deaths/100 episodes)	0.6

[a]80% in ≤2-year-olds.
Adapted from Snyder and Merson (88).

Table 17.2. Age-specific death rates for intestinal infectious diseases.[a]

Country	Year	Infants	Pre-school children	School children
Germany, Federal Republic of	1982	2.3	0.4	0.0
Japan	1982	3.3	0.4	0.0
United States	1980	3.5	0.2	0.0
France	1981	8.3	0.3	0.1
German Democratic Republic	1978	27.6	0.8	0.0
Bulgaria	1982	62.0	1.4	0.1
Romania	1982	134.2	4.6	0.1
Chile	1981	133.7	5.0	0.6
Kuwait	1982	219.3	6.2	0.8
Brazil	1980	612.2	36.2	1.9
Mexico	1982	717.3	59.6	4.5
El Salvador	1981	745.2	76.8	6.6
Paraguay	1980	1622.9	142.2	6.3

[a] Per 100,000 population.
From Kumate and Isibasi (58), with permission.

microscopy, and rotavirus by electron microscopic examination of thin sections of duodenal mucosa. It is of interest that Norwalk virus has not yet been propagated in tissue culture, whereas the cultivation of rotavirus directly from stool specimens into tissue culture was achieved only recently (81,92). It should be emphasized that the Norwalk virus is not implicated as a cause of severe gastroenteritis in infants and young children, although it has been associated with mild gastroenteritis in this age group (9,77). It is considered to be the cause of about 40% of community or institutional outbreaks of nonbacterial epidemic gastroenteritis in adults and school-age children (28,53). However, it is the rotavirus group that has consistently been shown to be the single most important etiologic agent of severe diarrhea of infants and young children in both developed and developing countries; it is associated with about one-third to one-half of such episodes (45).

For example, in a long-term cross-sectional study (covering more than 8 years) of children hospitalized with diarrheal illness in Washington, D.C., 34.5% of 1537 children shed rotavirus in feces (11) (Fig. 17.2). In a similar study in Japan (covering more than 6 years) 45% of 1910 hospitalized children were rotavirus-positive (55). Rotavirus diarrhea most frequently affects children 6 to 24 months of age. The next highest frequency is in infants less than 6 months of age.

Similarly, in studies in developing countries, rotaviruses are characteristically the major etiologic agents of severe diarrheal illness in infants and young children (45). For example, in a 1-year study of 6352 diarrhea patients observed at a treatment center in Bangladesh, 45% of the children

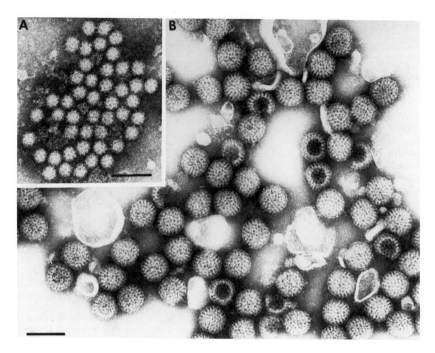

Fig. 17.1. (A) A group of Norwalk virus particles observed after incubation of 0.8 ml. of Norwalk stool filtrate (prepared from a stool of a volunteer administered the Norwalk agent) with 0.2 ml. of a 1:5 dilution of a volunteer's prechallenge serum and further preparation for EM. The quantity of antibody on these particles was rated as 1+. The bar = 100 nm. From: Kapikian, A.Z., et al., 1972 (bar added). (B) Human rotavirus particles observed in a stool filtrate (prepared from a stool of an infant with gastroenteritis) after incubation with PBS and further preparation for EM. The particles appear to have a double-shelled capsid. Occasional "empty" particles are seen. The bar = 100 rm. From: Kapikian, A.Z., et al., 1974, copyright by AAAS.

who were less than 5 years of age were rotavirus-positive (10). The next most commonly detected pathogens in this age group were the enterotoxigenic *Escherichia coli*, detected in 28%.

Compelling Need for a Rotavirus Vaccine

Although rotaviruses can cause mild diarrheal illness or subclinical infections, their capacity to induce severe dehydrating diarrheal illnesses with greater frequency than other agents makes the quest for a rotavirus vaccine especially compelling. The aim of a rotavirus vaccine is to prevent severe rotavirus diarrhea during the first 2 years of life, a period when

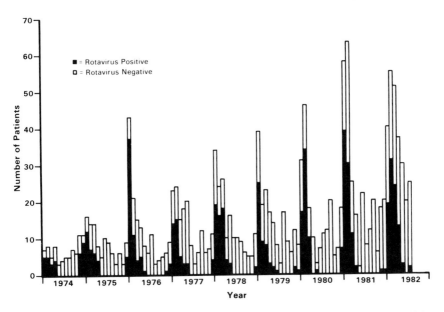

Fig. 17.2. Rotavirus infections in inpatients with gastroenteritis, mid-January 1974–July 1982 (as demonstrated by EM, IEM, and Rotavirus Confirmatory ELISA).

such illness is most serious. Because animal studies demonstrated that resistance to rotavirus disease was mediated primarily by local intestinal immunity, most efforts are directed at developing a live attenuated oral vaccine rather than a parenterally administered formulation (87).

It would be helpful, before discussing rotavirus vaccine strategies, to consider briefly some salient characteristics of the virus (41). Rotaviruses are classified as a new genus in the family Reoviridae. They cause diarrhea not only in humans but also in practically all animal species studied as well as in several avian species (Table 17.3). They are 70 nm in diameter with a double capsid (Fig. 17.1B). The inner capsid surrounds a core that contains the genome consisting of 11 segments of double-stranded RNA. The name rotavirus reflects its appearance, as the smooth outer margin gives the appearance of the rim of a wheel placed on short spokes radiating from a wide hub (24). The virion has a density of 1.36 g/cm^3 in cesium chloride (48).

Rotaviruses are antigenically distinct from the three reovirus serotypes (50). Rotaviruses possess three important antigenic specificities—group, subgroup, and serotype—which are mediated by three proteins: group and subgroup by protein VP6 (encoded by RNA segment 6) and serotype by proteins VP7 and VP3 (encoded by RNA segments 8 or 9, and 4, respectively) (41) (Fig. 17.3). Most animal and human rotaviruses share a common group antigen and are thus classified as group A rotaviruses (4,13).

Table 17.3. General characteristics of rotaviruses visualized in stools[a] of humans and animals.

Rotavirus host	Designation of viral agent	Clinical syndrome associated with agent	Ref.
Human infant	Reovirus-like (RVL) agent, rotavirus, duovirus, infantile gastroenteritis virus	Diarrheal illness	8, 22
Calf (bovine)	Nebraska calf diarrhea virus (NCDV), neonatal calf diarrheal reovirus (NCDR), rotavirus	Diarrheal illness	67
Infant mouse	Epizootic diarrhea of infant mice (EDIM), rotavirus	Diarrheal illness	70
Piglet	Rotavirus	Diarrheal illness	74
Foal	Rotavirus	Diarrheal illness	23
Lamb	Rotavirus	Diarrheal illness	86
Young rabbit	Rotavirus	Diarrheal illness	14
Monkey	Simian agent (SA)-11, S:USA:78:1; S:USA:79:2; rotavirus	None (SA-11); diarrheal illness	61, 89
Sheep and calves	Offal ("O") agent, rotavirus	(Derived from mixed intestinal washings from abattoir waste)	
Newborn deer	Rotavirus	Diarrheal illness	91
Newborn antelope	Rotavirus	Diarrheal illness	72
Young chimpanzee	Rotavirus	Diarrheal illness	3
Young gorilla	Rotavirus	Diarrheal illness	3
Young turkey	Rotavirus	Diarrheal illness	5, 66
Chicken	Rotavirus	Diarrheal illness	40
Young goat	Rotavirus	Diarrheal illness	83
Young kitten	Rotavirus	Diarrheal illness	85
Dog	Rotavirus	None	75
Newborn impala	Rotavirus	Pneumoenteric illness	20
Newborn addax	Rotavirus	Pneumoenteric illness	20
Newborn gazelle	Rotavirus	Pneumoenteric illness	20

[a]"O" agent derived from mixed intestinal washings of sheep and calves from abattoir waste, and SA-11 from rectal swab specimen; both agents were visualized initially in harvests of cell culture passaged material. In addition, serological evidence of rotavirus infection has been reported in goats.

ds RNA

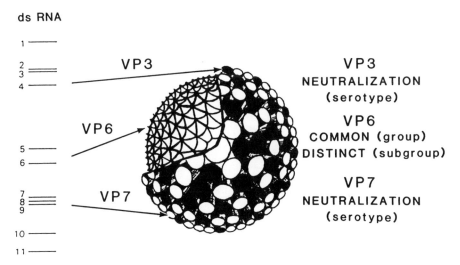

Fig. 17.3. Rotavirus gene coding assignments for antigenic specificities.

Several human and animal rotavirus strains have been described that do not share the common group antigen and are referred to as "pararotaviruses" or non-group A rotaviruses, which currently comprise groups B to F (13). Five human rotavirus serotypes are recognized, and they are further separated into two subgroups (1,37,62,102).

Jennerian Strategy for a Rotavirus Vaccine

The use of a live attenuated rotavirus strain from a nonhuman host—an approach pioneered by Edward Jenner's use of a cowpox strain to immunize humans against smallpox—holds great promise for vaccination against rotavirus diarrhea. This Jennerian approach was suggested by several early studies. It was discovered that human and animal rotaviruses share a common group antigen that makes them indistinguishable in certain serologic assays with hyperimmune animal sera (24,43,44,50,99) (Table 17.4). In addition, children infected with a human rotavirus develop a serologic response (by complement fixation) not only to the human rotavirus but also to animal rotaviruses such as bovine, simian, or murine strains (43) (Table 17.4). As noted above, this common antigen is shared by all group A rotaviruses and is located on the inner capsid.

Experimental studies in colostrum-deprived gnotobiotic calves demonstrated the feasibility of such a Jennerian approach. Table 17.5 shows that each of five calves inoculated in utero with a bovine rotavirus (NCDV) failed to develop illness following challenge at birth with a human rotavirus

Table 17.4. Antigenic relationships among a human rotavirus and various animal rotaviruses with hyperimmune animal and convalescent human sera by complement fixation.

Rotavirus (origin)	Reciprocal of serum antibody titer to indicated virus (source of serum)					Reciprocal of child's acute/convalescent serum			
	HRV (guinea pig)	NCDV (guinea pig)	EDIM (mice)	SA-11 (guinea pig)	"O" (guinea pig)	Child 1	Child 2	Child 3	Child 4
HRV (human)	640	1024	1280	512	2048	<4/128	<4/64	<4/128	<4/128
NCDV (calf)	1280	2048	1280	2048	8192	<4/32	<4/16	<4/32	<4/32
EDIM (mouse)	640	512	320	1024	4096	<4/32	<4/<4	<4/32	<4/16
SA-11 (monkey)	1280	2048	2560	2048	8192	<4/32	<4/8	<4/16	<4/16
"O" (sheep or cattle)	1280	2048	640	2048	4096	<4/32	<4/8	<4/32	<4/16

HRV = human rotavirus. NCDV = Nebraska calf diarrhea virus. EDIM = Epizootic diarrhea of infant mice.
From Kapikian et al. (43), with permission.

Table 17.5. Immunity to Nebraska calf diarrhea virus (NCDV) [serotype 6] and Human Rotavirus (HRV) D strain [serotype 1] in Gnotobiotic calves exposed to NCDV or veal infusion broth (VIB) or "nothing" *in utero*.

	Inoculum		No. of calves with diarrhea after postnatal challenge
No. of Calves	Prenatal (into amniotic sac)[a]	Postnatal (into duodenum)[b]	
2	VIB	NCDV	2 (100%)
2	NCDV	NCDV	0
8	VIB (4) or None (4)	HRV-"D"	7[c] (87.5%)
5	NCDV	HRV-"D"	0[c]

[a]2–24 Weeks before delivery by cesarean section.
[b]Within 2 days of birth.
[c]p = 0.005 Fisher exact test (two-tailed).
Adapted from Wyatt et al (105), with permission.

type 1 (D) strain, whereas seven of eight control animals that did not receive the virus in utero developed diarrhea following challenge with the type 1 strain (105). In addition, most calves inoculated with bovine rotavirus in utero developed neutralizing antibodies not only to the bovine rotavirus but also to human rotavirus serotypes 1, 2, and 3 (104). In later confirmatory studies, piglets inoculated with NCDV and challenged with various human rotaviruses shed significantly lower amounts of virus than did control animals (110).

Vaccination with an Attenuated Bovine Rotavirus Strain

The bovine rotavirus NCDV (Lincoln) strain was developed as a vaccine candidate by Smith-Kline-RIT and has been tested extensively in humans (93–95). This vaccine (designated RIT 4237) is a cold-adapted mutant that has undergone 147 passages in bovine embryonic kidney cells and seven in African green monkey kidney cells. In two placebo-controlled field trials in 178 and 331 Finnish infants 6 to 12 months of age, the vaccine was more than 80% effective in preventing clinically significant diarrheal illness (93,95). In the larger study the predominant infecting serotype was rotavirus type 1. The vaccine did not induce protection against milder rotavirus illnesses, which was not surprising because under natural conditions rotavirus reinfections occur frequently also (30,49,54). Although the RIT 4237 vaccine held great promise as a candidate rotavirus vaccine, its efficacy in developing countries has been disappointing. For example, in Rwanda and the Gambia its protective efficacy against rotavirus diarrhea was poor, and in Peru after a single dose it was marginal (18,32,58a). Thus Smith-Kline-RIT is currently studying a putative, less attenuated strain of bovine rotavirus that has undergone fewer cell culture passages and thus might be more antigenic (96a).

Vaccination with an Attenuated Rhesus Rotavirus (Strain MMU 18006) Vaccine

We have applied the "Jennerian" approach by exploring another animal rotavirus—rhesus rotavirus strain MMU 18006—as a vaccine candidate (47,51,103). This strain was obtained from the stool of a 3½-month-old rhesus monkey with diarrhea (89). The vaccine strain has undergone nine passages in primary or secondary MK cell cultures and seven in DBS-FRhL-2 cells, a semicontinuous diploid fetal rhesus monkey lung cell strain developed by the Office of Biologics, U.S. Food and Drug Administration, as a potential cell substrate for vaccine production (51,97) (Fig. 17.4). The

Passage No in indicated type of cell culture	Cell Culture
(1)	CMK (original isolation)
(2)	CMK (passage sent to NIH)
(1)	AGMK
(2)	AGMK (1st plaque purification)
(3)	AGMK (2nd plaque purification)
(4)	AGMK (3rd plaque purification)
(5)	AGMK (amplification in roller tube cultures)
(6)	AGMK (amplification in flask culture)
(7)	AGMK (Flow laboratories)
(1)	DBS-FRhL-$_2$ (rhesus monkey diploid cell strain)
(2)	DBS-FRhL-$_2$
(3)	DBS-FRhL-$_2$
(4)	DBS-FRhL-$_2$ (harvest ether treated)
(5)	DBS-FRhL-$_2$
(6)	DBS-FRhL-$_2$ (prevaccine seed [passage 22 of FRhL-$_2$ cells])
(7)	DBS-FRhL-$_2$ (vaccine lot designated RRV-1 [passage 24 of FRhL-$_2$ cells]

Summary: Passage history of RRV-1 vaccine candidate CMK2 AGMK7 DBS-FRhL-$_2$ 7.

Fig. 17.4. Passage history rhesus rotavirus (RRV) candidate vaccine lot RRV-1 (strain MMU18006). (CMK = cynomolgus monkey kidney; AGMK = African green monkey kidney.)

rhesus rotavirus was selected for clinical trials because it had never been isolated from humans under natural conditions, it shared neutralization specificity with human rotavirus serotype 3, and it grew efficiently in DBS-FRhL-2 diploid cells, which unlike primary African green MK cells would likely not contain adventitious agents (47).

Following in vitro and in vivo safety tests, we initiated phase 1 studies to determine the reactogenicity and antigenicity of this animal virus in adult volunteers and proceeded in stepwise fashion to children and infants. The chronology of the 21 phase 1 studies and one phase 2 study (study 21) is shown in Table 17.6 (2,16a,47,51,60,71,96,101). The painstaking, careful progression of clinical studies that lead up to field trials is clearly evident.

In the early phase 1 studies in the United States the vaccine was safe and antigenic when administered undiluted (10^6 PFU) or at a 1:10 dilution (10^5 PFU). The vaccine was given orally after the administration of buffer and/or formula to neutralize stomach acid, as rotaviruses are acid-labile (47,94). However, in study 20 in Finland, in which the RRV or RIT 4237 vaccines were given to 6- to 8-month-old infants, the RRV vaccine was associated with fever in 16 (64%) of 25 vaccinees and the RIT vaccine in only 4 (17%) of 24 vaccinees (96). Watery stools were observed in 20% of the RRV vaccinees also. Similar reactions were found in Umea, Maryland, and West Virginia (2,47,60). It should be noted however, that the RRV vaccine was more immunogenic than the RIT 4237 vaccine (96).

Because of such reactions in certain of the studies, the reactogenicity and immunogenicity of lowered doses (10^4 or 10^3 PFU) of RRV vaccine were evaluated in 4- to 10-month-old children in Venezuela (study 22) (71). Neither dose was reactogenic. Moreover, an antibody response was demonstrated in 82% of the children receiving the higher dose. The reason for the absence of reactions in this study and in the early U.S. studies (in which a higher dose was given) was unclear. However, analysis of the prevaccination neutralizing antibody titer in serum of selected vaccinees in the United States, Venezuela, and Finland suggested that the absence of reactions was related to the significantly higher prevaccination serum antibody titers in the U.S. and Venezuelan children (45,47). Thus preexisting antibody may have modified the clinical response to the vaccine without significantly affecting the "take" rate.

Because of this finding we evaluated the vaccine in less than 6-month-old infants in whom the presence of maternal antibody might prove beneficial in preventing vaccine reactions (71). It would be especially important in developing countries where vaccination at a very young age (e.g., birth to 2 months) may be necessary, as rotavirus diarrhea can occur at an early age and because it might be the only time when the child has contact with medical personnel (31). We therefore tried to identify a dose of RRV vaccine in the very young that was capable of stimulating a silent immunizing infection under the cover of passively acquired maternal rotavirus antibodies. In a keystone study, 1- to 4-month-old infants in Venezuela were

Table 17.6. Sequence of studies in humans with rhesus rotavirus vaccine candidate RRV-1 (MMU 18006) administered orally (1ml).

1. Jan. 20, 1984.	2 young adults with high levels of rotavirus serum antibody given vaccine (undiluted) at University of Maryland (Drs. Levine, Clements, Black).
2. Jan. 24, 1984.	9 young adults with lower levels of rotavirus serum antibody given vaccine (undiluted) at University of Maryland (Drs. Levine, Clements, Black).
3. May 9, 1984.	25 young adults with lowest available serum neutralizing antibody to RRV at University of Rochester (Drs. Dolin, Christy): 13 received vaccine (undiluted), 12 placebo.
4. May 19, 1984.	15 young adults with lowest available serum neutralizing antibody to RRV at University of Rochester (Drs. Dolin, Christy): 8 received vaccine (undiluted), 7 placebo.
5. June 21, 1984.	12 children 5–12 years of age with serum neutralizing antibody to RRV at Vanderbilt University (Dr. Wright): 6 received vaccine (undiluted), 6 placebo.
6. June 22, 1984.	6 young adults with lowest available serum neutralizing antibody to RRV at University of Maryland (Drs. Clements, Levine): 3 received vaccine diluted 1:10 and 3 diluted 1:100.
7. July 9, 1984.	6 children 35–61 months of age with serum neutralizing antibody to RRV at Marshall University (Drs. Belshe, Anderson): 3 received vaccine (undiluted), 3 placebo.
8. July 12, 1984.	8 children 5–12 years of age with serum neutralizing antibody to RRV at Vanderbilt University (Dr. Wright): 4 received vaccine (undiluted), 4 placebo.
9. August 1, 1984.	3 children 31–59 months of age with serum neutralizing antibody to RRV at Marshall University (Drs. Belshe, Anderson): 2 received vaccine (undiluted), 1 placebo.
10. August 6, 1984.	12 children 2–12 years of age with serum neutralizing antibody to RRV at Vanderbilt University (Dr. Wright): 6 received vaccine (undiluted), 6 placebo.
11. Sept. 6, 1984.	6 children 22 months to 4 years of age at Vanderbilt Univ. (Dr. Wright): 3 received 1/10 dilution of vaccine because some had low or undetectable neutralizing antibody to RRV, 3 placebo.
12. Sept. 17, 1984.	5 children 8–22 months of age with serum neutralizing antibody to RRV at Marshall Univ. (Drs. Belshe, Anderson): 3 received vaccine (undiluted), 2 placebo.

13. Sept. 27, 1984. 5 children 12–19 months of age with serum neutralizing antibody to RRV at Vanderbilt Univ. (Dr. Wright): 3 received vaccine (undiluted), 2 placebo.

14. Oct. 1, 1984. 9 children 6–22 months of age (not all sera prescreened for RV antibody) at Marshall Univ. (Drs. Belshe, Anderson): 5 received vaccine (1/10 dilution), 4 placebo.

15. Oct. 1, 1984. 8 children 4–19 months of age (serum not prescreened for RV antibody) at Univ. of Maryland (Drs. Losonsky, Levine): 3 received vaccine (1/10 dilution), 5 placebo [3 placebo recipients vaccinated Oct. 22, 1984].

16. Oct. 22, 1984. 7 children 8 months to 3 years of age (low or undetectable neutralizing antibody to RRV [not all screened]) at Vanderbilt Univ. (Dr. Wright): 4 received vaccine (undiluted), 3 placebo.

17. Nov.–Dec. 1984. 9 children 4–12 months of age (serum not prescreened for RV antibody) at Marshall Univ. (Drs. Belshe, Anderson): 4 received vaccine (1/10 dilution), 5 placebo.

18. Dec. 11–14, 1984. 13 children 6–24 months of age (serum not prescreened for RV antibody) at Vanderbilt Univ. (Dr. Wright): 7 received vaccine (1/10 dilution), 6 placebo.

19. Nov.–Dec. 1984. 19 children 5–19 months of age (serum not prescreened for RV antibody) at Univ. of Maryland (Drs. Levine, Rennels): 11 received vaccine (1/10 dilution), 8 placebo.

20. Dec. 11–14, 1984. 51 children 6–8 months of age (serum not prescreened for RV antibody) at Univ. of Tampere, Tampere, Finland (Dr. Vesikari): 26 received RIT 4237 vaccine, 25 RRV vaccine (1/10 dilution).

21. Jan.–Feb. 1, 1985. 106 children 4–12 months of age (serum not prescreened for RV antibody) at Univ. of Umea, Umea, Sweden (Dr. Gothefors): 54 received vaccine (1/10 dilution), 52 placebo.

22. Feb.–March, 1985. 54 children 4–10 months of age (serum not prescreened for RV antibody) at National Institute of Dermatology, Caracas, Venezuela (Drs. Perez-Schael and Flores): 17 received further diluted vaccine (1/100 dilution), 18 further diluted vaccine (1/1000 dilution), 19 placebo.

Note: [a]38 young adults received RRV vaccine and 19 placebo; [b]178 children received RRV vaccine and 155 placebo [or RIT 4237]: 140 of the 178 RRV vaccinees and 121 of the 155 placebo (or RIT 4237) recipients were in the 4–12 month age group.

Table 17.7. Clinical trials of RRV vaccine candidate.

Institution (investigators)	No. of infants enrolled	Age of enrollees (months)
University of Maryland (Rennels, Losonsky, Levine)	30	2–11
Institute of Dermatology, Caracas, Venezuela (Perez-Schael, Flores)	240	1–10
Vanderbilt University, Tennessee (Wright)	30	4–12
Marshall University, West Virginia (Anderson, Belshe)	30	4–12
University of Umea, Sweden (Gothefors, Wadell)	106[a]	4–12
University of Tampere, Finland (Vesikari)	200	2–5
Johns Hopkins University/Whiteriver and Tuba City, Arizona (Santosham, Sack)	210	2–5
University of Rochester, New York (Christy, Dolin)	176	2–4
King Edward Medical College, Pakistan[b] (Jalil)	74	1.5
University of Umea, Sweden (Gothefors, Wadell)	40	1–4

[a]10^5PFU/ml, all others 10^4PFU ml dose of vaccine.
[b]Phase 1 study.

given a 10^4 PFU dose of RRV in stepwise fashion in Venezuela. This dose was nonreactogenic but quite antigenic, as 75% of these infants developed a rotavirus antibody response (71). As a result of this study, field trials with this vaccine were initiated at various locations in the under 5-month-old group. [This study is not listed in Table 17.6 because it became part of the phase 2 study (as did study 22), shown in Table 17.7.]

RRV Vaccine Field Trials Evaluating Clinical Efficacy

Placebo-controlled field trials of a 10^4 PFU dose of RRV vaccine are currently in progress in approximately 600 infants under 5 months of age in the United States and overseas (46) (Table 17.7). Preliminary data on vaccine efficacy is now available. The vaccine is not currently being given

Table 17.8. Proposed NIH rating system (0–3) for estimating severity of gastroenteritis in rotavirus vaccine field trials.

Sign or symptom	Score
Duration of diarrhea ($\geq$ 3 loose or watery stools in 24 hours)	
< 2 days	1
2–3 days	2
$\geq$ 4 days	3
Maximum No. of diarrheal stools in 24 hours	
< 3	0
3	1
4–5	2
$\geq$ 6	3
Duration of vomiting	
No vomiting	0
1–2 days	2
$\geq$3 days	3
Maximum No. of vomiting episodes in 24 hours (an episode must be separated from another episode by $\geq$3 hours)	
1	1
2	2
$\geq$3	3
Dehydration	
None	0
$\leq$ 5%	2
> 5%	3
Maximum temperature (rectal)	
< 38.1°C (<100.6°F)	0
38.1–38.9°C (100.6–102°F)	1
> 38.9°C (>102°F)	2
Hospitalization necessary	
Yes	3
No	0

Note: Maximum severity score = 20.
Modified from Hjelt et al. (33)

to children over 5 months of age because higher reaction rates were observed above this age presumably due to the loss of most maternally derived serum antibodies to rotavirus (as noted above).

In order to standardize observations relating to the severity of clinical illness in field trials, we modified a scheme described by Hjelt et al (33) as a rating system of clinical manifestations, as shown in Table 17.8.

The code for the efficacy trial in Venezuela has been broken, and the results are encouraging (25). As shown in Table 17.9, 247 infants, aged 1 to 10 months, were followed for up to 1 year: 123 received the 10^4 PFU dose of RRV and 124 the placebo. Twenty-one episodes of rotavirus diarrhea were observed: 16 in the control group and only 5 in the vaccinees, for an efficacy rate of 68%. When vaccine efficacy was analyzed according to severity of illness by a slightly modified version of our rating system described above, the vaccine induced marked protection against severe rotavirus diarrhea (clinical score ≤ 8) for the entire study group (Table 17.10). Later analysis also revealed a high degree of efficacy (100%) in the youngest age group (1 to 4 months) against rotavirus diarrhea (Table 17.11). This finding was especially encouraging, as the vaccine should be given to the youngest age groups in developing countries.

In a small trial in Maryland (10^5 PFU dose) the RRV vaccine also appeared to induce resistance to diarrheal illness (73). In addition, in the Umea, Sweden and Tampere, Finland trials the vaccine appeared to induce a reduction in the number of clinically significant diarrheal episodes. In contrast, however, the vaccine failed to induce protection against diarrheal illness in the Rochester vaccine trial (46a).

The RRV and RIT 4237 vaccines are undergoing a clinical efficacy trial in a study at the American Indian reservation in Whiteriver, Arizona. The code for this trial is to be broken shortly.

Reassortant Vaccines Against Other Reassortant Serotypes

Because the efficacy of the RRV vaccine is variable, it appears that a multivalent vaccine with components to all four serotypes may be needed. It can be achieved by taking advantage of the property of rotaviruses to

Table 17.9. Protective Efficacy of RRV vaccine against RV diarrhea in children 1 to 10 months of age at time of vaccination in Caracas, Venezuela

Inoculum	No. in each group	No. with RV diarrhea[a]
RRV	123	5
Placebo	124	16

Protection rate = 68%, $p = 0.013$.
[a] $\geq$ Three loose or watery stools in 24 hours.
Data from Flores et al (25).

Table 17.10. Protective efficacy of RRV vaccine against RV diarrheal illness of varying severity in children 1 to 10 months of age at time of vaccination in Caracas, Venezuela.

RV diarrheal illness severity score	No. of episodes of RV diarrheal illness in indicated group		Protection rate (%)	p
	RRV ($n = 123$)	Placebo ($n = 124$)		
All RV diarrheal illness	5	16	68	0.013
RV diarrhea with rating $\geqq 7$	2	9	77	0.035
RV diarrhea with rating $\geqq 8$	1	7	86	0.04
RV diarrhea with rating $\geqq 9$	0	6	100	0.01
RV diarrhea with rating $\geqq 10$	0	5	100	0.03

Adapted from Flores et al (25).

Table 17.11. Protective efficacy of RRV vaccine against RV diarrhea in infants 1 to 4 months of age at time of vaccination in Venezuela.

Inoculum	No. in each group	No. with RRV diarrhea[a]
RRV vaccine	49	0
Placebo	54	12

Protection rate = 100%, $p < 0.001$.
[a]Three loose or watery stools in 24 hours.
Data from Flores et al. (125).

undergo genetic reassortment with high efficiency during co-infection. Such reassortants are made by co-infecting cell cultures with two rotavirus strains under selective pressure of antibody against one of them (Fig. 17.5). Thus the rhesus RV (or bovine RV) strain can be used as a donor of attenuating genes, and a human rotavirus of a specific serotype can be used as the donor of a single gene—that which codes for the major neutralization protein (VP7) (68,69). Such single gene substitution reassortants have been prepared for each of the four epidemiologically important human rotavirus serotypes, with the rhesus and/or bovine (UK) rotavirus as the donor of the other ten genes (68,69). VP3, another outer capsid protein (in addition to VP7), can also induce high levels of neutralizing antibody and such antibody can protect against RV disease experimentally (35,70b). Therefore it may be possible to construct a reassortant that has a broadly reactive VP3 as well as the VP7 of a different serotype for use as a vaccine candidate (35).

Phase 1 Trials with Serotype 1 Reassortant (D × RRV) and Serotype 2 Reassortant (DS1 × RRV)

We are in the midst of phase 1 clinical trials with two reassortant rotavirus vaccine candidates. One is the D × RRV reassortant, which has a single gene that encodes VP7 from the human RV parent (strain D) and the remaining ten genes from the animal RV parent (strain RRV). This reassortant has the neutralization specificity of serotype 1 via VP7. The other is the DS-1 × RRV reassortant, which has a single gene that codes for VP7 from the human rotavirus parent and the remaining ten genes from the rhesus rotavirus parent. This reassortant has the neutralization specificity of serotype 2 via VP7 (68).

Let us first consider our studies with the D × RRV reassortant. As noted earlier with the RRV vaccine, such studies move slowly, as they rely on the availability of human volunteers. In total, 12 studies have been carried out with the D × RRV candidate vaccine: 43 individuals received the reassortant and 25 a placebo, as shown chronologically in Table 17.12.

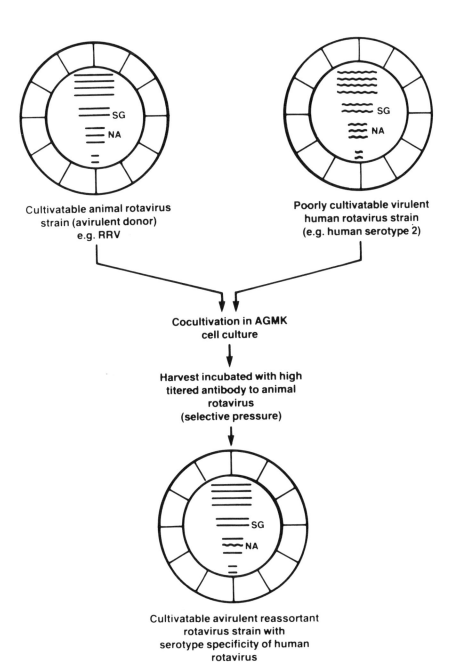

Fig. 17.5. Production of reassortant rotavirus vaccine. Adapted from Kapikian et al., 1986b (47).

Table 17.12. Sequence of Studies with D × RRV reassortant vaccine (serotype 1 [VP7] administered orally to adults and children.

Date	Study
4/13/86	2 adult volunteers with a high level (1:640) of serum antibody by PRN to D × RRV reassortant given 10^6 PFU of vaccine at Johns Hopkins University.
5/15/86	8 adult volunteers with lowest available level (1:160) of serum antibody by PRN to D × RRV reassortant given 10^6 PFU of vaccine at Johns Hopkins University.
6/10/86	3 children 3–10 years of age given 10^4 PFU of vaccine (2 children given placebo) at Vanderbilt University.
6/25/86	3 children 3–10 years of age given 10^5 PFU of vaccine (2 children given placebo) at Vanderbilt University.
8/7/86	4 children 6–30 months of age with a high level ($\geqq$ 1:40–1:1280) of serum antibody by PRN to D × RRV reassortant given 10^4 PFU of vaccine (3 children given placebo) at Johns Hopkins University.
8/22/86	2 children 14 and 26 months of age given 10^4 PFU of vaccine (2 children given placebo) at Vanderbilt University.
9/6/86	4 children 9–36 months of age with a high level (1:320–1:1280) of serum antibody by PRN to D × RRV reassortant given 10^4 PFU of vaccine (4 children given placebo) at Johns Hopkins University.
9/12/86	4 children 7–42 months of age given 10^4 PFU of vaccine (3 children given placebo) at Vanderbilt University.
10/2/86	3 children 5–15 months of age given 10^4 PFU of vaccine (3 children given placebo) at Vanderbilt University.
10/22/86	4 children 2–4 months of age with a high level (1:160–1:1280) of serum antibody by PRN to D × RRV reassortant given 10^4 PFU of vaccine (3 children given placebo) at Johns Hopkins University.
10/27/86	3 children 4–10 months of age given 10^4 PFU of vaccine (1 child 13 months of age given placebo) at Vanderbilt University.
11/18/86	3 children 2–5 months of age with varying levels (1:40, 1:80, 1:320) of serum antibody by PRN to D × RRV reassortant given 10^4 PFU of vaccine (2 children given placebo) at Johns Hopkins University.

Initially, two volunteers with a high level of PRN serum antibody (1:640) to the D × RRV reassortant were given 10^6 PFU of this vaccine orally after ingestion of buffer. Neither volunteer became ill. Subsequently, eight additional volunteers with the lowest available PRN serum neutralizing antibody (1:160) titers to the D × RRV reassortant were given this vaccine. None developed illness. We then began stepwise studies with this reassortant in 3- to 10-year-old children who were given 10^4 or 10^5 PFU after administration of buffer. Because illness was not observed, we studied the younger age groups, 6 months to 3 years of age, and currently 6 weeks to 6 months of age, with a 10^4 PFU dose of vaccine. A major difficulty when studying young children is the frequency of intercurrent infections, which makes it difficult to assess the reactogenicity of the vaccine.

How antigenic is the reassortant vaccine? In the adult volunteers the vaccine appeared to be antigenic, as more than 80% developed serum antibody responses. It appears that the vaccine is less antigenic in children (where studies are in progress). Perhaps the high levels of maternally acquired prevaccine serum antibodies suppress the "take" rate.

Similar studies have been initiated with the DS-1 × RRV reassortant vaccine (Table 17.13). A total of 22 individuals have been vaccinated, and 9 received the placebo. Initially, two adult volunteers with a high level of prechallenge PRN antibody (1:160 or 1:640) to this reassortant were given 10^6 PFU orally following ingestion of buffer. Neither volunteer developed illness. Subsequently, 14 volunteers with little, if any, prevaccination antibody (<1:80) to this reassortant were given the vaccine. None developed illness. We then began stepwise studies in children 3 to 10 years of age with 10^4 PFU of vaccine. Because illness was not observed we extended the studies to the 6-month to 3-year age group. If this phase is completed satisfactorily, children 6 weeks to 6 months of age will be enrolled for study.

How antigenic is this vaccine? In adult volunteers it was quite antigenic, as more than 90% developed an antibody response. In studies of four

Table 17.13. Sequence of studies with DS-1 × RRV reassortant vaccine (serotype 2 [VP7] administered orally to adults and children.

5/1/86	2 adult volunteers with a high level (1:160 or 1:640) of serum antibody by PRN to DS-1 × RRV reassortant given 10^4 PFU of vaccine at Johns Hopkins University.
5/15/86	14 adult volunteers without serum antibody (<1:80) by PRN to DS-1 × RRV reassortant given 10^4 PFU of vaccine at Johns Hopkins University.
8/21/86	4 children 6–9 years of age with low level of serum antibody (1:41–1:64) by PRN to DS-1 × RRV reassortant given 10^4 PFU of vaccine (7 children given placebo) at Marshall University.
10/22/86	2 children 9–20 months of age with low level of serum antibody (1:20) by PRN to DS–1 × RRV reassortant given 10^4 PFU of vaccine (2 children given placebo) at Marshall University.

children 3 to 10 years of age, each developed a neutralizing antibody response in preliminary data. The highly antigenic nature of this vaccine may reflect the low prevalence of serum PRN antibody to the VP7 of this reassortant in the general population. In addition to the low antibody prevalence in adults, it has been difficult to find children 6 months to 3 years of age who possess PRN serum antibody to this reassortant even at a level of 1:10. Why do we want vaccinees with serum antibody to the reassortant? We are aiming to vaccinate children under 5 months of age who have high levels of naturally acquired passive rotavirus antibody, as such antibody may modify the reactogenicity of the vaccine.

Do we need a polyvalent vaccine that is composed of each of the epidemiologically important serotypes to achieve maximum protection? This issue has not yet been resolved. It is conceivable that the heterotypic vaccines (RIT 4237 and RRV) may protect only against severe diarrhea and not against the milder forms of illness because the vaccine-induced heterotypic antibody titers may not be high enough to prevent *any* illness but may be adequate to protect against severe illness. The elegant studies of Chiba et al demonstrated that the mere presence of naturally occurring RV antibody was not sufficient to protect against even homotypically induced illness (16). A serum neutralization titer of at least 1:128 was needed for protection.

Thus a vaccine with a serotype 1 or 3 component would most likely also contain a serotype 2 component for maximum protection, as the latter appears to be least related to the other three serotypes (26). Of course, a polyvalent vaccine comprised of each of four serotypes, if necessary, can be formulated readily, as reassortants for each of the serotypes are available along with the serotype 3 component, RRV.

Conclusion

Other possible approaches to vaccine development include (1) molecular biologic techniques, (2) cold adaptation of rotaviruses, (3) the use of strains obtained from neonates undergoing asymptomatic infections that may represent naturally occurring attenuated strains, and (4) the use of such asymptomatic strains as a donor of VP3 in construction of reassortants with a VP7 from a virulent rotavirus (6,35,36,47,63). Currently, the use of conventional cell culture techniques for the growth of animal RV strains appears promising for the development of an effective RV vaccine. If the Jennerian approach does not induce immunity to rotavirus disease induced by each of the serotypes, single gene substitution reassortants for each of the serotypes are available. It should be noted that none of these vaccines would prevent illness due to the non-group A rotaviruses (also known as pararotaviruses) that have caused large outbreaks of gastroenteritis in China (13,38). Surveillance for these atypical rotaviruses, which do not share the common group antigen and which have been recovered only

Table 17.14. Unusual clinical manifestations associated temporally with rotavirus infections.

Manifestation	Refs.
Chronic or recurrent diarrhea (postgastroenteritis syndrome)	21, 84, 90
Abortion	15, 109
Intussusception	56, 57
Gastrointestinal bleeding	17, 65, 88a, 88b, 90a
Elevated serum transaminase	19a, 90
Henoch-Schönlein purpura	17
Reye's syndrome (fatal)	79
Encephalitis	79, 92a, 92c
Hemolytic uremic syndrome	21, 100a
Disseminated intravascular coagulation	100a
Febrile or afebrile convulsions with gastroenteritis	59, 64a, 70c, 79, 92b
Prolonged diarrhea in primary immunodeficiency	82
Exanthem subitum	78
Sudden infant death syndrome	107
Upper respiratory or middle ear symptoms	27, 30, 65, 73a
Severe gastroenteritis in pre- or post-bone marrow transplant patients (average age of study patients 20–21 years) (fatalities)	106
Necrotizing enterocolitis in neonates	16b, 76, 76a
Hemorrhagic gastroenteritis in neonates	16b, 76, 76a
Kawasaki syndrome	64
Serious disease in patients with severe combined immunodeficiency (fatalities)	39
Pneumonia or bronchitis	65, 80
Aseptic meningitis	100
Transient hyperphosphatasemia	34
Hepatic abscess	29
Laryngitis	70a
Colitis	26a
Rash	65, 79
Conjunctivitis	65

sporadically outside China, should be continued to determine their prevalence (13). Vaccine development, if needed, for these agents must await their growth in tissue culture.

The previous discussion has focused on the rationale for development of a rotavirus vaccine. It is clear that vaccine studies must proceed cautiously and carefully with constant vigilance for detection of any unexpected, unusual clinical manifestation that may be related temporally to vaccination. With such vigilance, a decision may have to be made whether an observed condition was associated etiologically with vaccination or was merely incidental to it. It is the most difficult area of vaccinology, as the decision to continue or to stop a vaccine study may hang precariously on a single observation in a single patient, and often the answer is not clear-cut. The presence of a control group in any vaccine trial aids immeasurably in this decision-making process and should be a *sine qua non* except under the most unusual circumstances. The dilemma of temporally associated events is real and poses serious issues even before initiating a vaccine trial. For example, the literature contains case reports or single outbreaks of unusual clinical conditions associated with naturally occurring rotavirus infections (Table 17.14). Are these conditions caused by rotavirus infections, or are they merely coincidental to them? Would a live rotavirus vaccine cause similar unusual conditions? At this time, because the frequency of such conditions is extremely rare in comparison to the number of rotavirus infections, it appears (with the exception of the severe manifestations in immunocompromised patients and in neonates with necrotizing enterocolitis and with hemorrhagic gastroenteritis) that most are temporally, but not etiologically, associated. It was on this basis that the vaccine studies were initiated, a decision that appears to have been correct. However, the final answer regarding such other relationships must await continued clinical observations and vigilance.

References

1. Albert MJ, Unicomb LE, Bishop RF: Cultivation and characterization of human rotaviruses with "supershort" RNA patterns. *J Clin Microbiol* 1987;25:1635–1640.
2. Anderson EL, Belshe RB, Bartram J, et al: Evaluation of rhesus rotavirus vaccine (MMU 18006) in infants and young children. *J Infect Dis* 1986;153:823–831.
3. Ashley CR, Caul EO, Clark SKR, et al: Rotavirus infections of apes. *Lancet* 1979;2:477.
4. Bachmann P, Bishop RF, Flewett TH, et al: Nomenclature of human rotaviruses: designation of subgroups and serotypes. *Bull WHO* 1984;62:501–503.
5. Berglund ME, McAdaragh JP, Stotz I: Proceedings of the 26th Western Poultry disease conference at University of California, Davis, 1986, pp 129–130.
6. Bishop RF, Barnes GL, Cipriani E, Lund JS: Clinical immunity after neonatal rotavirus infection: a prospective longitudinal study in young children. *N Engl J Med* 1983;309:72–76.
7. Bishop RF, Davidson GP, Holmes IH, Ruck BJ: Virus particles in epithelial cells of duodenal mucosa from children with viral gastroenteritis. *Lancet* 1973;2:1281–1283.
8. Bishop RF, Davidson GP, Holmes IH, Ruck BJ: Detection of a new virus by electron microscopy of fecal extracts from children with acute gastroenteritis. *Lancet* 1974;1:149–151.
9. Black RE, Greenberg HB, Kapikian AZ, et al: Acquisition of serum antibody to Norwalk virus and rotavirus and relation to diarrhea in a longitudinal study of young children in rural Bangladesh. *J Infect Dis* 1982;145:483–489.

10. Black RE, Merson MH, Mizanur Rahman ASM, et al: A two-year study of bacterial, viral and parasitic agents associated with diarrhea in rural Bangladesh. *J Infect Dis* 1980;142:660–664.

11. Brandt CD, Kim HW, Rodriguez WJ, et al: Pediatric viral gastroenteritis during eight years of study. *J Clin Microbiol* 1983;18:71–78.

12. Brandt CD, Parrot RH, Chandra R, et al: Diarrhea virus and sudden infant deaths. *Pediatr Res,* abstract 1068.

13. Bridger JC: Novel rotaviruses in animals and man, in *Ciba Foundation Symposium 128: Novel Diarrhoea Viruses.* New York, Wiley, 1987, pp 5–15.

14. Bryden AS, Thouless ME, Flewett TH: Rotavirus in rabbits. *Vet Rec* 1976;99:323.

15. Buffet-Janvresse C, Berhard E, Magard H: Responsabilite des rotavirus dans les diarrhees du nourissan. *Nouv Presse Med* 1976;5:1249–1251.

16. Chiba S, Yokoyama T, Nakata S, et al: Protective effect of naturally acquired hemotypic and hetertypic rotavirus antibodies. *Lancet* 1986;2:417–421.

16a. Christy C, Madore HP, Treanor JJ, Pray K, Kapikian AZ, Chanock RM, Dolin R: Safety and immunogenicity of live attenuated rhesus monkey rotavirus vaccine. *J Infect Dis* 1986;154:1045–1047.

16b. Deerlove J, Lathan P, Deerlove B, et al: Clinical range of neonatal rotavirus gastroenteritis. *Br Med J* 1983;286:1473–1475.

17. De Lage G, McLaughlin B, Berthiaume L: A clinical study of rotavirus gastroenteritis. *J Pediatr* 1978;93:455–457.

18. De Mol P, Zissis G, Butzler JP: Failure of live attenuated oral rotavirus vaccine. *Lancet* 1986;2:108.

19. Dingle JH, Badger GF, Jordan WS: *Illness in the Home: A Study of 25,000 Illnesses in a Group of Cleveland Families.* Cleveland, Western Reserve University Press, 1964, pp 19–32.

19a. Dominick HC, Maas G: Rotavirus infectionen im kindersalter. *Klin Paediatr* 1979;191:33–39.

20. Eugster AK, Strother J, Hartfield DA: Rotavirus (reovirus-like) infection of neonatal ruminants in a zoo nursery. *J Wild Dis* 1978;14:351–354.

21. Flewett TH: Clinical features of rotavirus infections, in Tyrell DAJ, Kapikian AZ (eds): *Virus Infections of the Gastrointestinal Tract.* New York, Marcel Dekker, 1982, pp 125–145.

22. Flewett TH, Bryden AS, Davies H: Virus particles in gastroenteritis. *Lancet* 1973;2:1497.

23. Flewett TH, Bryden AS, Davies H: Diarrhea in foals and other animals. *Vet Rec* 1975;97:477.

24. Flewett TH, Bryden AS, Davies H, et al: Relationship between virus from acute gastroenteritis of children and newborn calves. *Lancet* 1974;2:61–63.

25. Flores J, Perez-Schael I, Gonzalez M, et al: Protection against severe rotavirus diarrhea by rhesus rotavirus vaccine in Venezuelan infants. *Lancet* 1987;1:882–884.

26. Flores J, Perez I, White L, et al: Genetic relatedness among human rotaviruses as determined by RNA hybridization. *Infect Dis* 1982;37:648–655.

26a. Fujita Y, Hiyoshi K, Wakasugi N, et al: Transient improvement of the West syndrome in two cases following rotavirus colitis. *No To Hattutsu* 1988;20:59–63.

27. Gordon AG: Rotavirus infections and the sompe syndrome. *J Infect Dis* 1982;146:117–118.

28. Greenberg HB, Valdesuso J, Yolken RH, et al: Role of Norwalk virus in outbreaks of nonbacterial gastroenteritis. *J Infect Dis* 1979;139:564–568.

29. Grunow JE, Dunton SF, Waner JL: Human rotavirus-like particles in hepatic abscess. *J Pediatr* 1985;106:73–76.

30. Gurwith M, Wenman W, Hinde E, et al: A prospective study of rotavirus infection in infants and young children. *J Infect Dis* 1981;144:218–224.

31. Halsey N, Galazka A: The effectiveness of DPT and oral poliomyelitis immunization schedules initiated from birth to 12 weeks of age. *Bull WHO* 1985;63:1151–1169.

32. Hanlon P, Hanlon L, Marsh V, et al: Trial of an attenuated bovine rotavirus vaccine (RIT 4237) in Gambian infants. *Lancet* 1987;1:1342–1345.

33. Hjelt K, Graubelle PC, Anderson L, et al: Antibody response in serum and intestine in children up to six months after a naturally acquired rotavirus gastroenteritis. *J Pediatr Gastroenterol Nutr* 1986;5:74–80.

34. Holt PA, Steel AE, Armstrong AM: Transient hyperphosphatasaemia of infancy following rotavirus infection. *J Infect* 1985;9:283–285.

35. Hoshino Y, Sereno MM, Midthun K, et al: Independent segregation of two antigenic specificities (VP3 and VP7) involved in neutralization of rotavirus infectivity. *Proc Natl Acad Sci USA* 1985;82:8701–8704.

36. Hoshino Y, Wyatt RG, Flores J, et al: Serotypic characterization of rotaviruses derived from asymptomatic human neonatal infections. *J Clin Microbiol* 1985;21:425–430.

37. Hoshino Y, Wyatt RG, Greenberg HB, et al: Serotypic similarity and diversity of rotaviruses of mammalian and avian origins as studied by plaque reduction neutralization. *J Infect Dis* 1984;149:694–702.

38. Hung T, Chen G, Wang C, et al: Waterbourne outbreak of rotavirus diarrhea in adults in China caused by a novel rotavirus. *Lancet* 1984;1:1139–1142.

39. Jarvis WR, Middleton PJ, Gelfand EW: Significance of viral infections in severe combined immunodeficiency disease. *Pediatr Infect Dis* 1983;2:187–192.

40. Jones RC, Hughes CS, Henry RR: Rotavirus infection in commercial laying hens. *Vet Rec* 1979;104:22.

41. Kapikian AZ, Chanock RM: Rotaviruses, in Fields BN, et al (eds): *Virology*. New York, Raven Press, 1985, pp 863–906.

42. Kapikian AZ, Chanock RM: Norwalk group of viruses, in Fields BN, et al (eds): *Virology*. New York, Raven Press, 1985, pp 1495–1517.

43. Kapikian AZ, Cline WL, Kim HW, et al: Antigenic relationships among five reovirus-like (RVL) agents by complement fixation (CF) and development of a new substitute CF antigen for the human RVL agent of infantile gastroenteritis. *Proc Soc Exp Biol Med* 1976;152:535–539.

44. Kapikian AZ, Cline WL, Mebus CA, et al: New complement-fixation test for the human reovirus-like agent of infantile gastroenteritis: Nebraska calf diarrhoea virus used as antigen. *Lancet* 1975;1:1056–1061.

45. Kapikian AZ, Flores J, Hoshino Y, et al: Rotavirus: the major etiologic agent of severe infantile diarrhea may be controllable by a "jennerian" approach to vaccination. *J Infect Dis* 1986;153:815–822.

46. Kapikian AZ, Flores J, Hoshino Y, et al: Prospects for development of a rotavirus vaccine against rotavirus diarrhea of infants and young children. *Rev Infect Dis* (in press).

46a. Kapikian AZ, Flores J, Midthun K, et al: Development of a rotavirus vaccine by a "Jennerian" and modified "Jennerian" approach. *Vaccines 88* Cold Spring Harbor, New York, Cold Spring Harbor Laboratory, 1988, pp 151–158.

47. Kapikian AZ, Hoshino Y, Flores J, et al: Alternative approaches to the development of a rotavirus vaccine, in Holmgren J, Lindberg A, Mollby R (eds): *Development of Vaccines and Drugs Against Diarrhea.* 11th Nobel Conference, Stockholm, 1985. Lund, Studentlitteratur, 1986, pp 192–214.

48. Kapikian AZ, Kalica AR, Shih JW, et al: Buoyant density in cesium chloride of the human reovirus-like agent of infantile gastroenteritis by ultracentrifugation, electron microscopy, and complement-fixation. *Virology* 1976;70:564–569.

49. Kapikian AZ, Kim HW, Wyatt RG, et al: Human reovirus-like agent as the major pathogen associated with "winter" gastroenteritis in hospitalized infants and young children. *N Engl J Med* 1976;294:965–972.

50. Kapikian AZ, Kim HW, Wyatt RG, et al: Reovirus-like agent in stools: association with infantile diarrhea and development of serologic tests. *Science* 1974;185:1049–1053.

51. Kapikian AZ, Midthun K, Hoshino Y, et al: Rhesus rotavirus: a candidate vaccine for prevention of human reovirus disease, in Lerner RA, Chanock RM, Brown F (eds): *Vaccines 85: Molecular and Chemical Basis of Resistance to Parasitic, Bacterial, and Viral Diseases.* Cold Spring Harbor, New York, Cold Spring Harbor Laboratory, 1985, pp 357–367.

52. Kapikian AZ, Wyatt RG, Dolin R, et al: Visualization by immune electron microscopy of a 27nm particle associated with acute infectious non-bacterial gastroenteritis. *J Virol* 1972;10:1075–1081.

53. Kaplan JE, Gary GW, Baron RC, et al: Epidemiology of Norwalk gastroenteritis and the role of Norwalk virus in outbreaks of acute nonbacterial gastroenteritis. *Ann Intern Med* 1982;96:756–761.

54. Kim HW, Brandt CD, Kapikian AZ, et al: Human reovirus-like agent (HRULA) infection: occurrence in adult contacts of pediatric patients with gastroenteritis. *JAMA* 1977;238:404–407.

55. Konno T, Suzuki H, Kutsushima N, et al: Influence of temperature and relative humidity on human rotavirus infection in Japan. *J Infect Dis* 1983;147:125–128.

56. Konno T, Suzuki H, Kutsuzawa T, et al: Human rotavirus and intussusception. *N Engl J Med* 1977;297:945.

57. Konno T, Suzuki H, Kutsuzawa T, et al: Human rotavirus infection in infants and young children with intussusception. *J Med Virol* 1978;2:265–269.

58. Kumate J, Isibasi A: Pediatric diarrheal disease: a global perspective. *Pediatr Infect Dis* 1986;5:S21–S28.

58a. Lanata CF, Black RE, del Aguila, et al: Protection of Peruvian children against rotavirus diarrhea of specific serotypes by RIT 4237 attenuated bovine vaccine. *J Infect Dis* (in press).

59. Lewis HM, Parry JV, Davies HA, et al: A year's experience of the rotavirus system and its association with respiratory illness. *Arch Dis Child* 1979;54:339–346.

60. Losonsky GA, Rennels MB, Kapikian AZ, et al: Safety, infectivity, transmissibility, and immunogenicity of rhesus rotavirus vaccine (MMU 18006) in infants. *Pediatr Infect Dis* 1986;5:25–29.

61. Malherbe HH, Strickland-Cholmley M: Simian rotavirus SA-11 and the related "O" agent. *Arch Ges Virusforsch* 1967;22:235–245.
62. Matsuno S, Hasegawa A, Mukoyama A, Inouye S: A candidate for a new serotype of human rotavirus. *J Virol* 1985;54:623–624.
63. Matsuno S, Murakami S, Takagi M, et al: Cold adaption of human rotavirus. *Virus Res* 1987;7:273–280.
64. Matsuno S, Utagawa E, Sugiura A: Association of rotavirus infection with Kawasaki syndrome. *J Infect Dis* 1983;148:177.
65. McCormack JG: Clinical features of a rotavirus gastroenteritis. *J Infect Dis* 1982;4:167–174.
66. McNulty MS, Allan GM, Stuart JC: Rotavirus infection in avian species. *Vet Rec* 1978;103:319–320.
67. Mebus CA, Underdahl NR, Rhodes MG, Twiehaus MJ: Calf diarrhea (scours): reproduced with a virus from a field outbreak. *Univ Nebraska Res Bull* 1969;233:1–16.
68. Midthun K, Greenberg HB, Hoshino Y, et al: Reassortant rotaviruses as potential live rotavirus vaccine candidates. *J Virol* 1985;53:949–954.
69. Midthun K, Hoshino Y, Kapikian AZ, Chanock RM: Single gene substitution rotavirus reassortants containing the major neutralization protein (VP7) of human rotavirus serotype 4. *J Clin Microbiol* 1986;24:822–826.
70. Much D, Zajac I: Purification and characterization of epizootic diarrhea of infant mice virus. *Infect Immun* 1972;6:1019–1024.
70a. Nigro G and Midulla M: Acute laryngitis associated with rotavirus gastroenteritis. *J Infect* 1983;7:81–83.
70b. Offit PA, Clark HF, Blavat G, Greenberg HB: Reassortant rotaviruses containing structural proteins: VP3 and VP7 from different parents induce antibodies protective against each parental serotype. *J Virol* 1986;60:491–496.
70c. Ohno A, Taniguchi K, Sugimoto K, et al: Rotavirus gastroenteritis and afebrile infantile convulsion. *No To Hattatsu* 1982;14:520–521.
71. Perez-Schael I, Gonzalez M, Daoud N, et al: Reactogenicity and antigenicity of the rhesus rotavirus vaccine in Venezuelan children. *J Infect Dis* 1987;155:334–337.
72. Reed DE, Daley CA, Shave HJ: Reovirus-like agent associated with neonatal diarrhea in pronghorn antelope. *J Wild Dis* 1976;12:488–491.
73. Rennels MB, Losonsky GA, Levine MM, et al: Preliminary evaluation of the efficacy of rhesus rotavirus vaccine strain MMU 18006 in young children. *Pediatr Infect Dis* 1987;5:587–588.
73a. Rodriguez WJ, Kim HW, Arrobio JO, et al: Clinical features of acute gastroenteritis associated with human reovirus-like agent in infants and young children. *J Pediatr* 1977;91:188–193.
74. Roger SM, Craven JA, William I: Demonstration of reovirus-like particles in intestinal contents of piglets with diarrhea. *Austr Vet J* 1975;51:536.
75. Roseto A, Lema F, Sitbon M, et al: Detection of rotavirus in dogs. *Soc Occup Med* 1979;7:478.
76. Rotbart HA, Levin MJ, Yolken RH, et al: An outbreak of rotavirus-associated neonatal necrotizing enterocolitis. *J Pediatr* 1983;103:454–459.
76a. Rotbart HA, Nelson WL, Glode MP, et al: Neonatal rotavirus-associated necrotizing enterocolitis: case control study and prospective purveillance during an outbreak. *J Pediatr* 1988;112:87–93.
77. Ryder R, Singh N, Reeves WC, et al: Evidence of immunity induced by

naturally acquired rotavirus and Norwalk virus infection on two remote Panamanian islands. *J Infect Dis* 1985;151:99–105.

78. Saitoh Y, Matsuno S, Mukuyoma A: Exanthem subitum and rotavirus. *N Engl J Med* 1983;304:845.

79. Salmi TT, Arstila P, Koivikko A: Central nervous system invovlement in patients with rotavirus gastroenteritis. *Scand J Infect Dis* 1978;10:29–31.

80. Santosham M, Yolken RH, Quiroz E, et al: Detection of rotavirus in respiratory secretions of children with pneumonia. *J Pediatr* 1983;103:583–585.

81. Sato K, Inaba Y, Shinozaki T, et al: Isolation of human rotavirus in cell cultures. *Arch Virol* 1981;69:155–160.

82. Saulsbury FT, Winkelstein JA, Yolken RH: Chronic rotavirus infection in immunodeficiency. J Pediatr 1980;97:61–65.

83. Scott AC, Luddington J, Lucas M, Gilbert FR: Rotavirus in goats. *Vet Rec* 1978;103:145.

84. Shepherd RW, Truslow S, Walker-Smith JA, et al: Infantile gastroenteritis: a clinical study of reovirus-like agent infection. *Lancet* 1975;2:1082–1083.

85. Snodgrass DR, Angus KW, Gray EW: A rotavirus from kittens. *Vet Rec* 1979;104:222–223.

86. Snodgrass DR, Smith W, Gray EW, Herring JA: A rotavirus in lambs with diarrhea. *Res Vet Sci* 1976;20:113–114.

87. Snodgrass DR, Wells PW: Passive immunity in rotaviral infections. *J Am Vet Med Assoc* 1978;173:565–568.

88. Snyder JD, Merson MH: The magnitude of the global problem of acute diarrhoeal disease: a review of active surveillance data. *Bull WHO* 1982;60:605–613.

88a. Stoll BJ, Glass RI, Huq MI, et al: Surveillance of patients attending a diarrheal disease hospital in Bangladesh. *Br Med J* 1982;285:1185–1188.

88b. Stoll BJ, Glass RI, Banu H, et al: Value of stool examination in patients with diarrhea. *Br Med J* 1983;286:2037–2040.

89. Stuker G, Oshiro L, Schmidt NL: Antigenic comparisons of two new rotaviruses from rhesus monkeys. *J Clin Microbiol* 1980;11:202–203.

90. Tallett S, MacKenzie C, Middleton A, et al: Clinical, laboratory, epidemiological features of viral gastroenteritis in infants and children. *Pediatrics* 1977;60:217–222.

90a. Taylor PR, Merson MH, Black RE, et al: Oral rehydration therapy for treatment of rotavirus diarrhea in a rural treatment centre in Bangladesh. *Archiv Dis Childhood* 1980;55:376–379.

91. Tzipori S, Caple IW, Butler R: Isolation of a rotavirus from deer. *Vet Rec* 1976;99:398.

92. Urasawa T, Urasawa S, Taniguchi K: Sequential passages of human rotavirus in MA-104 cells. *Microbiol Immunol* 1981;25:1025–1035.

92a. Ushijima H, Bosu K, Abe T, Shinozaki T: Suspected rotavirus encephalitis. *Archiv Dis Childhood* 1986;61:692–694.

92b. Ushijima H, Tajima T, Tagaya M, et al: Rotavirus and central nervous system. *Brain and Development* 1984;6:138.

92c. Ushijima H, Araki K, Abe T, Shinozaki T: Anti-rotavirus antibody in cerebrospinal fluid. *Archiv Dis Childhood* 1987;62:298–299.

93. Vesikari T, Isolauri E, Delem A, et al: Clinical efficacy of the RIT 4237 live attenuated bovine rotavirus vaccine in infants vaccinated before a rotavirus epidemic. *J Pediatr* 1985;107:189–194.

94. Vesikari T, Isolauri E, D'Hondt E, et al: Increased take rate of oral rotavirus vaccine in infants after milk feeding. *Lancet* 1984;2:700.

95. Vesikari T, Isolauri E, D'Hondt E, et al: Protection of infants against rotavirus diarrhea by RIT 4237 attenuated bovine rotavirus strain vaccine. *Lancet* 1984;1:977–981.

96. Vesikari T, Kapikian AZ, Delem A, Zissis G: A comparative trial of rhesus monkey (RRV-1) and bovine (RIT 4237) oral rotavirus vaccines in young children. *J Infect Dis* 1986;153:832–839.

96a. Vesikari T, Rautanen T, Isolauri E, et al: Immunogenicity and safety of a low passage level bovine rotavirus candidate vaccine RIT 4256 in human adults and young infants. *Vaccine* 1987;5:105–108.

97. Wallace RE, Vasington PJ, Petricciani JC, et al: Development of a diploid cell line from fetal rhesus monkey lung for virus vaccine production. *In Vitro* 1973;8:323–332.

98. Walsh JA, Warren KS: Selective primary health care: an interim strategy for disease control in developing countries. *N Engl J Med* 1979;301:967–974.

99. Woode GN, Bridger JC, Jones JM, et al: Morphological and antigenic relationships between viruses (rotaviruses) from acute gastroenteritis of children, calves, piglets, mice and foals. *Infect Immun* 1976;14:804–810.

100. Wong CJ, Price Z, Bruckner DA: Aseptic meningitis in an infant with rotavirus gastroenteritis. *Pediatr Infect Dis* 1986;3:244–246.

100a. WHO Scientific Working Group: Rotavirus and other viral diarrhoeas. *Bull WHO* 1980;58:183–198.

101. Wright PF, Tajima T, Thompson J, et al: Candidate rotavirus vaccine (rhesus rotavirus strain) in children: an evaluation. *Pediatrics* 1987;80:473–480.

102. Wyatt RG, Greenberg HB, James WD, et al: Definition of human rotavirus serotypes by plaque reduction assay. *Infect Immun* 1984;37:110–115.

103. Wyatt RG, Kapikian AZ, Hoshino Y, et al: Development of rotavirus vaccines, in *Control and Eradication of Infectious Diseases. An International Symposium.* Washington, DC: Pan American Health Organization, 1985, pp 17–28.

104. Wyatt RG, Kapikian AZ, Mebus CA: Induction of cross-reactive serum neutralizing antibody to human rotavirus in calves after in utero administration of bovine rotavirus. *J Clin Microbiol* 1983;18:505–508.

105. Wyatt RG, Mebus CA, Yolken RH, et al: Rotaviral immunity in gnotobiotic calves: heterologous resistance to human virus induced by bovine virus. *Science* 1979;203:548–550.

106. Yolken RH, Bishop CA, Townsend TR, et al: Infectious gastroenteritis in bone-marrow transplant recipients. *N Engl J Med* 1982;306:1009–1012.

107. Yolken RH, Murphy M: Sudden infant death syndrome associated with rotavirus infection. *J Med Virol* 1982;10:291–296.

108. Yow MD, Melnick JL, Blattner RJ, et al: The association of viruses and bacteria with infantile diarrhea. *Am J Epidemiol* 1970;92:33–39.

109. Zissis G: Het belang van rotavirus in die etiologie van infantiele diaree. MD thesis. Vrije Universiteit, Brussels, 1979.

110. Zissis G, Lambert JP, Marbehant P, et al: Protection studies in colostrum-deprived piglets of a bovine rotavirus vaccine candidate using human rotavirus strains for challenge. *J Infect Dis* 1983;148:1061–1068.

CHAPTER 18

Mucosal Immunity in the Gastrointestinal Tract in Relation to ETEC Vaccine Development

Ann-Mari Svennerholm, Jan Holmgren,
Yolanda Lopez-Vidal, Joaquin Sanchez,
and Christina Åhrén

Introduction

Enteric infection with enterotoxinogenic *Escherichia coli* (ETEC) is one of the most important causes of diarrhea in developing countries and is also the most frequent cause of diarrhea among travelers. Illness resulting from ETEC infection is characterized by watery diarrhea, often accompanied by low grade fever, abdominal cramps, malaise, and vomiting. In its most severe form ETEC infection may result in cholera-like disease leading to severe dehydration and sometimes death.

Studies in animals and human volunteers have shown that ETEC infection may give rise to substantial immunity against rechallenge with the homologous organisms (2,21). The protection achieved may also extend to heterologous ETEC strains sharing critical protective antigens with the infecting strain (2). These findings suggest the possibility of inducing effective immunity against ETEC illness by vaccination.

The development of an effective vaccine against ETEC-induced diarrheal disease must be based on the use of immunogens which give rise to immune responses that would efficiently interfere with one or more of the major pathogenic events. These events include (a) the colonization and multiplication of ETEC bacteria in the small intestine, and (b) the elaboration and action of a heat-labile (LT) or heat-stable (ST) enterotoxin, or both, that are responsible for the increased water and electrolyte secretion in the intestinal lumen (29).

Another prerequisite for an effective ETEC vaccine is that it must be able to stimulate the local immune system in the gut to an efficient degree. Studies in experimental animals indicate that locally produced secretory immunoglobulin A (SIgA) antibodies and immunological memory for production of this class of antibodies are of prime importance for protection against enterotoxin-induced diarrheal disease. Such antibodies can be efficiently stimulated by oral antigen administration, whereas the parenteral route is usually inefficient. The encouraging results from an ongoing field trial of an oral B subunit–whole cell cholera vaccine in Bangladesh (see

Chap. 4), also support that oral vaccination may be efficient in stimulating protective immunity against noninvasive enteric pathogens such as ETEC.

In this review we will discuss important aspects for the development of vaccines against ETEC disease, including an account of mechanisms of disease and immunity in ETEC disease, identification of major protective antigens, and modes of stimulating effective immunity against these antigens locally in the intestine. We will also describe various approaches to constructing an oral ETEC vaccine.

Intestinal Immune Mechanisms

The gut has a complex system of immunological and nonimmunological host defense factors by which it can combat enteric infections. The function of these factors is to protect the body against harmful effects from various enteric pathogens and their secreted toxins and probably against allergy from food components as well (17). The best known entity providing specific immunoprotection in the gut lumen or on the mucosal surface is the SIgA system, which seems to be specifically designed to respond to locally applied antigens. By means of specialized epithelial cells, so-called M cells, that cover the Peyer's patches in the small intestine, antigens are efficiently transported from the gut lumen into the lymphoid follicles of the patches, where SIgA antibody responses originate. After stimulation, B cell blasts migrate via the lymph into the bloodstream and ultimately "home" back to the intestinal lamina propria, where they differentiate to become mainly IgA-secreting plasma cells. Studies suggest that different sets of T cells can direct the antigen-stimulated B cells from the Peyer's patches toward IgA production (18). The IgA dimers produced by the plasma cells in the lamina propria are efficiently transported across the intestinal epithelium into the gut lumen. This phenomenon is explained by the fact that the epithelial cells on their basolateral membrane contain a glycoprotein, the secretory component (SC), which functions as a specific receptor for the binding and transport of IgA dimers (and for locally produced IgM). The SC also functions as a protective shield against proteolytic degradation of the SIgA molecule after its release into the intestinal lumen.

By which mechanisms do secretory IgA antibodies confer protection against noninvasive enteric pathogens? Obviously one of the main functions of SIgA antibodies is to interfere with colonization of intestinal pathogens. It could be done by agglutinating microorganisms in the gut lumen, by interfering with motility and other means of microorganisms to penetrate the mucous coat, and by blocking bacterial surface structures responsible for adherence to the epithelium. The antibodies might also interfere with utilization of various growth factors in the intestinal environment such as iron. Another important effect of SIgA antibodies is their capacity to neutralize enterotoxin, particularly by preventing specific binding of the toxins to their specific receptors on the mucosal cells. SIgA antibodies may also augment the protective capacity of nonimmunological components of the mucosal barrier, e.g., by enhancing proteolysis and trapping bacteria in the mucous coat.

The intestinal SIgA response is of relatively short duration, lasting a few weeks up to a month or two, but it exhibits immunological memory and may thereby be progressively stimulated by repeated doses of antigen. Thus, as studied in rats (27) and mice (20), an increasing number of IgA antitoxin-containing cells in intestinal lamina propria and increased intestinal synthesis of antitoxic IgA antibodies have been obtained with each of several oral immunizations with cholera toxin (33). These IgA responses have also been closely associated with increased levels of protection against challenge with cholera toxin in intestinal loops (20,33). Vaccination studies in humans have also indicated that oral antigen administration may induce an immunological memory that can be boosted one or even several years after the initial antigen administration. Thus Bangladeshi volunteers who had initially been immunized with an oral cholera toxoid (cholera B subunit) and were reimmunized 15 months later responded to a booster dose that was too low to give any response in previously nonvaccinated Bangladeshi women; furthermore, the response in the primed women developed more quickly than in previously nonvaccinated volunteers, being manifest within 72 hours (35). In recent studies Jertborn et al (17b) have also shown that adult Swedes who had received two immunizations with an oral B subunit/whole-cell cholera vaccine 3 to 5 years previously responded considerably better with both antitoxic and antibacterial antibodies to a single oral booster with the same vaccine than did a matched group of previously nonvaccinated persons.

The induction of an immunological memory by intestinal antigen stimulation with bacterial as well as toxin antigens may explain the protection for several years, found in patients recovering from, for example, cholera or enterotoxin-induced diarrheal disease (21). This possibility gained additional support when Lycke and Holmgren (24) showed that after peroral immunizations of mice with cholera toxin there was a practically life-long immunological memory, which on renewed intestinal exposure to even minute amounts of antigen was triggered to provide an anamnestic-type mucosal antitoxic response within a few hours. Thus the immunological memory induced by an infection might then be boosted so efficiently and rapidly by a reinfection that it could abort the new infection at a stage where it would still be asymptomatic.

Although the major immunological protection against noninvasive enteric pathogens seems to be due to antibodies, and then mainly to SIgA, an additional role for cell-mediated immune mechanisms cannot be excluded. Thus the gut is loaded with T cells being present in the lamina propria and intraepithelially. Particularly the intraepithelial T cells are able to react with antigen on the luminal side of the epithelium. The function and properties of gut T cells have attracted increased interest during recent years. T-cell-mediated immune reactions as well as natural killer cell activity and antibody-dependent cellular cytotoxicity may contribute to the protection against various enteric pathogens as suggested by in vitro findings (25,26).

Pathogenic Mechanisms and Virulence Factors of ETEC

The most important virulence factors of ETEC are plasmid-mediated. These factors include the enterotoxins (LT and ST) and the fimbrial colonization factors. Whereas the enterotoxins are responsible for the water and electrolyte losses resulting in diarrhea, the colonization factors are important in allowing the bacteria to avoid being expelled from the intestine by peristalsis and other host defense mechanisms.

Escherichia coli LT is similar to cholera toxin in structure and function, having a toxic-active ADP-ribosylating A subunit (16) attached to five B subunits that mediate binding to cell membrane receptors consisting of GM_1 ganglioside and a closely related glycoprotein (15). Different from cholera toxin, however, which is actively secreted by *Vibrio cholerae* during cell growth (14), LT is accumulated in the periplasm of the ETEC cell and is not released until the organism dies and lyses.

The ST toxins can be divided into two subclasses, STa and STb, according to their host specificity (5), and these toxins have now also been found to differ substantially in primary structure. Because production of STb is rare among human ETEC isolates, only STa needs to be considered as a candidate antigen in a human ETEC vaccine. The STa enterotoxins are small polypeptides of either 18 or 19 amino acids, including six cysteine residues, and they stimulate guanylate cyclase activity in intestinal cells after binding to brush border membrane receptors (28). Different from LT, which is a good immunogen, STa is a nonimmunogenic unless coupled to a carrier protein (12). DNA sequence studies indicate slight molecular heterogeneity among STas (STaI or ST_P and STaII or ST_h), but the antigen epitope(s) of these toxins seems to be identical or closely similar (36). At variance with LT, which is trapped in the periplasm, ST is exported from the cell during growth in vivo as well as in vitro.

The capacity to colonize the small intestine is also a prerequisite for ETEC bacteria to produce disease. In many ETEC strains adhesion to intestinal mucosa is mediated by fimbriae, which are specific antigens. In strains pathogenic for man three distinct types of putative adhesion fimbriae have been described: the colonization factor antigens I and II (CFA/I and CFA/II) (8,9) and the PCF8775 antigen (39). However, additional adhesion fimbriae on human ETEC strains are likely to be recognized.

CFA/I is a single, homogeneous fimbrial antigen that causes mannose-resistant hemagglutination (MRHA) of human and bovine erythrocytes. CFA/II and PCF8775 antigen, on the other hand, are composed of more than one antigenic component each. The CFA/II system comprises the coli surface-associated antigens CS1, CS2, and CS3; and the PCF8775 system comprises the CS4, CS5, and CS6 antigens (6,31,41). CFA/II-positive strains usually carry CS3 together with either CS1 or CS2, all of which have a fimbrial structure and cause MRHA of bovine but not human erythrocytes (6,31). PCF8775-positive strains of serogroup 025 usually possess CS4 and CS6, whereas those of serogroups 0115 or 0167 have the combination of CS5 and CS6 (41). Whereas CS4 and CS5 are fimbriae and give MRHA of various species of erythrocytes, CS6 has a nonfimbrial structure and does not cause MRHA.

Epidemiology of ETEC

Enterotoxigenic *E. coli* has been shown to cause diarrhea worldwide but is much more common in developing countries. In hospital-based studies from various developing countries the percentage of cases in which ETEC was found ranged from 10 to 50%; an average figure for children under 5 years of age would probably be between 20 and 25% (4). It has been estimated that in the developing countries of Asia, Africa, and Latin America there may be as many as 650 million cases of diarrhea due to ETEC, resulting in nearly 800,000 deaths annually. Because of this high incidence, especially in younger children diarrhea due to these organisms is of major importance for morbidity and mortality (4).

ETEC has also been shown to be the most frequent cause of travelers' diarrhea. Among short-term travelers to Latin America the median attack rate of diarrhea was 52% and for travelers to Africa 54%; ETEC was associated with 36% of all the diarrheal episodes (4).

The relative frequency with which *E. coli* produces LT, ST, or both varies in different regions of the world. Analyses of ETEC from various areas suggest that strains producing both toxins are largely restricted to a small number of serotypes, which are entirely unrelated to the so-called enteropathogenic O groups. The ability to produce only ST or only LT seems to occur in a much broader range of serotypes. However, in addition to the known geographic variations in serotypes, there are other variations as well.

The reported prevalence of the different colonization factor antigens differs considerably in studies from various parts of the world. The discrepancies may partly be explained by technical problems residing in both the toxin detection methods and the loss of CFA-encoding plasmids during laboratory storage and passage of the strains (3). However, it may also reflect true variations in enterotoxin and serotype profiles of ETEC isolates in different settings, as these properties appear closely associated with

CFA expression. In a prospective study in Bangladesh we found that as many as 75% of ETEC strains isolated from patients with diarrhea carried either CFA/I or CFA/II, whereas these adhesins were never identified on *E. coli* strains from healthy controls (13). In a study from Mexico, Evans et al (10) showed that 86% of ETEC strains isolated from travelers' diarrhea carried CFA/I. However, in a retrospective study encompassing diarrhea-provoking strains of *E. coli* from 14 countries, only 7% of the strains carried CFA/I (40). Whereas CFA/I seem to be the most prevalent adhesin in, for example, Bangladesh (13), studies from Japan (40) suggest that CFA/II is more common than CFA/I, and ongoing studies in Mexico suggest that PCF8775 may be the most prevalent of the hitherto identified adhesins (Y. Lopez-Vidal, unpublished, A. Cravioto, personal communication).

Against this background additional epidemiological studies of the enterotoxin profile as well as the prevalence of different adhesins on ETEC isolates should be carried out in various geographic areas to provide a better basis for the design of a future ETEC vaccine.

Protective Antigens and Synergism

A number of studies in animals and humans have shown that ETEC infections evoke significant antitoxin as well as antibacterial antibody responses in intestine and in serum. The antitoxic immune response is against LT because, as mentioned, ST is not immunogenic in its natural state. Because LTs produced by different human ETEC isolates appear to be immunologically similar (34), immunizations with one human LT-derived immunogen should probably provide effective cross-protection against a wide range of LT-producing human ETEC strains. Similar to cholera, antitoxic immunity against ETEC disease is mainly directed against the B subunit portion of the toxin molecule, although studies with monoclonal antibodies have shown that, different from cholera toxin, *E. coli* LT can also be neutralized by specific anti-A antibodies (37). Because elimination of the A subunit from LT results in a completely nontoxic molecule the LT B pentamer seems to be an ideal component in an ETEC vaccine for stimulating anti-LT immunity.

Immunization of experimental animals with STa coupled to an immunogenic protein, e.g., bovine serum albumin or the B subunit pentamer of cholera toxin or LT, have resulted in antibody formation not only against the carrier proteins but also against STa (12,19,36). We have found that polyclonal antisera raised against such ST conjugates, as well as monoclonal antibodies against specific epitopes on STa are capable of neutralizing ST toxic activity in intestine. This finding suggests that anti-ST immunity is conceivable and might be a worth-while goal to achieve by a vaccine against ETEC. Klipstein et al (19) have reported that a conjugate of synthetic ST and LTB was effective in inducing protective immunity against ST/LT-producing *E. coli* strains in a rat model. Results in our own laboratory, however, suggest the presence of residual ST toxic activity in such ST–B subunit conjugates, and this problem must of course be overcome before an ST-containing toxoid could be considered.

Antibacterial immunity against ETEC may to a great extent be ascribed to immunity against the various colonization factors even though anti-O,H immunity may also play a role for ETEC of homologous O and H types. In rabbits we have shown that antibodies against CFA/I or CFA/II conferred passive protection against ETEC strains carrying the homologous adhesins (1). This protection could most likely be ascribed to an antiadhesive effect of the antibodies, as isolated Fab fragments were also effective (1). We have also found that antibodies against one subtype of CFA/II (CS1 + CS3) protected well also against a heterologous subtype of CFA/II (CS2 + CS3), suggesting protective immunogenicity of the shared CS3 component of the CFA/II complex. Evans et al (11) reported that in human volunteers a sequence of parenteral and oral immunizations with purified CFA/I conferred a significant degree of protection against subsequent challenge with CFA/I-carrying ETEC bacteria, and Levine et al (22) found that intestinal administration of purified CFA/II by a gastric tube resulted in substantial specific antibody formation in the gut.

In collaboration with B. Rowe and colleagues (Public Health Laboratories, Colindale, London) we have evaluated the role of the various CS components of PCF8775 as adhesins and protective antigens. It was done by comparing the colonizing ability and protective immunogenicity in rabbits (the RITARD model) of different mutant strains being devoid of one or two of the CS components. Data from these studies suggest that CS6 is a colonization factor in rabbits and is capable of inducing protective immunity against ETEC-carrying CS6 in combination with CS4 or CS5 (37b).

When grown in vitro in liquid culture, most ETEC strains may express type 1 (or common) pili, which cause mannose-sensitive HA (7). However, type 1 pili have been found in the same frequency on nonpathogenic *E. coli* and on clinical ETEC isolates, and there is evidence that ETEC may not express the type 1 pili in vivo in the intestine (23,42). Therefore there is little reason to believe that type 1 pili are important in ETEC pathogenesis or immunity. At variance with the relatively limited number of toxins and colonization factor antigens hitherto associated with ETEC, these bacteria represent a large number of O:K:H serotypes. This heterogeneity suggests that O antigens should not be considered primary protective antigens in a future ETEC vaccine. However, a close association between a few O groups and enterotoxin production as well as the presence of certain fimbriae have been described. Thus CFA/I is most commonly found on 015, 025, 063, 078, and 0128 strains, CFA/II on 06 and 08 strains, and PCF8775 fimbriae on 025, 0115, and 0167 strains (6,9,41). Studies in experimental animals have also shown that antibodies against either the homologous complete LPS or against broadly cross-reactive LPS structures (Re-type LPS or 014 Common antigen) can protect well against intestinal infection with serogroup heterologous ETEC bacteria (34).

Though in principle either antibacterial or antitoxic immunity alone could provide protection against ETEC disease, an ETEC vaccine should most likely provide both antibacterial and antitoxic immunity in order to be optimally effective. It would enable the vaccine-induced immunity to simultaneously interfere with both of the main separate pathogenic events, i.e. colonization and toxin action. Furthermore, we have found that antibodies against CFA/I or CFA/II do in fact cooperate synergistically with anti-LT antibodies in protecting experimental animals against challenge with ETEC bacteria expressing the homologous CFA and toxin (1). This finding is in accordance with our findings in a nonligated intestine animal model, the RITARD model, that the protection induced by an initial ETEC infection against rechallenge was stronger when the immunizing strain had both the CFA type and LT in common with the challenge strain than when it had only one of these antigens, CFA or LT, in common (2).

Intestinal Immune Responses to ETEC Disease

Using the RITARD model we have shown that recovery from intestinal infection with CFA-carrying ETEC is associated with highly significant protection against rechallenge with homologous or O-,K-H-heterologous strains carrying the same CFA (2). We have also shown that intestinal immunization with bacteria carrying the CS4 and CS6 components of PCF8775 or the CS6 component alone were considerably more effective than CS-deficient mutants of the same strain for inducing protection against challenge with serotype homologous bacteria carrying the two CS components (37). These results suggest that intestinal administration of various colonization factor antigens, at least as presented on live bacteria, can induce substantial protective immunity in the gut. There is also evidence for protective anti-LT immunity in the RITARD model. Thus intestinal immunization with CFA/II-carrying ST/LT-producing strains conferred significant protection against subsequent challenge with CFA/I-positive, LT-producing *E. coli* bacteria of a different serotype (2).

Although normal fecal *E. coli* bacteria as well as PCF8775 antigen-deficient mutants are poorly effective in protecting against fully virulent CFA-positive strains, we recently found that a CFA/I-deficient mutant derived from a clinical CFA/I-positive ST/LT isolate protected effectively against colonization as well as disease by the original strain used for rechallenge (23). These results were interpreted to suggest that the mutant during growth in the intestine may have expressed an additional adhesive factor(s) that induced a local immune response that protected against bacterial colonization as well as disease by the challenge strain. Taken together, all these studies in rabbits suggest that different colonization factor antigens or adhesive factors are important protective antigens that can stimulate an effective immune response in intestine.

In order to provide basic information for subsequent vaccination studies, the intestinal immune response in patients convalescing from moderate to severe ETEC disease was studied for intestinal immune responses against several of the major protective antigens on ETEC bacteria (32). Fifteen adult patients were admitted to the Dhaka Hospital of the International Centre for Diarrhoeal Disease Research, Bangladesh (ICDDR,B) with acute watery diarrhea, moderate to severe dehydration, and ETEC isolated from stool samples were studied. Ten patients had disease from *E. coli* producing LT alone or in combination with ST, and five were infected with ST-only strains. Of the 15 patients, 10 were infected with CFA/I- or CFA/II-carrying strains. Intestinal immune responses against LT, CFA/I, CFA/II, and lipopolysaccharide (LPS) of the homologous infecting strain were studied in intestinal lavage fluid by different ELISA techniques.

Following infection, most of the patients had significant rises in local intestinal IgA antibodies against these antigens if present on the infecting strains. As shown in Table 18.1, 80% of the patients infected with LT-positive strains responded to LT, 63% of the patients infected with CFA-carrying strains responded to CFA, and 78% of all the patients responded to the LPS of their infecting ETEC strain. In most patients, local intestinal responses of IgA antibodies to LT, CFA, and LPS were elevated within 9 days after infection and had decreased significantly by 4 weeks after onset of disease. In these patients we also found evidence that ETEC disease can "prime" the gut immune system to respond more efficiently to immunization with an enterotoxin antigen. Thus oral immunization of the ten patients convalescing from LT disease with cholera B subunit 1 month after onset of diarrhea resulted in early and considerably higher

Table 18.1. Magnitude and duration of local antibody responses in intestinal lavage fluid to LT, CFA/I, or CFA/II and homologous LPS in patients with natural ETEC disease.

Days after coming to hospital	Geometric mean IgA antibody titer total IgA (U/mg)[a]		
	Anti-LT ($n = 10$)	Anti-CFA ($n = 8$)	Anti-LPS ($n = 9$)
"0"	≤0.9	≤0.5	≤0.9
9	4.4	4.0	12.2
28	1.1	0.5	1.1
Responders	80%	63%	78%

[a]Antibody responses were determined only if the antigen (LT or CFA/I or CFA/II) was expressed on the infecting strain.

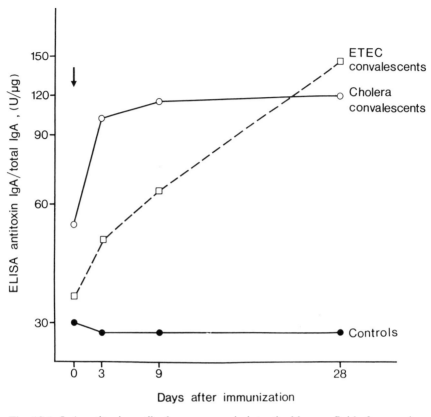

Fig. 18.1. IgA antitoxin antibody responses in intestinal lavage fluid after one immunization with cholera B subunit (0.5 mg) of convalescents from moderate to severe cholera or ETEC disease (LT or LT/ST-mediated) and of nonimmunized matched controls.

antitoxin responses in intestine than seen in previously nonimmunized controls who were given the same oral vaccination concurrently (Fig. 18.1).

Vaccine Development

Based on the identification of the key protective antigens of ETEC bacteria and of the immune mechanisms operating against ETEC infections, different aspects for the development of an effective ETEC vaccine should be considered. First, an ETEC vaccine should be intended for oral administration to allow optimal stimulation of protective antibodies locally in the gut. Second, the vaccine should induce antibacterial as well as antitoxic immune responses in the intestine to enable synergistic cooperation of these antibody specificities for protection against ETEC disease, i.e., be based on some form of combination of bacterial cell- and toxin-derived antigens. This goal could be achieved through alternative means. Three possible ETEC vaccine candidates that might also be combined to some extent are as follows.

1. *A mixture of the major protective antigens in purified form.* The most important of the somatic antigens are those colonization factor antigens that in epidemiological studies in different geographic areas have been shown to have the highest prevalence on ETEC strains. These antigens are CFA/I, the CFA/II complex (or maybe some of its subcomponents), and the PCF8775 antigen (or one or two of its subcomponents). Although probably of less importance than the above fimbrial antigens, the most common O antigens that are associated with ETEC in different areas (e.g., O15, O25, O63, O78, O128, O6, O8, O115, and O167) might also be considered for inclusion in an ETEC vaccine. Most of these antigens—the colonization factor antigens and the O antigens—can be purified relatively simply by standard methods (32), although simple purification methods for individual outer membrane proteins, such as the CS6 component, are as yet lacking.

With regard to the toxoid component of such a combined vaccine, it may consist of the B subunit pentamer of LT. Such LTB can either be isolated from holotoxin by separation techniques (38) or be purified from an *E. coli* strain producing LTB only. Several plasmids are now available that produce large quantities of LTB both when expressed in *E. coli* cells and when transferred to and expressed in *V. cholerae* organisms (from which cholera toxin production has been deleted) (14). The use of *V. cholerae* as an expression system has the interesting feature that the LTB is then secreted to the extracellular milieu (in *E. coli* it is periplasmic), which simplifies isolation of the LTB protein, e.g., by means of affinity chromatography on GM_1 ganglioside columns (38).

Since a nontoxic but antibody-binding polypeptide of STa has recently been identified (37c), an LTB–ST conjugate may also be considered. As previously mentioned, we have raised polyclonal as well as monoclonal antibodies against STa (with ST neutralizing capacity) by immunizing experimental animals with conjugates of STa and BSA or STa and cholera B subunit (36). Unfortunately, however, these latter conjugates have retained some ST toxic activity.

An alternative approach to producing such combined LT–ST conjugates would be to try to fuse the gene encoding for ST expression with one of those encoding for LTA or LTB expression, thereby hoping to achieve production of a chimeric protein. For vaccine purposes, such a protein should obviously be completely nontoxic as well as exhibiting protective ST as well as LT antigenicity. Gene fusions of ST to LTB would clearly avoid any inherent LT toxic activity in the hybrid proteins, and very recently we could genetically fuse a nontoxic, antibody-binding ST-related decapeptide to the amino-end of the B-subunit of cholera toxin (Sanchez J, Svennerholm A-M, and Holmgren J, to be published). This fusion protein was completely nontoxic and gave rise to significant antibody levels against STa and LTB. It has also been possible to fuse the gene-encoding ST to the carboxy end of the gene coding for the LTA subunit (30). The hybrid gene directed expression of chimeric LTA-STa protein. The simultaneous presence in *E. coli* bacteria of a plasmid directing the expression of this fusion protein with another plasmid directing the expression of the LTB subunit resulted in a protein complex that bound to GM_1 ganglioside and reacted with monoclonal antibodies against either LTA, LTB, or STa, indicating that the STa and the LT epitopes remain immunologically intact after fusion.

2. *Live bacteria expressing the major colonization factor antigens and producing enterotoxoid.* A live vaccine with these properties would have the main attraction to being able to provide the local immune system of the gut with sustained antigen stimulation. Because the various colonization factors are not normally expressed on the same bacteria, such vaccines probably have to consist of several strains. One or more of these strains could then be genetically manipulated to produce LTB or even chimeric LT–ST conjugates. An inherited problem of such a live vaccine is the risk of overgrowth of one or more of the included vaccine strains. Other problems of a live enteric vaccine that should be considered are the risk of reversion to toxicity, low production of enterotoxoid during in vivo growth, and death of most organisms during storage.

3. *Killed ETEC bacteria representing the major O groups, expressing the key colonization factor antigens and containing enterotoxoid.* Development of such a vaccine requires identification of suitable *E. coli* strains representing the major O groups and expressing the different colonization factor antigens in high concentration on their surfaces. It also requires elaboration of suitable methods for killing the bacteria with retained antigenicity of the critical surface antigens. Strains harboring LT plasmids can probably not be used as potential vaccine strains, as biologically active LT might be released from the periplasm of the killed bacteria when placed in the intestine and cause side effects.

The toxoid component of such a whole-cell killed vaccine could either be periplasmic LTB or chimeric LT–ST conjugates produced by one or more of the vaccine strains (as discussed above). Such toxoids can be expected to be released from the inactivated organisms during the passage through the small intestine. Alternatively, purified LTB or a nontoxic LTB–ST conjugate could be added to a cocktail of killed bacteria.

Of these alternatives we regard the third approach as being the currently most practical one. In preliminary studies we have developed useful methods, based on mild formalin treatment, for killing candidate vaccine strains with preservation of both the CFA and O antigens. Suitable enterotoxoids consisting of B subunit in combination with synthetic nontoxic ST peptides are under development, and chimeric ST–LT proteins have been prepared by chemical coupling (Svennerholm A-M, to be published) or gene fusion (30).

Acknowledgments. Financial support for the studies from our laboratory was obtained from the Swedish Medical Research Council (grant 16X-3382), the Swedish Agency for Research Cooperation with Developing Countries (SAREC), the World Health Organization, and The Faculty of Medicine, University of Göteborg, Göteborg, Sweden.

The skillful technical assistance of Ms. Kerstin Andersson, Christina Olbing, and Gudrun Wiklund is gratefully acknowledged.

References

1. Åhrén C, Svennerholm A-M: Synergistic protective effect of antibodies against Escherichia coli enterotoxin and colonization factor antigens. *Infect Immun* 1982;38:74–79.
2. Åhrén C, Svennerholm A-M: Experimental enterotoxin-induced Escherichia coli diarrhea and protection induced by previous infection with bacteria of the same adhesin or enterotoxin type. *Infect Immun* 1985;50:225–261.
3. Åhrén C, Gothefors L, Stoll B, et al: Comparison of methods for detection of colonization factor antigens (CFAs) on enterotoxigenic Escherichia coli. *J Clin Microbiol* 1986;23:586–591.
4. Black RE: The epidemiology of cholera and enterotoxigenic E. coli diarrheal disease, in Holmgren J, Lindberg A, Möllby R (eds): *Development of Vaccines and Drugs Against Diarrhea.* 11th Nobel Conference, Stockholm. Lund, Studentlitteratur, 1986, pp 23–32.
5. Burgess MN, Bywater RJ, Cowley CM, et al: Biological evaluation of a methanol-soluble, heat-stable Escherichia coli enterotoxin in infant mice, pigs, rabbits and calves. *Infect Immun* 1978;21:526–531.
6. Cravioto A, Scotland SM, Rowe B: Hemagglutination activity and colonization factor antigens I and II in enterotoxigenic and nonenterotoxigenic strains of Escherichia coli isolated from humans. *Infect Immun* 1982;36:189–197.
7. Duguid JP, Smith IW, Dempster G, Edmunds PN: Nonflagellar filamentous appendages ("fimbriae") and haemagglutinating activity in Bacterium coli. *J Pathol Bacteriol* 1955;70:335–348.
8. Evans DG, Silver RP, Evans DJ Jr, et al: Plasmid-controlled colonization factor associated with virulence in Escherichia coli enterotoxigenic for humans. *Infect Immun* 1975;12:656–667.
9. Evans DG, Evans DJ Jr: New surface-associated heat-labile colonization factor antigen (CFA/II) produced by enterotoxigenic Escherichia coli of serogroups 06 and 08. *Infect Immun* 1978;21:638–647.
10. Evans DG, Evans DJ Jr, Tjoa WS, DuPont HL: Detection and characteri-

zation of colonization factor of enterotoxigenic Escherichia coli isolated from adults with diarrhea. *Infect Immun* 1978;19:727–736.

11. Evans DG, Graham DY, Evans DJ Jr, Opekun A: Administration of purified colonization factor antigens (CFA/I, CFA/II) of enterotoxigenic Escherichia coli to volunteers: response to challenge with virulent enterotoxigenic Escherichia coli. *Gastroenterology* 1984;87:934–940.

12. Frantz JC, Robertson DC: Immunological properties of Escherichia coli heat-stable enterotoxins: development of a radioimmunoassay specific for heat-stable enterotoxins with suckling mouse activity. *Infect Immun* 1981;33:193–198.

13. Gothefors L, Åhrén C, Stoll B, et al: Presence of colonization factor antigens on fresh isolates of fecal Escherichia coli: a prospective study. *J Infect Dis* 1985;152:1128–1133.

14. Hirst TR, Sanchez J, Kaper JB, et al: Mechanism of toxin secretion by Vibrio cholerae investigated in strains harboring plasmids that encode heat-labile enterotoxins of Escherichia coli. *Proc Natl Acad Sci USA* 1984;81:7752–7756.

15. Holmgren J, Fredman P, Lindblad M, et al: Rabbit intestinal glycoprotein receptor for Escherichia coli heat-labile enterotoxin lacking affinity for cholera toxin. *Infect Immun* 1982;38:424–433.

16. Holmgren J: Toxins affecting intestinal transport processes, in Sussman M (ed): *The Virulence of Escherichia coli*. London, Academic Press, 1985, pp 177–191.

17. Holmgren J, Lycke N: Immune mechanisms in enteric infections, in Holmgren J, Lindberg A, Möllby R (eds): *Development of Vaccines and Drugs Against Diarrhea*. 11th Nobel Conference, Stockholm. Lund, Studentlitteratur, 1986, pp 9–22.

17b. Jertborn M, Svennerholm A-M, Holmgren J: Five-year immunological memory after oral cholera vaccination. *J Infect Dis* 1988;157:374–377.

18. Kawanishi H, Strober W: T cell regulation of IgA immunoglobulin production in gut-associated lymphoid tissues. *Mol Immunol* 1983;20:917–930.

19. Klipstein FA, Engert RF, Clements JD, Houghton RA: Protection against human and porcine enterotoxinogenic Escherichia coli in rats immunized with a cross-linked toxoid vaccine. *Infect Immun* 1983;40:924–929.

20. Lange S, Nygren H, Svennerholm A-M, Holmgren J: Antitoxic cholera immunity in mice: influence of antigen deposition on antitoxin-containing cells and protective immunity in different parts of the intestine. *Infect Immun* 1980;28:17–23.

21. Levine MM: Immunity to cholera as evaluated in volunteers, in Ouchterlony Ö, Holmgren J (eds): *Cholera and Related Diarrheas*. 43rd Nobel Symposium, Stockholm. Basel, Karger, 1980, pp 195–203.

22. Levine M, Morris JG, Losonsky G, et al: *Fimbrial (Pili) Adhesins as Vaccines: Protein–Carbohydrate Interactions in Biological Systems*. FEMS Symposium No. 31. London, Academic Press, 1986, pp 143–145.

23. Lopez-Vidal Y, Åhrén C, Svennerholm A-M: Colonization diarrhoea and protective immunogenicity of a CFA-deficient enterotoxin-producing Escherichia coli mutant in a non-ligated intestine experimental model. *Acta Pathol Microbiol Scand* Sect. B. 1987;95:123–130.

24. Lycke N, Holmgren J: Intestinal mucosal memory and presence of memory cells in lamina propria and Peyer's patches in mice 2 years after oral immunization with cholera toxin. *Scand J Immunol* 1986;23:611–616.

25. Morgan DR, Dupont HL, Wood LD, Kohl S: Cytotoxicity of leukocytes from normal and Shigella-susceptible (opium-treated) guinea pigs against virulent Shigella sonnei. *Infect Immun* 1984;46:22–24.

26. Nencioni L, Villa D, Boraschi D, et al: Natural and antibody-dependent cell-mediated activity against Salmonella typhimurium by peripheral and intestinal lymphoid cells in mice. *J Immunol* 1983;130:903–907.

27. Pierce NF, Cray WC Jr: Determinants of the localization, magnitude and duration of a specific mucosal IgA plasma cell response in enterically immunized rats. *J Immunol* 1982;128:1311–1315.

28. Rao MC: Toxins which activate guanylate cyclase: heat-stable enterotoxins, in *Microbial Toxins and Diarrheal Disease*. Ciba Foundation Symposium 112. London, Pitman, 1985, pp 74–93.

29. Sack R: Enterotoxigenic Escherichia coli: identification and characterization. *J Infect Dis* 1980;142:279–286.

30. Sanchez J, Uhlin B-E, Grundström T, et al: Immunoactive chimeric ST-LT enterotoxins of Escherichia coli generated by in vitro gene fusion. *FEBS Lett* 1986;208:194–198.

31. Smyth CJ: Two mannose-resistant haemagglutinins on enterotoxigenic Escherichia coli of serotype 06:K15:H16 or H- isolated from travellers' and infantile diarrhoea. *J Gen Microbiol* 1982;128:2081–2096.

32. Stoll BJ, Svennerholm A-M, Gothefors L, et al: Local and systemic antibody responses to naturally acquired enterotoxinogenic Escherichia coli diarrhea in an endemic area. *J Infect Dis* 1986;153:527–534.

33. Svennerholm A-M, Lange S, Holmgren J: Correlation between intestinal synthesis of specific immunoglobulin A and protection against experimental cholera in mice. *Infect Immunol* 1978;21:1–6.

34. Svennerholm A-M, Åhrén C: Immune protection against enterotoxigenic E. coli: search for synergy between antibodies to enterotoxin and somatic antigens. *Acta Pathol Microbiol Immunol Scand [C]* 1982;90:1–6.

35. Svennerholm A-M, Gothefors L, Sack DA, et al: Local and systemic antibody responses and immunological memory in humans after immunization with cholera B subunit by different routes. *Bull WHO* 1985;62:909–918.

36. Svennerholm A-M, Wikström M, Lindblad M, Holmgren J: Monoclonal antibodies against Escherichia coli heat-stable toxin (STa) and their use in a diagnostic ST ganglioside GM_1 enzyme-linked immunosorbent assay. *J Clin Microbiol* 1986;24:585–590.

37. Svennerholm A-M, Wikström M, Lindblad M, Holmgren J: Monoclonal antibodies to Escherichia coli heat-labile enterotoxins: neutralizing activity and differentiation of human and porcine LTs and cholera toxin. *Med Biol* 1986;64:23–30.

37b. Svennerholm A-M, Lopez-Vidal Y, Holmgren J, et al: Role of PCF8775 antigen and its coli surface subcomponents for colonization, disease and protective immunogenicity of enterotoxigenic Escherichia coli in rabbits. *Infect Immun* 1988;56:523–528.

37c. Svennerholm A-M, Lindblad M, Svennerholm B, Holmgren J: Synthesis of nontoxic, antibody-binding Escherichia coli heat-stable enterotoxin (ST_a) peptides. *FEMS Microbiol Lett* (in press).

38. Tayot J-L, Holmgren J, Svennerholm L, et al: Receptor-specific large scale purification of cholera toxin on silica beads derivatized with lyso-GM_1-ganglioside. *Eur J Biochem* 1981;113:249–258.

39. Thomas LV, Cravioto A, Scotland SM, Rowe B: New fimbrial antigenic type (E8775) that may represent a colonization factor in enterotoxigenic Escherichia coli in humans. *Infect Immun* 1982;35:1119–1124.
40. Thomas LV, Rowe B: The occurrence of colonization factors (CFA/I, CFA/II and E8775) in enterotoxigenic Escherichia coli from various countries in South East Asia. *Med Microbiol Immunol* 1982;171:85–90.
41. Thomas LV, McConnell MM, Rowe B, Field AM: The possession of three novel coli surface antigens by enterotoxigenic Escherichia coli strains positive for the putative colonization factor PCF8775. *J Gen Microbiol* 1985;131:2317–2326.
42. To SC-M, Moon HW, Runnels PL: Type 1 pili (F1) of porcine enterotoxigenic Escherichia coli: vaccine trial and tests for production in the small intestine during disease. *Infect Immun* 1984;43:1–5.

CHAPTER 19

Shigella Vaccines

Samuel B. Formal, Thomas L. Hale, and Christine Kapfer

Shigellosis (bacillary dysentery) is endemic throughout the world. Because the infectious dose is low (ten organisms), the disease can be spread not only by infected food or water but also by person-to-person contact. Shigellosis is of special concern in developing countries where conditions of sanitation are poor, personal hygiene practices often primitive, and malnutrition common. In endemic areas shigellosis is a disease of childhood. The incidence is low in breast-fed infants, but in weaned infants it may reach a peak of 200,000 cases per 100,000 population per year. In contrast, the incidence of disease in industrialized nations is low (50 cases per 100,000 population per year). The severity of illness may vary from mild diarrhea to severe dysentery (stools of small volume that contain blood, mucous, and inflammatory cells). In extreme cases, which occur most commonly in children, shigellosis can result in hypotensive shock and death. Inflammation and ulcerative lesions of the colon occur in moderate and severe cases. Other signs of illness include fever, cramps, and tenesmus.

Although there are more than 30 serotypes of shigellae, usually only two or three predominate in any given geographic area. In industrialized countries *Shigella sonnei* is predominant, whereas *S. flexneri* is most frequently seen in less developed nations. *S. sonnei, S. flexneri* 2a, and *S. flexneri* 3 are responsible for 99% of shigellosis in the United States. In addition to the common serotypes, *S. dysenteriae* 1 (Shiga) is of special interest because this species can cause epidemic disease. Shiga epidemics were rare until a serious outbreak occurred in Central America in 1969. This epidemic involved a half million people of all ages, and it was accompanied by higher than usual mortality. Since the Central American outbreak, epidemic Shiga dysentery has been seen in Bangladesh, India, Nepal, Burma, Sri Lanka, and Africa. Considering the epidemic potential

---

*The views of the authors do not purport to reflect the position of the Department of the Army or the Department of Defense.

of *S. dysenteriae* 1, vaccines against this species, in addition to the more common *S. sonnei* and *S. flexneri* species, are required.

Pathogenesis

Shigellosis occurs after ingestion of virulent organisms, and an important step in the pathogenesis of this disease is invasion of the colonic epithelium (14,27). Mutant *Shigella* strains that do not invade colonic epithelial cells are uniformly avirulent. Following penetration, the organisms are initially contained within vacuoles derived from the membrane of the epithelial cell. The organisms lyse these vacuoles, multiply in the cytoplasm, and invade adjacent cells (22). This process causes the death of epithelial cells and results in the formation of ulcerative lesions in the colon. Organisms that reach the lamina propria evoke an intense inflammatory reaction that apparently kills the extracellular bacteria because bacteremia due to shigellae is rare.

A number of laboratory models can be used to evaluate the invasive potential of shigella strains, including: (a) the Sereney test, which assesses the ability to produce keratoconjunctivitis in the eyes of rabbits or guinea pigs (23) and which, in turn, is a reflection of the penetration of the corneal epithelial cells by shigellae (17); (b) a test employing cultured mammalian cells as serogates of the colonic epithelium and which demonstrates the capacity of shigellae to invade these cells in vitro (14); and (c) the histological examination of the intestinal epithelium for intracellular organisms following oral challenge in animal models (14,24).

Genetic studies have indicated that both plasmid and chromosomal genes are involved in the virulence of shigellae. When chromosomal genes were transferred by conjugation or transduction from an *E. coli* K-12 donor to virulent *S. flexneri* recipients, three regions were found to affect the virulence of the hybrid strains in laboratory models. Transfer of a locus designated *kcp*, which is co-transducible with the *pur*E locus, resulted in loss of the ability to cause either keratoconjunctivitis or a fatal infection in starved guinea pigs (8). *S. flexneri* hybrids that incorporated the *xyl-rha* chromosomal region of *E. coli* K-12 into their genomes lost the capacity to cause a fatal infection in starved guinea pigs but retained the ability to evoke keratoconjunctivitis (3). *S. flexneri* hybrids that inherited an *E. coli* K-12 locus co-transducible with the His[+] marker could not express the O antigenic determinant (Table 19.1), and they were uniformly avirulent in animal models (7).

Each of these hybrid classes retained the ability to invade cultured epithelial cells. Those that inherited either the *E. coli xyl-rha* region or the *pur*E locus could elicit an acute inflammatory reaction in the intestinal mucosa of laboratory animals but could not kill these animals. In contrast, the rough hybrids that had inherited the His[+] marker produced no significant inflammatory reaction in the bowel. These observations indicate that chromosomal genes express virulence determinants that are necessary for survival of the pathogen in tissues. It is not known what virulence factor is encoded by the *kcp* locus, but it has been shown that the *xyl-rha* region of *S. flexneri* encodes an aerobactin iron-binding system (11). It has also been reported that this segment of the chromosome of *S. dysenteriae* 1 encodes the production of Shiga toxin (25).

Genes located on 120 or 140 megadalton (mDa) plasmids are necessary for expression of the invasive phenotype. Sansonetti et al (20) demonstrated that loss of a 140-mDa plasmid from *S. flexneri* correlates with loss of the invasive phenotype. Transfer of this plasmid to an avirulent *S. flexneri* strain that had lost a 140-mDa plasmid restored virulence (20). Furthermore, transfer of the *S. flexneri* plasmid to an avirulent mutant of invasive *Escherichia coli* also reinstated its ability to invade epithelial cells (19). Similar results have been obtained with the 120-mDa plasmid of *S. sonnei*. Studies in minicells have demonstrated that plasmid-containing invasive strains express seven outer-membrane polypeptides (OMPs) with molecular weights ranging from 20 to 78 kDa. Non-plasmid-containing strains do not express these OMPs nor do genotypically virulent strains grown at 30°C (12). The role of these plasmid-encoded OMPs in invasion is presently being investigated. The 140-mDa *S. flexneri* plasmid has been shown also to be responsible for the survival and growth of shigellae within mammalian cells. This capacity to multiply correlates with the ability to lyse endocytic vacuoles, which, in turn, correlates with the temperature-regulated expression of a "contact hemolysin" (22).

In addition to plasmid genes that encode the invasive phenotype, genes situated on a 120-mDa plasmid in *S. sonnei* are also necessary for expression of the form I somatic antigen (Table 19.1) (18). This 120-mDa plasmid is lost at high frequency, and the resulting form II clones are avirulent owing to loss of the invasive phenotype and to the inability to express a smooth somatic antigen. Indeed, the 120-mDa plasmid is lost at such a high frequency that it is difficult to explain how *S. sonnei* continues to persist as a pathogen. In contrast to *S. sonnei,* genes on a 6-mDa plasmid are necessary for expression of the somatic antigen of *S. dysenteriae* 1 (29). Loss of this plasmid results in an avirulent, rough phenotype. Transfer of the 6-mDa plasmid from *S. dysenteriae* 1 to *E. coli* K-12, however, is

Table 19.1. Genetic control of the expression of O-specific oligosaccharide by *Shigella* spp.

Serotype	Genetic loci	Ref.
S. dysenteriae 1	6-mDa plasmid and *his*	29
	chromosomal loci	13
S. flexneri 2a	*his* and *pro* chromosomal loci	7
S. sonnei	120 mDa plasmid	18

not sufficient to allow the recipient *E. coli* to express the *S. dysenteriae* 1 somatic antigen. The *his*$^+$ locus from the *S. dysenteriae* chromosome together with the 6-mDa plasmid is required for expression of the complete *S. dysenteriae* 1 antigen (Table 19.1) (13).

The behavior of E. coli K-12 recombinants that have inherited large segments of *S. flexneri* chromosome is not altered in any of the model systems used to assess virulence (21), and *E. coli* K-12 transconjugants that have inherited the *S. flexneri* 140-mDa plasmid are able to invade HeLa cells but do not cause significant changes in animal models. Transfer of *S. flexneri* chromosomal segments to the later plasmid-containing transconjugant strain can alter its biological activity. For example, transfer of the *S. flexneri* 2a His$^+$ marker, which is linked to genes encoding O-antigen expression in this serotype (Table 20.1) resulted in an invasive transconjugant strain that could evoke an inflammatory reaction in the ileal mucosa of a ligated rabbit intestinal loop. This strain did not produce fluid in the ileal loop, nor did it cause a positive Serney test. Transfer of the *S. flexneri* *kcp*$^+$ locus and the *xyl-rha* segment to the latter hybrid produced recombinants that could evoke a positive Serney test. Some of these recombinants also elicited a fluid response in the ligated rabbit ileal loop assay (21).

Attenuated Vaccines

Perhaps because shigellosis is a superficial or local infection of the colonic mucosa, immunization by the parenteral route using killed vaccines has not been successful. Even living, virulent *Shigella flexneri* 2a organisms administered subcutaneously in monkeys fail to protect these animals against oral challenge with homologous organisms (6). However, Dzhikidze et al (2), have reported promising results using parenterally administered preparations of ribosomes from *S. sonnei* to protect monkeys against experimental oral challenge with *S. sonnei*. None of the ten immunized animals became ill, whereas all five controls exhibited disease of varying severity following challenge. These observations must obviously be investigated further.

Epidemiological data suggest that prior infection with shigellae confers some degree of resistance to subsequent illness caused by organisms of the same serotype. These observations further suggest that attenuated oral vaccines that would stimulate local antibody and/or a cellular immune response would confer protection against clinical illness. For example, spontaneously arising avirulent strains of *S. flexneri* 2a, which lack the ability to invade the intestinal mucosa, have been shown to be safe and effective oral vaccines in monkeys. However, multiple doses of large numbers of organisms were required to induce resistance to subsequent challenge with virulent organisms of the same serotype (4). Another spontaneous avirulent mutant strain isolated by Istrati has been reported to be safe and effective in large Rumanian field trials, but it requires multiple inoculations (15). Attenuated vaccines prepared from streptomycin-dependent mutant strains have been widely tested in Yugoslavia. Several oral doses conferred a significant degree of protection in field studies; however, these vaccines were somewhat unstable (16).

In contrast to the noninvasive attenuated vaccine strains described above, a single dose of a genetically attenuated *S. flexneri* 2a strain, which was constructed by incorporating the *xyl-rha* region of *E. coli* K-12 into the *S. flexneri* 2a chromosome, was sufficient to produce resistance to homologous challenge in monkeys (4,5). This vaccine was able to invade the intestinal mucosa, but it was unable to multiply within mucosal epithelial cells (3). The ability of an invasive vaccine to deliver antigen in a predictable fashion to antibody-forming cells in the lamina propria and to simulate cellular immunity could account for the effectiveness of this strain. Even though it produced no apparent disease in monkeys, the invasive *S. flexneri* hybrid caused diarrhea when fed in high doses to volunteers.

Results from field studies of human populations and monkey vaccinations indicate that protection is associated with the shigella somatic antigen. If this case holds true, it is not likely that we shall be able to protect against all shigellosis because of the more than 30 serotypes. This antigenic heterogeneity may not be a practical problem, however, because only two or three serotypes usually predominate in any given geographic area. Immunization of monkeys with a polyvalent vaccine consisting of four serotypes protected them against challenge with any of these serotypes (5), so a polyvalent vaccine composed of attenuated strains of *S. sonnei* and two or three serotypes of *S. flexneri* could substantially reduce endemic shigellosis. It would also be desirable to include a vaccine strain that protects against epidemic *S. dysenteriae* 1.

Hybrid Vaccines Using Carrier Strains

Experience summarized above suggests that the somatic antigen is associated with protection against shigellosis and that the most effective attenuated vaccines are those that transport this antigen into the mucosal tissue. Therefore an ideal vaccine would be invasive and express shigella somatic antigens, but it would lack the potential to revert to virulence. One approach to this problem is to transfer only the genes encoding the shigella somatic antigen (summarized in Table 19.1) to an invasive but avirulent carrier strain. One such carrier strain is *Salmonella typhi* 21a, which is presently in use as a living oral vaccine (28). There is no direct evidence that this strain invades the intestinal mucosa, but the fact that three doses produce significant immunity against typhoid fever suggests that it may be the case. The 120-mDa plasmid, which encodes *S. sonnei* form I somatic antigen synthesis, has been conjugally transferred into *S. typhi* 21a, and the resulting transconjugant expresses the somatic antigens of both *S. typhi* and *S. sonnei*. Following parenteral immunization, this vaccine protects mice against parenteral challenge with either *S. typhi* or *S. sonnei* (9). Doses of up to 1×10^{10} cells have been fed to a limited number of human volunteers, and no unacceptable reactions were observed (26). Three doses of 1×10^9 cells, administered orally within a period of 7 days, have yielded evidence of protection against dysentery in volunteers challenged with *S. sonnei*. However, lot-to-lot variation in the efficacy of this product has so far precluded its use in field studies (1).

A second carrier-based vaccine strain was constructed by the conjugal transfer of the His$^+$ and Pro$^+$ chromosomal markers from *S. flexneri* 2a to an *E. coli* K-12 strain that carried the *S. flexneri* 140-mDa plasmid. In the case of *S. flexneri*, the group somatic antigen is linked to the His$^+$ marker, and the type-specific antigens are linked to the Pro$^+$ marker. Therefore a serotype 2a vaccine could be selected by isolation of transconjugants that repaired histidine and proline auxotropies in the *E. coli* K-12 recipient. Because it carries the 140-mDa plasmid, the hybrid can invade HeLa cells in vitro and produce a mild inflammatory response in the lamina propria when injected into rabbit ileal loops. The pathology produced in the loop mucosa is suggestive of transient mucosal invasion. This vaccine caused no apparent adverse reaction in monkeys when fed in doses as high as 1×10^{11} cells, and the animals shed the organism for fewer than 4 days. Animals that received the vaccine were significantly more resistant to experimental challenge with *S. flexneri* than were control monkeys (10). The safety and efficacy of invasive *E. coli* hybrid vaccines in humans must now be determined.

References

1. Black RE, Levine MM, Clements ML, et al: Prevention of shigellosis by a Salmonella typhi–Shigella sonnei bivalent vaccine. *J Infect Dis* 1987;155:1260–1265.
2. Dzhikidze EK, Kavtaadze KN, Levenson VI, et al: Ribosomal dysentery vaccine. IV. Study of the reactogenicity, antigenic and protective activity of Shigella sonnei ribosomes in experiments on monkeys. *Z. Microbiol* 1981;7:53–59.
3. Formal SB, LaBrec EH, Kent TH, Falkow S: Abortive intestinal infection with an Escherichia coli–Shigella flexneri hybrid strain. *J Bacteriol* 1965;89:1374–1382.
4. Formal SB, LaBrec EH, Palmer A, Falkow S: Protection of monkeys against experimental shigellosis with attenuated vaccines. *J Bacteriol* 1965;90:63–68.
5. Formal SB, Kent TH, May HC, et al: Protection of monkeys against experimental shigellosis with a living attenuated oral polyvalent dysentery vaccine. *J Bacteriol* 1966;92:17–22.

6. Formal SB, Maenza RM, Austin S, LaBrec EH: Failure of parenteral vaccines to protect monkeys against experimental shigellosis. *Proc Soc Exp Biol Med* 1967;125:347–349.

7. Formal SB, Gemski P Jr, Baron LS, LaBrec EH: Genetic transfer of Shigella flexneri 2a antigens to Escherichia coli K-12. *Infect Immun* 1970;1:279–287.

8. Formal SB, Gemski P Jr, Baron LS, LaBrec EH: A chromosomal locus which controls the ability of Shigella flexneri to evoke kerato-conjunctivitis. *Infect Immun* 1971;3:73–79.

9. Formal SB, Baron LS, Kopecko DJ, et al: Construction of a potential bivalent vaccine strain: introduction of Shigella sonnei form I antigen genes into the gal E Salmonella typhi Ty21a typhoid vaccine strain. *Infect Immun* 1981;34:746–750.

10. Formal SB, Hale TL, Kapfer C, et al: Oral vaccination of monkeys with an invasive Escherichia coli K-12 hybrid expressing Shigella flexneri 2a somatic antigen. *Infect Immun* 1984;46:465–469.

11. Griffiths E, Stevenson P, Hale TL, Formal SB: Synthesis of aerobactin and a 76,000-dalton iron-regulated outer membrane protein by Escherichia coli K-12–Shigella flexneri hybrids and by enteroinvasive strains of Escherichia coli. *Infect Immun* 1985;49:61–71.

12. Hale TL, Sansonetti PJ, Schad PA, et al: Characterization of virulence plasmids and plasmid-associated outer membrane proteins in Shigella flexneri, Shigella sonnei and Escherichia coli. *Infect Immun* 1983;40:340–350.

13. Hale TL, Guerry P, Seid RC Jr, et al: Expression of lipopolysaccharide O antigen in Escherichia coli K-12 hybrids containing plasmid and chromosomal genes from Shigella dysenteriae 1. *Infect Immun* 1984;46:470–475.

14. LaBrec EH, Schneider H, Magnani TJ, Formal SB: Epithelial cell penetration as essential step in the pathogenesis of bacillary dysentery. *J Bacteriol* 1964;88:1503–1518.

15. Meitert T, Pencu E, Cuidin L, Tonciu M: Vaccine strain Sh flexneri T32–Istrati: studies in animals and in volunteers; antidysentery immunoprophylaxis and immunotherapy by live vaccine Vadizen (Sh flexneri T32–Istrati). *Arch Roum Pathol Exp Microbiol* 1984;43:251–278.

16. Mel DM, Gangarosa EJ, Radovanovic ML: Studies on vaccination against bacillary dysentery. 6. Protection of children with oral immunization using streptomycin-dependent Shigella strains. *Bull WHO* 1971;45:457–464.

17. Piechaud MS, Szturm-Rubensten S, Piechaud D: Evolution histologique de la kerato-conjuctivite a bacillus dysenteriques du cobaye. *Ann Inst Pasteur* 1958;94:298–309.

18. Sansonetti PJ, Kopecko DJ, Formal SB: Shigella sonnei plasmids: evidence that a large plasmid is necessary for virulence. *Infect Immun* 1981;34:75–83.

19. Sansonetti PJ, D'Hauteville H, Formal SB, Taucas M: Plasmid mediated invasiveness in "shigella-like" Escherichia coli. *Ann Inst Pasteur Microbiol* 1982;1324:351–355.

20. Sansonetti PJ, Kopecko DJ, Formal SB: Involvement of a plasmid in the invasive ability of Shigella flexneri. *Infect Immun* 1982;35:852–860.

21. Sansonetti PJ, Hale TL, Dammin GJ, et al: Alterations in the pathogenicity of Escherichia coli K-12 after transfer of plasmid and chromosomal genes from Shigella flexneri. *Infect Immun* 1983;39:1392–1402.

22. Sansonetti PJ, Ryter A, Clerc P, et al: Multiplication of Shigella flexneri within

HeLa cells: lysis of the phagocytic vacuole and plasmid-mediated contact hemolysis. *Infect Immun* 1986;51:461–469.

23. Sereney B: Experimental kerato-conjunctivitis Shigellosa. *Acta Microbiol Hung* 1957;4:367–376.

24. Takeuchi A, Sprinz, H, LaBrec EH, Formal SB: Experimental bacillary dysentery: an electron microscopic study of the response of the mucosa to bacterial invasion. *Am J Pathol* 1965;47:1011–1044.

25. Timmis KN, Clayton CL, Sikizaki T: Localization of Shiga toxin gene in the region of Shigella dysenteriae chromosome specifying virulence functions. *FEMS Microbiol Lett* 1985;30:301–305.

26. Tramont EC, Chung R, Berman S, et al: Safety and antigenicity of typhoid-Shigella sonnei vaccine (strain 5076-1C). *J Infect Dis* 1984;149:133–136.

27. Voino-Vasenetsky MV, Khavkin TN: A study of intraepithelial localization of dysentery causative agents with the aid of fluorescent antibodies. *J Microbiol* 1964;12:98–100.

28. Wahdan MH, Serie C, Germaniner R, et al: A controlled field trial of a live oral typhoid vaccine. *Bull WHO* 1980;58:469–474.

29. Watanabe H, Timmis KN: A small plasmid in Shigella dysenteriae 1 specifies one or more functions essential for O antigen production and bacterial virulence. *Infect Immun* 1984;43:391–396.

CHAPTER 20

DNA Sequence Homology Among *ipa* Genes of *Shigella* spp. and Enteroinvasive *Escherichia coli*

Malabi M. Venkatesan, Jerry M. Buysse, and Dennis J. Kopecko

Bacillary dysentery, caused by *Shigella* and enteroinvasive *Escherichia coli* (EIEC), is endemic throughout the world, and it accounts for 10 to 20% of acute diarrheal disease worldwide (11). Clinical symptoms of the disease range from a mild diarrhea to a severe dysenteric syndrome characterized by fever, abdominal cramps, frequent defecation of blood and mucus, and, in extreme cases, hypotensive shock. In developing countries inadequate sanitation and malnutrition contribute to the prevalence of the disease, which can be caused by ingesting as few as 10 to 100 bacteria (10). Shigellosis is restricted to primate hosts and all serotypes of *Shigella*, as well as nine serotypes of EIEC, can cause the disease. Understanding the mechanism of dysentery pathogenesis should facilitate the construction of rapid diagnostic probes for *Shigella* and EIEC and will help to identify genes and gene products that may be essential in devising an effective dysentery vaccine.

Analysis of infected colons in monkeys and humans has shown that the colonic mucosa is the primary site of bacterial infection. The colonic mucosal surface is covered with an epithelial cell monolayer that overlays connective tissue containing blood and lympathic vessels, collectively known as the lamina propria. The ability to invade epithelial cells and multiply intracelluarly is the distinctive pathogenic feature of *Shigella* and EIEC strains. Once internalized, the bacteria disseminate laterally to adjacent epithelial cells and to the lamina propria. Unlike some *Salmonella*, however, the bacteria rarely enter the circulatory system. The process of epithelial cell necrosis results in an acute inflammatory response (6) involving polymorphonuclear leukocytes from the lamina propria, which limits the infection to the superficial layers of the colon. Epithelial cell necrosis is apparently due to the production of bacterial enterotoxins that inhibit protein synthesis in eukaryotic cells; this necrosis leads to focal microulceration of the colon (18,20). The small-volume rectal discharges are due to fluid malabsorption in the colon and sloughing of necrotic tissue and mucus.

Genetic studies have revealed that products of several regions of the

Shigella chromosome contribute to the overall virulent phenotype (5,9,12,24). The ability to invade epithelial cells, which is the first step in the pathogenesis of the disease, is determined by the presence of a 120- to 140-megadalton (mDa) nonconjugative invasion plasmid (Figs. 20.1 and 20.2A) (22,23). Loss of this plasmid results in loss of the invasive phenotype, and such plasmid-cured bacteria are avirulent (9,24). Transfer of the invasion plasmid to avirulent plasmid-cured strains restores the ability to invade epithelial cells. In the laboratory, the invasive phenotype is measured by centrifuging the bacteria onto a monolayer of HeLa cells in tissue culture and monitoring microscopically the presence of bacteria inside the cell (13). This assay does not require a smooth lipopolysaccharide (LPS) bacterial cell surface and does not measure other essential virulence properties such as intracellular multiplication. The ability of shigellae or EIEC to bind Congo red dye correlates with the presence of the invasion plasmid (3,15,25). Congo red dye agar is therefore used as an initial screen to detect invasive plasmid-containing bacteria. Finally, a commonly used animal model for assessing *Shigella* virulence is the elicitation of a keratoconjunctivitis reaction in rabbits, mice, or guinea pigs (27). This assay mimics an intestinal infection in that the bacteria invade and multiply within the corneal epithelial cells.

Molecular characterization of the invasion plasmid has shown that at least four regions of the plasmid are required for maintenance of the virulent phenotype (Fig. 20.1) (1,12,17,26,28). Two of these regions are encoded within a 35- to 37-kilobase (kb) stretch of DNA and, along with the *virF* locus (21), are essential for three linked phenotypes (i.e., epithelial cell invasive ability, Congo red dye binding, and inhibition of bacterial growth). In addition, a positive Sereny reaction requires the *virG* locus, which apparently encodes a function that results in lysis of the endocytic vacuole and rapid intracellular multiplication (14).

The large invasion plasmid encodes the synthesis of several outer membrane polypeptides (7,8,14,19). At least six of these proteins—140, 78, 62, 60, 42, and 37 kilodaltons (kDa)—are important immunogens in that convalescent sera from monkeys and humans contain high titers of antibodies to these proteins (Fig. 20.2B). The synthesis of these polypeptides is temperature-regulated (expressed at 37°C and repressed at 30°C) and correlates with the requirement for growth at 37°C for HeLa cell invasion and virulence (16). *Shigella* and EIEC strains that do not synthesize some or all of these immunogenic polypeptides are noninvasive and avirulent. Using rabbit antisera specific for the invasion plasmid polypeptides and the λgt11 expression vector, we have cloned and characterized the invasion plasmid antigen *(ipa)* genes from *S. flexneri* 5, strain M90T-W, that code for the 62-kDa *(ipaB)*, 42 kDa *(ipaC)*, and 37-kDa *(ipad)* proteins (1). More recently, we have obtained plasmid recombinants that express the 78-kDa protein *(ipaA)* and cosmid recombinants that synthesize five of these immunogenic polypeptides (unpublished observations). The *ipaA, ipaB, ipaC,* and *ipaD* genes have been mapped to a contiguous DNA segment con-

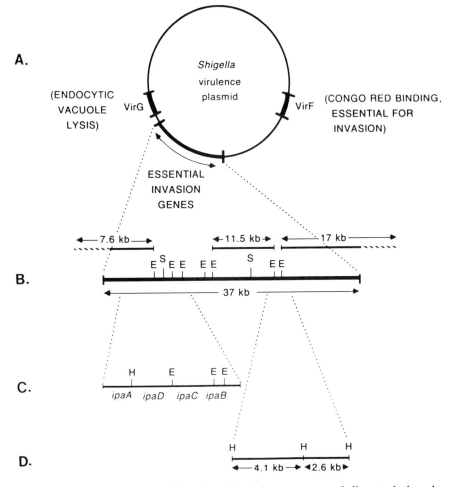

Fig. 20.1. Composite map of the plasmid virulence genes and dissected cloned regions based on interpretation of data from several published reports (1, 17, 25, 26, 28). **A:** Relative spatial relations among identified virulence genes of a hypothetical *Shigella* virulence plasmid approximately 210 kb in size. The *virG* region may encode endocytic vacuole lysis. (14), the *virF* locus encodes Congo red binding ability and is essential for virulence (21), and a region of about 35 kb encodes a number of essential invasion genes (17, 26). **B:** Linear map of the 37-kb minimal essential invasion gene region, isolated by Maurelli et al (17), is shown with restriction endonuclease sites marked for *Eco*R1 (E) and *Sal*1 (S). The locations of three large *Eco*R1 fragments, associated with invasive ability, are denoted above this cloned region. The orientation of the 37-kb region on the plasmid map is unknown. **C:** Linear map of the cloned invasion plasmid antigen (*ipa*) genes (*ipaA*), *iapB*, *ipaC*, and *ipaD*, which encode cell surface polypeptides of 37 to 78 kDa and are directly associated with invasive ability (1). Restriction endonuclease sites for *Eco*R1 (E) and *Hind*III (H) are shown, as is the proposed relation of these genes with the sequences cloned by Maurelli et al (17). **D:** Linear map of two *Hind*III (H) fragments cloned by Watanabe and Nakamura (28) and shown to encode the synthesis of four peptides of 38, 41, 47, and 80 kDa. The speculated location of these genes on the map of Maurelli et al (17) is shown. [From Kopecko et al (12), with permission.

(A)

(B)

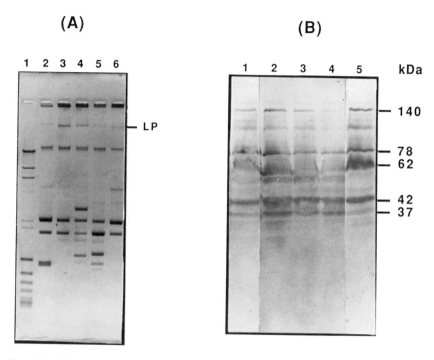

Fig. 20.2. Electrophoretic analysis of the large invasion plasmid (**A**) and its protein products (**B**). **A:** Plasmid DNA was isolated essentially by the method of Cassie-Denarie et al (2). DNA from *S. flexneri* 1 strain M25-8 (lane 2) *S. flexneri* 2 strain 2457T (lane 3) *S. flexneri* 3 strain J17 (lane 4), *S. flexneri* 4 strain M76-39 (lane 5) and *S. flexneri* 5 strain M90T-W (lane 6) was electrophoresed on 0.8% agarose gels in 20 m*M* Tris–HCl pH 7.9 20 m*M* sodium acetate, and 1 m*M* EDTA. Lane 1 contains a mixture of a *Hind*III digest of λ DNA and *Hinc*II digest of φx174 DNA. LP = large invasion plasmid. The band close to the largest λ *Hind*III fragment (~23 kb) is mostly fragmented large plasmid DNA. The two Col El-like small plasmids (~2 kb) are commonly present in all *S. flexneri* strains, and similar 2-4-kb plasmids are also seen in other *Shigella* spp. and EIEC. **B:** Whole-cell SDS lysates of *S. flexneri* 1 (lane 1) *S. flexneri* 2 (lane 2), *S. flexneri* 3 (lane 3), *S. flexneri* 4 (lane 4), and *S. flexneri* 5 (lane 5) were electrophoresed on polyacryl-amide–SDS gels, transferred onto nitrocellulose (Western blot), and reacted with mokey sera containing antibodies specific for invasion plasmid antigens. The strains used are as in **A**. The five invasion-associated antigens are clearly seen, and their approximate molecular weights are given in kilodaltons (kDa). A number of high-molecular-weight bands are also seen between the 140-kDa and 78-kDa proteins. At present, their identity remains unclear, although they are more evident when cells are grown in a minimal medium than in a nutrient broth.

tained at one end of the 35- to 37-kb invasion-essential region of the virulence plasmid (Fig. 20.1C) (1). A newly defined antigen, *ipaH*, coding for the synthesis of a 60-kDa protein, was also isolated from the λgt11 library (1). Although *ipaB* and *ipaH* proteins could not be separated readily on one-dimensional SDS—polyacrylamide gels, these proteins were found to be distinct immunologically; in addition, the *ipaB* and *ipaH* genes have been shown to be nonhomologous by DNA hybridization analysis (1). Genetic mapping of the *ipaH* locus showed that it was unlinked to the *ipaBCDA* region (Fig. 20.3).

The immunodominant nature of these surface-expressed outer membrane proteins, coupled with the fact that they are required for the expression of the invasive phenotype and virulence, indicate that they might be important triggers in the elicitation of a protective immune response against bacillary dysentery in primates and humans. Considering the most prev-

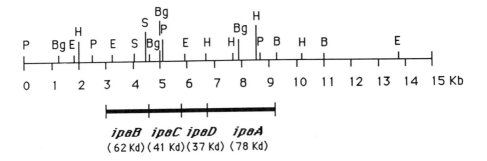

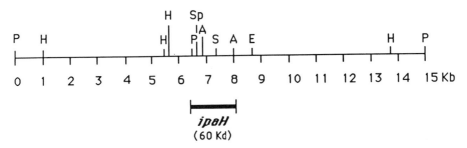

Fig. 20.3. Genetic map of the *ipaBCDA* region (1) and *ipaH*. The two loci are unlinked. This map was generated by hybridizing purified insert DNA from λgt11 recombinants to restriction enzyme digested invasion plasmid DNA (1). The DNA probes used in these studies were 470 bp (*ipaB*), 1750 bp (*ipaC*), 870 bp (*ipaD*), and 3.0 kb (*ipaH*). Restriction endonucleases used for the mapping studies included *Bam*HI (B), *Bgl*II(Bg), *Eco*RI(E), *Hind*III(H), *Pst*I(P), *Sal*I(S), *Sph*I(Sp), and *Ava*I(A).

alent serotypes of *Shigella* responsible for 95% of the shigellosis encountered worldwide (11), it appears that four oral vaccines, directed against *S. flexneri* serotypes 2a and 3, *S. sonnei,* and *S. dysenteriae* serotype 1, would be most efficacious. Identification of common genetic sequences encoding essential virulence proteins among these strains would be useful as rapid diagnostic probes and for the construction of dysentery vaccines. Toward that end, we isolated large plasmid DNA from several invasive, Sereny-positive strains of *Shigella* (Fig. 20.2A) and EIEC, as well as from an avirulent strain of *S. flexneri* serotype 5, strain M90T-A₃ (see below). These plasmids were digested with *Hind*III and electrophoresed on 0.8% agarose gels (Fig. 20.4). The general patterns of restriction fragments were different among the various species of *Shigella* and EIEC, a feature that has been observed previously (24). The plasmid DNA digests were transferred to nitrocellulose membranes and hybridized with radiolabeled *ipaB, ipaC, ipaD,* and *ipaH* sequences. Hybridization with *ipaB, ipaC,* and *ipaD* under high stringency reaction conditions revealed a 4.6-kb *Hind*III band common to all of the strains (Fig. 20.4B). Only in the case of one *S. dysenteriae* strain was a slightly larger band seen. No hybridization was observed between these *ipa* probes and DNA from *S. flexneri* 5, M90T-A₃, an isogenic avirulent strain derived from M90T-W, which contained an invasion plasmid that migrated slightly ahead of M90T-W. Our hybridization data show that a specific deletion that includes the entire *ipa* region has occurred in this stable but noninvasive strain.

Hybridization of *Shigella* and EIEC invasion plasmid DNA with *ipaH* sequences revealed a much more complex pattern. In each of the strains tested, several restriction endonuclease-generated bands reacted with the probes. Although common restriction fragments could be observed among some of the strains (Fig. 20.4), no two strains had identical patterns. Probes encoding the *ipaH* sequence hybridized to five distinct *Hind*III bands in the virulent M90T-W strain. Four of these bands could also be seen in M90T-A₃, but a 9.6-kb band found in the virulent strain was missing in M90T-A₃. The relation of *ipaH* sequences and the virulence phenotype is unclear at present. Explanations to account for the pattern of multiple bands seen with the *ipaH* probe include multiple copies of *ipaH* on the plasmid molecule or repeated elements within *ipaH* that are present elsewhere on the invasion plasmid. The heterogeneous banding pattern observed with the *ipaH* probe when various strains of *Shigella* spp. are examined is currently being exploited to develop an epidemiologic probe for bacillary dysentery.

The above studies clearly demonstrated that the plasmid-borne *ipaBCD* genes of all shigellae and EIEC are highly homologous, suggesting similarities in the molecular mechanisms involved in epithelial cell invasion. In order to evaluate this hypothesis, the probes were used to determine if other strains of enteric bacteria carried analogous invasion-associated

A.

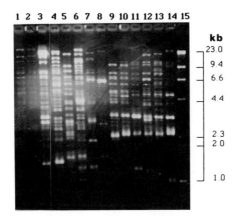

B.   C.

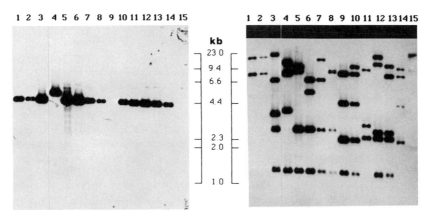

Fig. 20.4. Hybridization pattern of invasion plasmid DNA probed with *ipaC* and *ipaH*. **A:** Plasmid DNA from EIEC strain M41-63T (lane 1), *S. dysenteriae* 1 strain JVA/70 (lane 2) *S. dysenteriae* 1 strain 1617 (lane 3), *S. dysenteriae* strain CG1314 (lane 4), *S. boydii* strain CG1159 (lane 5), *S. sonnei* strain PA325 (lane 6), *S. sonnei* strain 53GI (lane 7), *S. flexneri* 6 CDC strain (lane 8), *S. flexneri* 5 M90T-A₃ (lane 9), *S. flexneri* 5 M90T-W (lane 10), *S. flexneri* 4 M76-39 (lane 11), *S. flexneri* 3 J17 (lane 12), *S. flexneri* 2 strain 2457T (lane 13), and *S. flexneri* 1 strain M25-8 (lane 14) was digested with *Hind*III and stained with ethidium bromide. **B,C:** The DNA was transferred to nitrocellulose and hybridized with *ipaC* (**B**) and *ipaH* (**C**) probes. The identical pattern of hybridization seen with *ipaC* (**B**) was also observed with *ipaB* and *ipaD* DNA sequences. Lane 15 has DNA molecular weight markers, which are expressed in kilobases at the side of the figure.

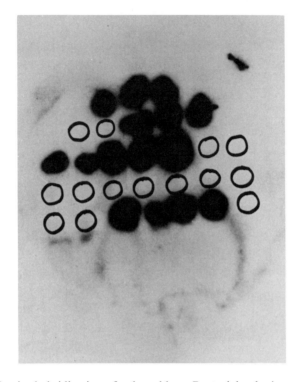

Fig. 20.5. In situ hybridization of colony blots. Bacterial colonies were streaked onto nylon filters (NEN Gene Screen) and hybridized with the *ipaC* DNA probe. The bacteria are labeled from left to right. **Top row:** *S. flexneri* 1 strain M25-8, *S. flexneri* 2 strain 2457T, and *S. flexneri* 3 strain J17. **Second row:** *E. coli* HB101, *E. coli* JM107, *S. dysenteriae* 1 strain 3818, *S. dysenteriae* 1 strain JVA/70 and *S. dysenteriae* strain CG1314. **Third row:** *S. dysenteriae* strain CG1768, *S. boydii* strain CG1159, *S. boydii* strain CG705-C, *S. sonnei* strain PA232, *S. sonnei* strain R071A, hemorrhagic *E. coli* (0157-H7) strain 1600-82, and (0157-H7) strain 562-84. **Fourth row:** EPEC (055-H7) strain 5380-66, EPEC (0125:H21) strain 1318-59, EPEC (0127:H−) strain 1322-54, EPEC (0111:H−) strain 4869-54, ETEC (LT 088:H21) strain 2185, ETEC (LT 0159:H21) strain 2094, and ETEC (ST 083:H17) strain 2021. **Fifth row:** ETEC (LT 0159:H21) strain 496, ETEC (ST 0153:H45) strain 1989, *S. dysenteriae* 1 strain 1617, *S. dysenteriae* 1 strain 3818, *S. dysenteriae* 1 strain JVA/70, and EIEC strain M4163T.

genes. Virulent and avirulent *Shigella* and EIEC as well as enteropathogenic *E. coli* (EPEC), enterotoxigenic *E. coli* (ETEC), enterohemorrhagic *E. coli* (EHEC), *Salmonella, Campylobacter, Citrobacter, Vibrio, Enterobacter, Klebsiella, Yersinia, Edwardsiella,* and *Aeromonas* cultures were used in colony blots and hybridized with the *ipaBCD* probes. A positive signal was observed only with laboratory-tested invasive, Sereny-positive strains of *Shigella* and EIEC. An example of such a colony blot is seen in Figure 20.5. No reaction was seen with any other enteric organism, suggesting that the invasion-associated *ipa* genes are unique to shigellae. Although the *ipa* gene sequences are intimately associated with the virulent phenotype, other plasmid as well as chromosomal genes are also involved in determining virulence (Fig. 20.1). Mutations in these loci would render the bacteria avirulent while retaining the *ipa* genes. To what extent these types of mutation occur in nature and would constribute to false positives is currently under investigation using the *ipa* probes in clinical studies of stool blots obtained from acute and convalescent dysentery patients.

Future efforts will be directed toward characterizing the *ipa* genes and their products structurally and functionally. We are also analyzing other virulence-associated genes in an effort to elucidate the interactions and control mechanisms that result in the virulent, invasive phenotype. Transcriptional studies will clarify mechanisms of gene regulation and control. Determinants of local and cell-mediated immunity (which certainly play a major role in protection) will also be studied so that an effective and inexpensive vaccine can be engineered.

References

1. Buysse JM, Stover CK, Oaks EV, et al: Molecular cloning of invasion plasmid antigen (ipa) genes from Shigella flexneri: analysis of ipa gene products and genetic mapping. J Bacteriol 1987;169:2561–2569.
2. Casse FC, Boucher C, Julliot JS, et al: Identification and characterization of large plasmid in Rhizobium meliloti using gel electrophoresis. *J Gen Microbiol* 1979;113:229–242.
3. Daskaleros PA, Payne SM: Characterization of Shigella flexneri sequences encoding Congo red binding (Crb): conservation of multiple (crb) sequences and role of IS1 in loss of the Crb+ phenotype. *Infect Immun* 1985;54:435–443.
4. Dinari G, Hale TH, Austin SW, Formal SB: Local and systemic antibody responses to Shigella infection in rhesus monkeys. *J Infect Dis* 1987;155:1065–1069.
5. Formal SB, Gemski P Jr, Baron LS, LaBrec EH: Genetic transfer of Shigella flexneri antigens to Escherichia coli K-12. *Infect Immun* 1970;1:279–287.
6. Formal SB, Hale TL, Boedeker EC: Interactions of enteric pathogens and the intestinal mucosa. *Proc R Soc Lond* 1983;B303:65–73.
7. Hale TL, Sansonetti PJ, Schad PA, et al: Characterization of virulence plasmids and plasmid associated outer membrane proteins in S. flexneri, S. sonnei and Escherichia coli. *Infect Immun* 1983;40:340–350.
8. Hale TL, Oaks EV, Formal SB: Identification and antigenic characterization of virulence associated, plasmid coded protein of Shigella spps. and enteroinvasive Escherichia coli. *Infect Immun* 1985;50:620–629.
9. Hale TL, Formal SB: Genetics of virulence in Shigella. *Microbiol Pathogenet* 1986;1:511–518.

10. Hornick RB: Bacterial infections of the intestine, in Weinstein L, Fields BN (eds): *Seminars in Infectious Disease,* Vol. 1. New York, Stratton, 1978, pp 68–96.

11. Kopecko DJ, Baron LS, Buysse J: Genetic determinants of virulence in Shigella and dysenteric strains of Escherichia coli: their involvement in the pathogenesis of dysentery. *Curr Top Microbiol Immunol* 1985;118:71–95.

12. Kopecko DJ, Venkatesan M, Buysse JB: Invasion: basic mechanisms and genetic control, in *Enteric Infections, Mechanisms, Manifestations and Management.* London, Chapman & Hall, (in press).

13. LaBrec EH, Schneider H, Magnani TJ, Formal SB: Epithelial cell penetration as an essential step in the pathogenesis of bacillary dysentery. *J Bacteriol* 1964;88:1503–1518.

14. Makino S, Sasakawa C, Kamata K, et al: A genetic determinant required for continuous reinfection of adjacent cells on large plasmid in S. flexneri 2a. *Cell* 1986;46:551–555.

15. Maurelli AT, Blackmon B, Curtiss R III: Loss of pigmentation in Shigella flexneria 2a is correlated with loss of virulence and virulence-associated plasmid. *Infect Immun* 1984;43:397–401.

16. Maurelli AT, Blackmon B, Curtiss R III: Temperature dependent expression of virulence genes in Shigella species. *Infect Immun* 1984;43:195–201.

17. Maurelli AT, Baudry B, deHauteville H, et al: Cloning of plasmid DNA sequences involved in invasion of HeLa cells by S. flexneri. *Infect Immun* 1985;49:164–171.

18. Meyer MP, Dixon PS, Rothman SW, Brown JE: Cytoxicity of Shiga toxin for primary cultures of human clonic and ileal epithelial cells. *Infect Immun* 1987;55:1533–1535.

19. Oaks EV, Hale TL, Formal SB: Serum immune response to Shigella protein antigens in rhesus monkeys and humans infected with Shigella spps. *Infect Immun* 1986;53:57–63.

20. O'Brien AD, Thompson MR, Gemski P, et al: Biological properties of Shigella dysenteriae 1 toxin. *Infect Immun* 1977;15:796–798.

21. Sakai T, Sasakawa C, Makino S, Yoshikawa M: DNA sequence and products analysis of the vir F locus responsible for E. coli binding and cell invasion of Shigella flexneri 2a. *Infect Immun* 1986;54:395–402.

22. Sansonetti PJ, Kopecko DJ, Formal SB: Shigella sonnei plasmids: evidence that a large plasmid is necessary for virulence. *Infect Immun* 1981;34:75–83.

23. Sansonetti PJ, Kopecko DJ, Formal SB: Involvement of a plasmid in the invasive ability of Shigella flexneri. *Infect Immun* 1982;35:852–860.

24. Sansonetti PJ, Hale TL, Oaks EV: Genetic of virulence in enteroinvasive Escherichia coli, in Schlessinger D (ed): *Microbiology—1984.* Washington, DC, American Society for Microbiology, 1985, pp 74–77.

25. Sasakawa CK, Kamata K, Sakai T, et al: Molecular alteration of the 140-Md. plasmid associated with loss of virulence and Congo red binding activity in Shigella flexneri. *Infect Immun* 1986;51:470–475.

26. Sasakawa C, Makino S, Kamata K, Yoshikawa M: Isolation, characterization and mapping of Tn5 insertions into the 140-megadalton invasion plasmid defective in the mouse Sereny test in Shigella flexneri 2a. *Infect Immun* 1986;54:32–36.

27. Sereny B: Experimental Shigella keratoconjunctivitis. *Acta Microbiol Acad Sci Hung* 1955;2:293–296.
28. Watanabe H, Nakamura A: Identification of Shigella sonnei form I plasmid genes necessary for cell invasion and their conversation among Shigella species and enteroinvasive Escherichia coli. *Infect Immun* 1986;53:352–358.

CHAPTER 21

Mucosal and Acquired Immunity in Giardiasis and Its Relation to Diarrhea and Malabsorption

N.K. Ganguly and R.C. Mahajan

Giardiasis is a relatively common intestinal protozoal infection in man and animals (2,46). Several reports indicate that the infection is more common in immunodeficient (3,47) and malnourished (44) hosts and in individuals with a tendency to feco-oral transmission, such as infants in day-care nurseries (8) or homosexuals (52). The disease may occur in both epidemic and endemic forms (40,64). The clinical spectrum of the disease ranges from an entirely asymptomatic infection to a mild, self-limiting illness, or chronic diarrhea with or without malabsorption (34,65). Despite the ubiquitous prevalence of parasites, the pathophysiology of the disease remains obscure. Therefore this chapter reviews the possible mechanism of diarrhea and malabsorption in this increasingly important infection.

Malabsorption may occur in up to 60% of giardiasis patients who have diarrhea (27) and may involve one or more nutrients. The nutrient most frequently malabsorbed is fat; however, the absorption of lactose (3,10), D-xylose (36,56), and vitamin B_{12} (13,27) may also be impaired. Vitamin and iron deficiencies have been reported to be associated with giardiasis (14,38). In experimental studies, reduced uptake of glucose, L-alanine, and glycine have been observed (5,6,23,60).

Various theories have been put forth to explain the possible mechanism of malabsorption, but none is complete satisfactory and acceptable. It is possible that more than one mechanism may be involved.

Interaction of *Giardia* and Epithelial Cells

Giardia adheres to the epithelial microvillous surface with the help of a sucking disc. Morphologic evidence from both light microscopic (29,67,68) and electron microscopic (17,43,45) studies suggest a direct interaction between the parasite and the epithelial cells. The mechanism of this interaction is not clear. Pictorial representations of electron micrographs reveal that the disc of the organism digs into the epithelial microvillous layer. Scanning electron micrographs show circular footprints of the formerly adherent trophozoites on the epithelial surface (17). Anand et al (4) demonstrated the direct damaging effects of *Giardia* trophozoites on the intestinal cells. For such type of damage a large number of parasites are needed, whereas sometimes smaller number of trophozoites have been shown to produce malabsorption (9).

Penetration by *Giardia* Trophozoites

Trophozoites have also been observed moving about in a random fashion between epithelial cells (9,18). What pathogenic role, if any, these invaders play is not clear. Whether invasion reflects the occasional opportunistic transgression between loosely adherent and dying epithelial cells or is the result of the penetrating skill of the parasite is also not clear.

Physical Interference

The absorptive surface areas may be mechanically blocked by a large number of attached trophozoites and thus may act as a mechanical barrier to absorption of nutrients (7,42,62,63). There are, however, no quantitative data to support this theory. Moreover, the large absorptive surface area of the small intestine would seem to compensate readily for such a possibility. We observed that a different quantum of parasite burden in the intestine did not affect the uptake of nutrients linearly (Table 21.1). The difficulty in demonstrating either cyst or trophozoites in the stool and duodenal contents of some symptomatic cases (11) suggests that, at least in some cases, only a small number of trophozoites may be needed to cause disease.

Table 21.1. Uptake of nutrients in brush border membrane vesicles prepared from *G. lamblia*-infected intestine (mean ± se).[a]

Group	*D*-Glucose	*L*-Phenylalanine	*L*-Lysine	*L*-Aspartic acid
Control	99.9±9.0	214.7±8.0	62.2±0.3	78.9±8.3
Group I (infected with 10,000 cysts/ animal)	60.5±4.02 (p <0.05)	140.7±6.0 (p <0.01)	41.1±3.3 (p < 0.01)	46.27±7.9 (p <0.05)
Group II (1000 cysts/ animal)	60.4±3.8 (p <0.05)	151.7±17.3 (p <0.05)	45.5±0.8 (p <0.001)	46.2±13.8 (p <0.05)
Group III (100 cysts/ animal)	59.3±5.5 (p <0.05)	158.1±8.3 (p <0.01)	41.7±1.6 (p <0.001)	48.7±8.2 (p <0.05)

[a]Units are picomoles per milligram of protein; incubation time was 30 seconds at 37°C.

Toxin Production

The production of certain toxin substances by *G. lamblia* that directly interact with the absorptive cells has been reported (1). This hypothesis has been tested in in vitro experiments, and it was found that cultured trophozoites exerted a toxic effect on fibroblast culture (49). However, no such toxin has been isolated so far, and no in vivo evidence is available for such a toxin.

Nutrient Competition

Competition for nutrients between *Giardia* and its host has been suggested as a factor in the pathogenesis of malabsorption (7,13,29,69). There is no obvious mechanism, however, by which such passive competition would result in morphological changes in the epithelium.

Bile Salt Deconjugation

Malabsorption observed in giardiasis may be caused by bacterial over-growth and bile salt deconjugation by *Giardia* trophozoites (57). However, other workers could not demonstrate bile salt deconjugation by *Giardia* trophozoites in their in vitro experiments (41,54,55).

Secondary Infections

The role of altered bacterial flora in the pathogenesis of giardiasis has been suggested (67,68). This hypothesis was sustained by observations that tetracycline sometimes cures diarrhea while the infection is still present (35). Tomkins et al made qualitative and quantitative studies of small intestinal bacteria present in *Giardia* infections and suggested that they might be contributing to the pathogenesis of disease (58,59).

Depression in Pancreatic Brush Border Enzymes

Reduced duodenal tryptic activity was observed in giardiasis patients (12,26). The precise mechanism by which it occurred is not clear.

In experimentally infected animals depression in brush border membrane enzymes has also been reported (5,6,20,22,35). In addition, a change in phospholipid content of the brush border membrane of infected animals was also noted (48). All these changes may be additive, causative factors in the malabsorption syndrome.

Immune Mediated Damage

An inflammatory response involving polymorphonuclear and mononuclear cells in the epithelium occurs in giardiasis (66,68). Such cells conceivably secrete inflammatory mediators such as kinins, which could alter epithelial structure and function. The intensity of the cellular response appears to be proportional to the intensity of the infection and/or the extent of ob-served epithelial change (66). This response is in all respects similar to that seen in other diseases of the small intestine associated with shortening of villi and elongation of crypts (e.g., gluten enteropathy). Whether this response is protective is not yet known.

Evidence obtained from the murine model of giardiasis suggests that an intact host immune response, particularly the cellular immune response, may be necessary for clinical disease (50). Increased intestinal lymphocyte counts were observed in infected animals similar to that reported in patients (25,28,31,37). These lymphocytes are mainly T cells that showed increased cytotoxic responses against parasite on the fifteenth postinfection day in infected animals (32,61). A significantly elevated cytotoxic response against enterocytes ($35.18 \pm 6.36\%$ in infected animals versus $16.80 \pm 5.3\%$ in con-trols) was observed during peak infection (tenth postinfection day) in these animals that gradually subsided (33). The transient nature of cytotoxic responses generated against enterocytes suggests that once the protective responses have appeared they somehow modulate the cytotoxic response. The reason(s) for this modulation could not be ascertained from present observations. In addition, these responses were detected in in vitro studies so the extent of the responses in vivo could not be ascertained from our study. With the availability of pure *Giardia* antigen, it would be interesting to challenge the lymphocytes in the gut mucosa with an oral antigen to see if they are as cytopathic in giardiasis as they are in graft-versus-host reactions.

Diarrhea

Although the mechanism of bacterial diarrhea is largely understood, that of parasitic diarrhea remains unknown. To date, no enterotoxin has been isolated from *G. lamblia* trophozoites (54). Tissue invasion has been dem-onstrated in humans (9,51) and experimental animals (21,24,45). It might bring out certain biochemical changes in absorptive cells and thus con-tribute to the diarrheogenic mechanism.

Normal to subtotal villous atrophy has been examined histologically in specimens obtained from patients with giardiasis (16). The atrophy was directly proportional to the severity of the diarrhea. Repeated biopsy after treatment demonstrated an improvement in the histologic picture and thus correlated with the improvement in diarrhea.

Table 21.2. Levels of cyclic AMP and prostaglandins E and F in *G. lamblia*-infected and uninfected mice.

	Postinfection days				
Group	3	6	9	12	15
Cyclic AMP (pmol/100 mg tissue)					
Control	36.6±6.7	—	38.3±6.1	—	37.0±3.3
Infected	73.3±9.8	93.24±16.7	106.6±23.3	—	40.0±10.0
p value	0.01	0.001	0.001	—	0.05
PGF (ng/100 mg tissue)					
Control	241.12±13.4	240.3±17.5	240.9±14.6	—	241.2±17.9
Infected	259.0±11.2	288.3±9.8	539.0±18.4	—	321.3±16.4
p value	—	0.01	0.001	—	0.001
PGF (ng/100 mg tissue)					
Control	33.05±1.9	—	33.1±1.4	—	32.8±2.35
Infected	39.3±4.5	53.3±2.0	58.6±2.8	119.6±5.3	84.6±2.0
p value	—	0.05	0.01	0.001	0.001

Data obtained from refs. 20 and 23.

Authors have observed a transient increase in cyclic AMP and prostaglandin E and F levels in the small intestine of *G. lamblia*-infected animals (Table 21.2) (19,20,22). The possibility that the immune system of the host might be playing some role in altering the level of these inflammatory substances has been raised (21,24,53). The cyclic nucleotide-mediated diarrhea in this infection appears to be an intriguing possibility that warrants further investigation. Additional studies are needed to estimate the level of cyclic nucleotides in the diarrheic stools of patients suffering from giardiasis alone.

A significantly higher uptake of Ca^{2+} was observed in the *G. lamblia*-infected animals (632 ± 40 pmol/mg protein) compared to the controls (557 ± 27 pmol/mg protein) (20,22). This increased intracellular calcium level is known to stimulate secretory activity (15). Calcium is also the mediator for cyclic AMP-induced secretion. This possibility is strengthened by the report of Ilundain and Naftalin that stelazine blocks chloride secretion induced by cholera enterotoxin (30). These studies suggest that when intracellular cytosolic Ca^{2+} ion concentration is increased, it binds to and activates a calcium-dependent regulatory protein, calmodulin. The activated calmodulin then stimulates electrolyte secretion by an as yet undetermined pathway. In our infected animals the activity of calmodulin in the microvillas core was found to be increased (19). The presence of serotonin (a calcium influx increasing agent) in *Entamoeba histolytic* trophozoites has been reported, and the role of this component in inducing diarrhea in amebiasis has been suggested (39). In the case of giardiasis, whether increased influx of Ca^{2+} or activation of calmodulin is by serotonin or through some other pathway needs evaluation.

References

1. Alp MH, Hislop IG: The effect of Giardia lamblia infestation on the gastrointestinal tract. *Aust Ann Med* 1969;18:232–237.
2. Ament ME, Rubin SE: Relation of giardiasis to abnormal intestinal structure and function in gastrointestinal immunodeficiency syndromes. *Gastroenterology* 1972;62:216–226.
3. Ament ME, Ocha MD, Davis BL: Structure and function of the gastrointestinal tract in primary immunodeficiency syndromes: a study of 39 patients. *Medicine (Baltimore)* 1973;52:227–248.
4. Anand BS, Chaudhary R, Jyothi A, et al: Experimental examination of the direct damaging effects of Giardia lamblia on intestinal mucosal scrappings of mice. *Trans R Soc Trop Med Hyg* 1985;79:613–617.
5. Anand BS, Kumar M, Chakravarti RN, et al: Pathogenesis of malabsorption in Giardia infection: an experimental study in rats. *Trans R Soc Trop Med Hyg* 1980;74:565–569.
6. Anand BS, Mahmood A, Ganguly NK, et al: Transport studies and enzyme assays in mice infected with human Giardia lamblia. *R Soc Trop Med Hyg* 1982;76:616–619.

7. Barleri D, DeBrito T, Hoshino S, et al: Giardiasis in childhood: absorption tests and biochemistry, histochemistry, light and electron-microscopy of jejunal mucosa. *Arch Dis Child* 1970;45:466–472.
8. Black RE, Dykes AC, Sinclair SP, Wells JG: Giardiasis in day care centres; evidence of person to person transmission. *Paediatrics* 1977;60:486–491.
9. Brandborg LL, Tamkersley CB, Gottleib S, et al: Histological demonstration of mucosal invasion by Giardia lamblia in man. *Gastroenterology* 1967;52:143–150.
10. Brown WR, Butterfield D, Savage D, Tada T: Clinical, microbiological and immunobiological studies in patients with immunoglobulin deficiencies and gastrointestinal disorders. *Gut* 1972;13:441–449.
11. Burke JA: Giardiasis in childhood. *Am J Dis Child* 1975;129:1304–1310.
12. Chawla LS, Sehgal AK, Broor SL, et al: Tryptic activity in the duodenal aspirate following a standard test meal in giardiasis. *Scand J Gastroenterol* 1975;10:445–447.
13. Cowen AE, Campbell CB: Giardiasis—a cause of vitamin B_{12} malabsorption. *Am J Dig Dis* 1973;18:384–390.
14. Devizia B, Poggi V, Vajro P, et al: Iron malabsorption in giardiasis. *J Pediatr* 1983;107:75–78.
15. Dobbins JW, Binder HJ: Pathophysiology of diarrhoea: alterations in fluid and electrolyte transport. *Clin Gastroenterol* 1981;10:605–625.
16. Duncombe VM, Bolin TD, Davis AE, et al: Histopathology in giardiasis: a correlation with diarrhoea. *Aust NZJ Med* 1978;8:392–396.
17. Erlandsen SL, Chase DG: Morphological alterations in the microvillous border of villous epithelial cells produced by intestinal microorganisms. *Am J Clin Nutr* 1974;27:1277–1286.
18. Fleck SL, Hames SE, Warhurst DC: Detection of Giardia in human jejunum by the immunoperoxidase method: specific and non-specific results. *Trans R Soc Trop Med Hyg* 1985;79:110–113.
19. Ganguly NK, Garg UC, Mahajan RC, et al: Calmodulin: role in the regulation of NaCl transport in Giardia lamblia infected mice. *Biochem Int* 1987;14:249–256.
20. Ganguly NK, Garg SK, Vasudeva V, et al: Prostaglandins E and F levels in mice infected with Giardia lamblia. *Indian J Med Res* 1984;79:755–759.
21. Ganguly NK, Mahajan RC, Kanwar SS, Walia BNS: Immunologic modulations of diarrhoegenic mechanism in Giardia lamblia infected mice. Presented at the IIIrd Asian Conference on Diarrhoeal Diseases, Bangkok, Thailand, 1985.
22. Ganguly NK, Mahajan RC, Radhakrishna V, et al: Effect of Giardia lamblia on the intestinal cyclic AMP level in mice. *J Diarrh Dis Res* 1984;2:69–72.
23. Ganguly NK, Mahajan RC, Vasudeva V, et al: Intestinal uptake of nutrients and brush border enzymes in normal and thymectomised Giardia lamblia infected mice. *Indian J Med Res* 1982;75:33–39.
24. Ganguly NK, Radhakrishna V, Bhagwat, AG, Mahajan RC: Electron microscopic studies of jejunum of mice infected with Giardia lamblia. *Indian J Med Res* 1985;81:102–110.
25. Gillon J, Thamery DAL, Ferguson A: Features of small intestinal pathology (epithelial cell kinetics, intraepithelial lymphocytes, disaccharidases) in a primary Giardia muris infection. *Gut* 1982;23:498–506.

26. Gupta RK, Mehta S: Giardiasis in children: a study of pancreatic functions. *Indian J Med Res* 1973;61:743–748.
27. Hartong WA, Gourley WK, Arvanitakis C: Giardiasis: clinical spectrum and functional structural abnormalities of the small intestinal mucosa. *Gastroenterology* 1979;77:161–169.
28. Heyworth MF, Owen RL, Seaman WE, et al: Harvesting of leukocytes from intestinal lumen in murine giardiasis and preliminary characterisation of these cells. *Dig Dis Sci* 1985;30:149–153.
29. Hoskins LC, Winawer SJ, Broitman SA, et al: Clinical giardiasis and intestinal malabsorption. *Gastroenterology* 1967;53:265–279.
30. Ilundain A, Naftalin RJ: Role of Ca^{+2} dependent regulatory protein in intestinal secretion. *Nature* 1979;279:446–447.
31. Kanwar SS, Ganguly NK, Mahajan RC, Walia BNS: Enumeration of gut lymphocytes population of Giardia lamblia infected mice. *J Diarrh Dis Res* 1984;2:243–248.
32. Kanwar SS, Ganguly NK, Walia BNS, Mahajan RC: Direct and antibody dependency cell mediated cytotoxicity against Giardia lamblia by splenic and intestinal lymphoid cells in mice. *Gut* 1986;27:73–77.
33. Kanwar SS, Ganguly NK, Walia BNS, Mahajan RC: The macrophages as an effects cell in Giardia lambia infection. *Med Micro Immun* 1986;176:83–86.
34. Knight R: Giardiasis, iso-poriasis and balantidiasis. *Clin Gastroenterol* 1978;7:31–47.
35. Leon-Barua R: The possible role of intestinal bacterial flora in the genesis of diarrhoea and malabsorption associated with parasitosis. *Gastroenterology* 1968;55:559.
36. Levinson JD, Nastro LJ: Giardiasis with total villous atrophy. *Gastroenterology* 1978;74:271–275.
37. MacDonald TT, Perfuson A: Small intestinal epithelial cell kinetics and protozoal infection in mice. *Gastroenterology* 1978;74:496–500.
38. Mahalanabis D, Simpson TW, Chakraborty ML, et al: Malabsorption of water miscible vitamin A in children with giardiasis and ascariasis. *Am J Clin Nutr* 1979;32:313–318.
39. McGowan K, Kane A, Asarkof N: Entamoeba histolytica causes intestinal secretion: role of serotonin. *Science* 1983;221:762–764.
40. Meyer EA, Jarroll EL: Giardiasis. *J Epidemiol* 1980;111:1–12.
41. Meyer EA, Radulescu S: Giardias and giardiasis. *Adv Parasitol* 1979;17:1–47.
42. Morecki R, Parker JG: Ultrastructural studies of the human Giardia lamblia and subjacent jejunal mocusa in a subject with steatorrhoea. *Gastroenterology* 1967;52:151–164.
43. Mueller JC, Jones AL, Brandborg LL: Scanning electron microscope observation in human giardiasis, in Jones O (ed): *Proceedings of the Workshop on Scanning Electron Microscopy in Pathology*. Chicago, 1977, pp 557–564.
44. Okeahialam TC: Giardiasis in protein energy malnutrition. *E Afr Med J* 1982;59:765–770.
45. Owen RL, Nemanic PC, Stevens DP: Ultrastructural observations on giardiasis in a murine model. I. Intestinal distribution, attachment and relationship to the murine system of Giardia muris. *Gastroenterology* 1979;76:757–769.

46. Petersen H: Giardiasis (lambliasis). *Scand J Gastroenterol* 1972;14:1–44.
47. Popovic O, Pendic B, Paljm A, et al: Giardiasis: local immune defense and responses. *Eur J Clin Invest* 1974;4:380–386.
48. Radhakrishna V, Ganguly NK, Mahajan RC, Majumdar S: Lipid profile of the brush border membrane of Giardia infected mice. *Indian J Med Res* 1983;78:503–508.
49. Radulescu S, Rau C, Iosif V, Meyer EA: Contribution to the study of the mechanisms of pathogenesis of Giardia infection. Presented at the Fifth International Congress on Protozoology, New York, 1977, p 125.
50. Roberts-Thomsonn IC, Mitchell FG: Prolonged infections in certain mouse strains and hypothymic (nude) mice. *Gastroenterology* 1978;75:42–46.
51. Saha TK, Ghosh TK: Invasion of small intestinal mucosa by Giardia lamblia in man. *Gastroenterology* 1977;72:402–405.
52. Schmerin MJ, Jones TC, Klein H: Giardiasis: association with homosexuality. *Ann Intern Med* 1978;88:801–803.
53. Smith PD: Pathophysiology and immunology of giardiasis. *Annu Rev Med* 1985;36:295–307.
54. Smith PD, Gillin FD, Sipra WM, Nash TE: Chronic giardiasis: Studies on drug sensitivity, toxin production and host immune response. *Gastroenterology* 1982;83:797–803.
55. Smith PD, Horburgh CR, Brown WR: In vitro studies on bile acid deconjugation and lipolysis inhibition by Giardia lamblia. *Dig Dis Sci* 1981;26:700–704.
56. Takano J, Yardley JH: Jejunal lesions in patients with giardiasis and malabsorption: an electron microscopic study. *Bull Johns Hopkins Hosp* 1964;116:413–429.
57. Tandon BN, Tandon RK, Satpathy BK, Shriniwas S: Mechanisms of malabsorption in giardiasis: a study of bacterial flora and bile salt deconjugation in upper jejunum. *Gut* 1977;18:176–181.
58. Tomkins AM, Wright SG, Drasar BS, James WPT: Colonization of jejunum by enterobacteria and malabsorption in patients with giardiasis. *Gut* 1976;17:397–402.
59. Tomkins AM, Wright SG, Drasar BS, James WPT: Bacterial colonization of jejunal mucosa in giardiasis. *Trans R Soc Trop Med Hyg* 1978;72:33–36.
60. Upadhyay P, Ganguly NK, Mahajan RC, Walia BNS: Intestinal uptake of nutrients in normal and malnourished animals infected with Giardia lamblia. *Digestion* 1985;32:243–248.
61. Upadhyay P, Ganguly NK, Walia BNS, Mahajan RC: Kinetics of lymphocytes sub-population in intestinal mucosa of protein deficient Giardia lamblia infected mice. *Gut* 1986;27:386–391.
62. Veghelyi PV: Coeliac disease initiated by giardiasis. *Am J Dis Child* 1939;57:894–899.
63. Veghelyi PV: Giardiasis. *Am J Dis Child* 1940;59:793–804.
64. Walia BNS, Ganguly NK, Mahajan RC, et al: Morbidity in preschool Giardia cyst excretors. *Trop Geogr Med* 1986;38:367–372.
65. Wolf MS: Giardiasis. *N Engl J Med* 1978;298:319.
66. Wright SG, Tomkins AM: Quantification of the lymphocytic infiltrate in jejunal epithelium in giardiasis. *Clin Exp Immunol* 1977;29:408–412.
67. Yardley JH, Takano J, Hendrix TR: Epithelial and other mucosal lesions of

the jejunum in giardiasis: jejunal biopsy studies. *Bull Johns Hopkins Hosp* 1964;115:389–406.
68. Yardley JM, Bayless TM: Giardiasis. *Gastroenterol* 1967;52:301–304.
69. Zamcheck N, Hoskins LC, Winawer SJ, et al: Histology and ultrastructure of the parasite and the intestinal mucosa in human giardiasis: effects of Atabrine therapy. *Gastroenterology* 1963;44:860–863.

CHAPTER 22

Immunology of *Entamoeba histolytica* in Human and Experimental Hosts

V.K. Vinayak

Amebiasis, an invasive enteric protozoal infection that can spread to multiple organ systems, is a major public health problem. It is caused by the cytolytic parasite *Entamoeba histolytica*. It has a worldwide distribution, one estimate indicating that 480 million people carry the parasite in their intestinal tracts. Serological investigations suggest that in approximately one-tenth of the total infected population, i.e., approximately 48 million annually, the parasite has invaded the intestinal mucosa or liver (122). The great interest of the parasitologists, microbiologists, immunologists, pathologists, clinicians, and biochemists has resulted in a voluminous literature. Despite such interest, the complicated host–parasite relation, immunological responses, and effective immunoprophylaxis have not yet been elucidated.

The parasite, *E. histolytica,* exists in two forms: (a) trophozoites, which dwell in the lumen of large intestine, multiply, and invade mucosa to produce typical amebic ulcers in cecum and colon; and (b) resistant "cysts," which are protected by a relatively rigid cell wall containing chitin (6). The pathogenic *E. histolytica* is characterized by its potential to produce disease in experimental animals (54,106,115), characteristic zymodeme (71,102) increased susceptibility to agglutinate concanavalin A (Con A) (77,98), increased phagocytic index (97), lack of negative surface charge (98), and contact-dependent cell lysis (66,76,117). It is, however, absolutely clear that pathogenic, attenuated, or nonpathogenic *E. histolytica* organisms are morphologically similar as seen by either light or transmission electron microscopy. The physicochemical environment in the gut, the level of innate resistance, the specific immunological resistance because of prior infection, the virulence of the parasite, and concomitant infections are the major contributory factors for the final outcome of the disease. The antigenic analysis of amebae is important with a view to (a) understanding host–parasite relations and their modulation by immune responses, (b) investigating immune responses elicited by the host, and (c) developing effective immunoprophylactic agents. Earlier investigations aimed at understanding modulation of the disease by immune responses of the host were basically hindered by nonavailability of pure amebic antigen. However, with the development of axenic cultures (pure growth) of *E. histolytica* (25), in-depth immunological investigations have been initiated in various laboratories.

Antigens of *E. Histolytica*

The intense immune responses, both humoral and cellular, in amebic patients indicate that the antigens of *E. histolytica* are highly immunogenic. Basically, there are two sources of antigen: cultured amebic trophozoites and amebic cysts.

Trophozoite Antigen

The *E. histolytica* trophozoite is a complex mosaic of antigens. The surface-associated antigens have been demonstrated by immobilization of trophozoites by immune sera (20,64) and surface labeling of fluorescence-labeled immune sera (10), abrogation of surface binding after absorption of immune sera with trophozoites (12), and antibody-mediated lysis of trophozoites by complement (41,58). By labeling the surface membrane of intact amebae with ^{125}I, Aust Kettis et al (8) identified 12 polypeptides on autoradiography after sodium dodecyl sulfate–polyacrylamide gel electrophoresis (SDS–PAGE) of the cell homogenate. The molecular weights of nine major antigens ranged from more than 150,000 to 9,000 daltons. Aley et al (5) isolated plasma membrane from axenic *E. histolytica* (HK 9) and obtained 12 major polypeptides with molecular weights ranging from 12 to 200 kilodaltons (kDa). These proteins were found to be glycoprotein in nature and tightly bound to the cell surface. Vinayak et al (119) also identified 12 polypeptides in isolated plasma membrane of axenic *E. histolytica* (NIH:200) with an immunodominant antigen having a molecular weight of 29 kDa. The surface-associated antigens are undoubtedly important in modulating the immune responses and in turn the disease process. Such investigations in relation to amebic infection are lacking.

The homogenate or crude extract of amebic trophozoites is complex. Krupp (47) found a basic pattern of 14 antigens by immunoelectrophoresis, whereas Chang et al (18) observed 32 precipitin bands by employing two-dimensional cross-immunoelectrophoresis. We observed that sonicated extracts of trophozoites have more than 50 antigenic determinants as demonstrated by SDS–PAGE. By molecular exclusion chromatography, three to five peaks (Fig. 22.1) have been reported (8,13,48,74,112). The hemagglutinating, complement-fixation, and precipitin activities reside mainly in the high-molecular-weight fraction (3,74,94). Ribonucleic acid (RNA) protein (84,85,116) and lysosomal fractions (6) from amebic trophozoites have been shown to be immunogenic. The excretory antigens of live amebae have not yet been isolated. However, Bos et al (14) isolated

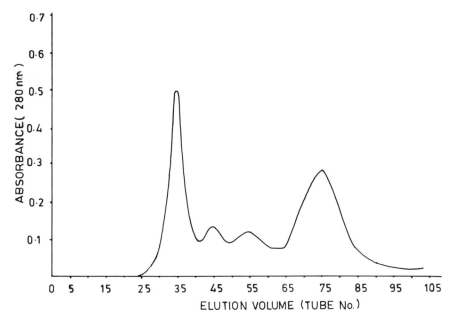

Fig. 22.1. Evolution profile of a sonacited extract of axenic *E. histolytica* (NIH:200) on a Sephadex G-200 column (2.5 × 45 cm, 0.05 Tris pH 7.4). fraction 1 (F-1) tubes 27 to 37; fraction 2 (F-2), tubes 42 to 60; fraction 3 (F-3) tubes, 65 to 84.

a cytotoxic agent and Lundblad et al (51) a glycosidase, which have been found to be antigenic in nature. It appears that such antigenic material is released upon disintegration of amebic trophozoites.

Cyst Antigens

Entamoeba histolytica cyst antigens have not yet been well characterized. The basic reason is nonavailability of pure and mature cysts of *E. histolytica* in the culture system. However, Yap et al (123) demonstrated that the cyst antigen of *E. invades* (reptile ameba) cross-reacts with *E. histolytica* antibodies.

Immune Responses

The clinical patients and the experimentally infected animals elicit strong humoral and cellular responses during the course of the disease.

Humoral Immune Responses

The humoral immune response is characterized by the appearance of antiamebic antibodies in the patients. Several types of antibody can be demonstrated by commonly used serological techniques (7,37,45,47, 55,93,100,101,103–108). The antiamebic antibodies are regularly produced in symptomatic and invasive amebic infections but with variable frequencies in asymptomatic and "noninvasive" amebic patients. The levels of antiamebic antibodies in hepatic amebic infection are relatively higher, though identical levels are also elicited in the "invasive" intestinal amebic patients. The levels of the antiamebic antibodies do not necessarily indicate the severity of infection (113). It is generally believed, however, that antiamebic antibodies appear after the invasion of gut mucosa by the parasite (9,100). The individuals may be high, moderate, low, or no responders. Therefore a mild degree of infection in a high responder would result in high antiamebic antibodies levels, whereas in low responders significantly low levels are elicited. The overall assessment of these antibodies in clinical patients indicate that the active disease seems to provoke a high antibody response, and asymptomatic or previously infected patients may produce relatively low titers. The antibodies are known to persist for years (40). The rise and fall in antibody levels, as seen in many bacterial and viral infections, does not take place in amebic patients, though seronegative conversion to seropositive results may be documented. Copro antibodies have been demonstrated by several workers (53,83,86). However, no attempt has been made to investigate the functional aspects in term of clearance of the trophozoites/cysts from the intestinal tract of the host.

In areas endemic for amebic infections, extensive studies have been carried out to determine the levels of major immunoglobulin classes in infected amebic patients (1,21,28). Though high levels of immunozlobrilin G (IgG) have been reported, it is not certain if the total rise in IgG levels is due to amebic invasion alone or to other, concomitant parasitic infections. However, at least part of the rise in IgG can be attributed specifically to amebic invasion. The total IgM and IgA levels are less affected, though relatively high levels of IgA have also been noted in amebic patients (43,60). Immediate-type hypersensitivity to *E. histolytica* antigen occurs frequently in sensitized individuals (52,72,107). Although IgE-type antibodies have been documented, a major problem arises because of the presence of helminthic infections. Helminths are known not only to induce high levels of specific and nonspecific IgE but also to potentiate the IgE response to unrelated antigen (59,99). There is a considerable debate and controversy over the role of antiamebic IgE (95) in the clearance of the parasite from the gut.

Cellular Immune Responses

The cell-mediated immune responses in patients have been studied in vivo by the delayed hypersensitivity (DTH) reaction to amebic antigen. The DTH responses have been demonstrated in previously infected or treated amebic patients (50,57). The acutely ill individuals do not show DTH responses (46,107). However, streptokinase and streptodornase produce positive DTH reactions, indicating that the immunosuppression is of a specific nature. The studies employing the migration inhibition test indicated transient cellular sensitization to amebic antigen (111), which tends to revert to normal 2 to 4 weeks after cessation of chemotherapy. Reduced blastogenic responses to phytohemagglutinin (PHA) and Con A have been reported by several workers (36,73,57). The blastogenic transformation of lymphocytes from patients with amebic liver abscess caused Segovia et al (78) to suggest that T lynmphocytes become sensitized to amebic antigen. However, in contrast to hepatic amebiasis, invasive colitis alone is not associated in a significant way with the development of a blastogenic response by lymphocytes or to migration inhibition factors to amebic antigen (111). Such observations indicate limited antigenic exposure in intestinal amebic patients. The observed lowered cell-mediated immune response was thought to be due to decreased T cell counts or to impaired function as a result of *E. histolytica* infection (29). However, other investigators did not make the same observations (50,78). The immunosuppression seen with antilymphocytic serum, corticosteroids, or irradiation favor the invasion by amebae (92,110,14) in experimental animals. Clinical reports by several workers subscribe to the fact that patients with fulminent amebic liver abscess have a history of cortisone therapy (26,89).

Immune Complex Reactions

The antibodies to colonic antigen in amebic patients indicate the development of immune complex reactions (49,69). It appears that prolonged exposure to parasite may provoke this reaction. Antibodies to human liver cells during hepatic amebiasis have been reported (27). Desimone et al (24) tested sera from patients of *E. histolytica* infection for anti-T-cell antibodies by assaying cross-reacting specificity with antigen defined by anti-Ia; they showed that activity was reduced when tested in an autologous mixed lymphocyte reaction. Amebic antigen has been detected in the circulation of amebic patients (62). In a study by Vinayak et al (118), specific immune complexes could be precipitated from the sera of patients of hepatic amebiasis by polyethylene glycol, and the specific amebic antigen could be demonstrated by sandwitch micro-ELISA. The precise role of immune complexes in pathogenic or immunomodulation need further elucidation.

Immunoprotection

The prior eradication of intestinal amebiasis seems to provide no observable protection to the subsequent exposure to amebic infection, even though a previous infection results in the development of humoral and cellular immune responses. Nevertheless, recurrence of amebic liver abscess is rare (4,44). Moreover, clearance of severe forms of amebic colitis also leads to some level of resistance (80). Consequently, on the basis of clinical evidence, it is believed that partial or complete resistance can be achieved by the host after eradication of the prior infection. The various observations in clinical patients seem to be attributed to varied immunological memory. In contrast to the unclear status of immunity acquisition in the human host, protection of various animals against *E. histolytica* has been documented unequivocally. In earlier experiments, dogs infected intracecally became refractory to reinfection (90). Immunity to reinfection was induced in 72% of hamsters by prior immunization with axenic amebic extract given along with Freund's complete adjuvant (82) or with live amebae at unusual sites (30). The whole amebic extract or its Sephadex-chromatographed fractions provide protection to cecal infection in guinea pigs (48,112,115). Protective immunity has also been achieved by exposure of the guinea pigs to repeated low doses of live amebae via the mesenteric vein or intracecally (42).

The partially purified amebic extract on Sephadex G-200 revealed three to five fractions (13,48,74) of high-molecular-weight protein (fraction I), resulting in a high degree of protection in guinea pigs (2,48,112,115), whereas fractions 2 and 3 failed to provide any significant level of protection in guinea pigs (Table 22.1). Fraction I (amebic protein F-I) also

Table 22.1. Immunoprophylactic potential of immunogens from *E. histolytica:* protection against cecal challenges with virulent isolates of *E. histolytica*.

Immunogen[a]	Immunized animals challenged (No.)	Animals with cecal lesions at sacrifice		Protection (%)
		No.	%	
Unimmunized (control)	14	13	92.8	7.2
FCA (control)	24	24	100.0	0
MMA + FCA	10	10	100.0	0
BAA + FCA	10	10	100.0	0
AAA + FCA	24	16	66.6	33.4
Chromatographed F-1 + FCA	20	1	5.0	95.0
F_2 + FCA	12	12	100.0	0
F_3 + FCA	9	7	77.7	23.3
Amebic RNA	18	1	5.5	94.5
Amebic RNA–RNase	13	12	92.3	7.7
F-1 RNA	11	1	9.9	91.1
F-1 RNA–RNase	7	7	100.0	0
Y-RNA	6	6	100.0	0

[a]FCA = Freund's complete adjuvant; MAA = monoaxenic amebic antigen; BAA = bacteria-associated antigen; AAA = axenic (NIH:200) amebic antigen; F-1, F-2, F-3 = amebic proteins of *E. histolytica* (NIH: 200); RNA = ribonucleic acid proteins; Y-RNA = yeast RNA.

provided protection to hamsters (33,65) upon intrahepatic challenge (Fig. 22.2). The RNA protein fraction obtained from amebic trophozoites stimulated amebic-specific cellular sensitization, which led to a high degree of protection (Table 22.1; Fig. 22.2) to subsequent intracecal challenge (85,116). The immunity was found to be specifically due to amebal RNA or RNA isolated from F-I amebic proteins. The immunization with yeast RNA (Y-RNA) failed to afford any degree of protection (Table 22.1). The small quantity (approximately 9%) of protein attached to the amebal RNA molecule did not participate in the protection, as treatment of RNA with pronase or trypsin did not abrogate the protection afforded by amebal RNA (115). Experimental vaccination of nonhuman primates with amebic extract, with glucan as adjuvant, provided protection against the intrahepatic challenge (2). In addition, immunity by lysosomal antigen has also been achieved in subhuman primates (79,80). Vinayak et al (119) have achieved success in providing protection to hamsters against intrahepatic challenge by immunization of animals with surface-associated amebic proteins (Fig. 22.2). Further immunization of animals with amebic protein F-I or plasma membrane antigens of *E. histolytica* entrapped in multilamellar phosphatidylcholine liposomes potentiated the immune system, which in turn resulted in protection of a high degree upon intrahepatic challenge (Fig. 22.2). These investigations indicate that immunity to amebic

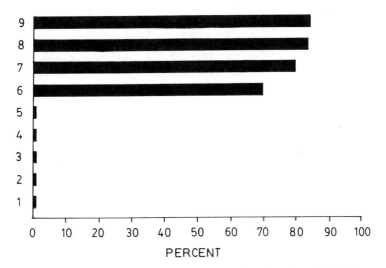

Fig. 22.2. Protection potentials of immunogens (1–9) to intrahepatic challenge with a virulent subline of axenic *E. histolytica* (NIH:200 V) in hamsters. Bars: (1) PBS - pH 7.2 buffer alone. (2) buffer + FCA. (3) empty multilamellar phosphatidylcholine liposomes. (4) intradermal TPS-1 medium. (5) intradermal 4 × 10⁶ live amebic trophozoites (NIH:200 V). (6) F-1 amebic proteins + FCA, (7) plasma membrane (PM) amebic proteins + FCA. (8) F-I entrapped in multilamellar phosphatidylcholine liposomes (MPL). (9) PM entrapped in MPL

infection can be induced with no undesirable reactions provided an appropriate antigen is employed for prophylaxis.

Mechanisms of Immunoprotection

The human host and experimental animals develop both humoral and cellular immune responses, and the precise role played by them in relation to partial or complete protection in clinical patients and experimental hosts are as follows.

Role of Anti-Amebic Antibodies

Entamoeba histolytica, like many other parasites, has been shown to active complement (70). Normal human serum inhibits growth of amebic trophozoites in culture and can kill the trophozoites in in vitro (22,56). These effects are attributable to the activation of alternate pathways of complement (41,58). In the presence of antiamebic antibodies, complement is activated by the classical pathway, resulting in the death of the parasite (41). The cytotoxicity is abrogated by adsorption of antiamebic antibodies with live trophozoites or amebic extract (81). The amebic trophozoites become immobile in the presence of heat-inactivated antisera but regain their motility after 45 to 60 minutes (20,64). Cross-linking agents, e.g., antibodies or Con A, induce redistribution of surface components into patches (63,98). The complexes are released as membrane "cups" containing amorphous sheets of soluble material (15,96). However, some part of membrane may be internalized (8). Repeated exposure to antiserum results in recapping (15). Amebae grown in the presence of antiserum lose their susceptibility to complement-mediated lysis (35). The shedding of antigen–antibody complexes may result in parasitic evasion of complement-mediated lysis.

The ineffectiveness of antiamebic antibodies as detected by serological methods does not rule out the possibility of the existence of protective antibodies, which might exist in the form of antibodies to the masked antigen or to antigens that are not expressed regularly. *E. histolytica* has a high plasma membrane (PM) turnover rate (35), and anti-PM antibodies should have some protective role. Our studies have revealed that the animals immunized with high-molecular-weight amebic protein F-I developed high anti-F-I antiamebic antibodies, and 95 to 100% protection was achieved to intracecal challenge at the same time. Immunization with crude amebic extract, amebic protein, F_2, or F_3 did not elicit high anti-F-I antibody levels (115,119). The F-I-immunized hamsters also had significantly higher anti-F-I antibodies at the time of challenge as well as at the time of sacrifice, which correlated well with the level of protection to intrahepatic challenge (Fig. 22.3). The most significant were the anti-PM antiamebic antibody levels, which appeared to provide protection against

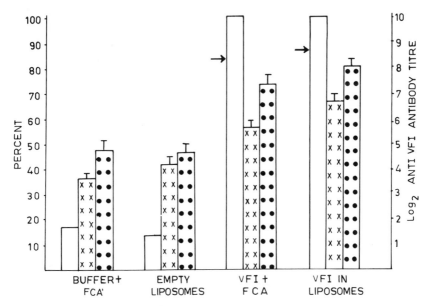

Fig. 22.3. Immune status at sacrifice of animals immunzied with F-1 amebic proteins (VF I) obtained from a virulent subline of *E. histolytica* (NIH:200 V) followed by challenge with virulent amebic trophozoites. ⬜ No amebic liver abscess. ⊠ Leukocyte migration inhibition (LMI) indices with specific amebic antigen, ⚫⚫ Log of anti-VF 1 anti-amebic indirect hemagglutination antibodies. Arrow = actual level of protection. I indicates the mean ± SD.

the establishment of amebae in the liver (Fig. 22.4). There appears to be a critical level of anti-PM antibodies that is responsible for prevention of the establishment of amebae in the liver. The buffer-immunized or empty-liposomes-immunized animals developed anti-PM antibodies at sacrifice, but the levels of such antigens were so low as to be effective in restricting the size of lesions (119). It appears worthwhile to investigate the levels of anti-PM antibodies in clinically acute cases of amebiasis.

Cell-Mediated Responses and Protective Immunity

Cell-mediated immunity plays a significant role in recovery from amebic infection, especially in patients with an amebic liver abscess (57,72,73,87,111). Asymptomatic undiagnosed amebiasis is dramatically exacerbated during corticosteroid treatment (26). Treatment of experimental animals led to the development of severe amebic pathology (110), and 20% of these treated animals developed hepatic lesions (114). Passive transfer of immune spleen cells has been shown to provide protection to hamsters (32). The treatment of spleen cells with anti-T-cell serum abol-

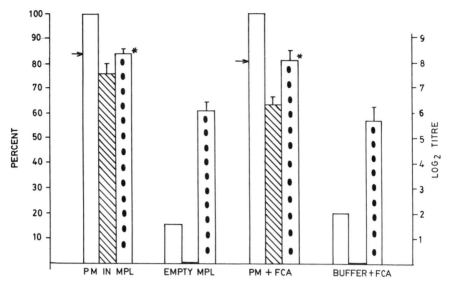

Fig. 22.4. Assessment of anti-PM antiamebic indirect glutinating antibodies and the degree of protection in immunized animals. ☐ No amebic liver abscess. ◸ Antibodies at challenge. ⊡ Antibodies at sacrifice. Arrow = actual protection. *$p<0.0001$.

ished the protective immunity by transferred spleen cells. Splenectomy of hamsters led to a significant increase in the mean weight of metastatic foci and liver abscesses (31). Our own observations have indicated that animals protected by amebic protein F-I have significantly high levels of cellular sensitization (Fig. 22.3). The level of cellular sensitization by amebic plasma membrane proteins is significantly correlated (<0.05) with the development of protection in hamsters (Fig. 22.5). The RNA stimulates the cellular system with no development of antiamebic RNA antibodies (85). The stimulation of cell-mediated immune responses by amebal RNA or RNA extracted from amebic protein F-I (Fig. 22.6) provided protection to intracecal amebic challenges in guinea pigs (116). In one of the investigations by Stern et al (88) it was reported that Nu^+/Nu^+ mice or mice treated with antithymocytes did not develop hepatic abscess. However, treatment of nu/nu or Nu^+ mice with silica (a selective macrophage toxin) increased the susceptibility of mice to hepatic amebic infection. Similarly, Ghadirian et al (34) observed that treatment of animals with antimacrophage serum resulted in a significant reduction in the number of metastatic amebic lesions. However, transfer of peritoneal cells from vaccine-protected hamsters could transfer the immunity to clean animals upon challenge. These investigations therefore highlighted the fact that cellular immune responses indeed have a significant role in affording protection at least against hepatic amebic infection.

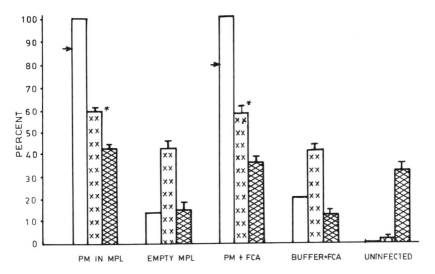

Fig. 22.5. Correlation of cellular sensitization and the level of protection in immunized hamsters. ☐ No amebic liver abscess, ☒ Percent LMI with PM amebic proteins. ▨ LMI with PHA. Arrow = actual protection. *$p<0.001$.

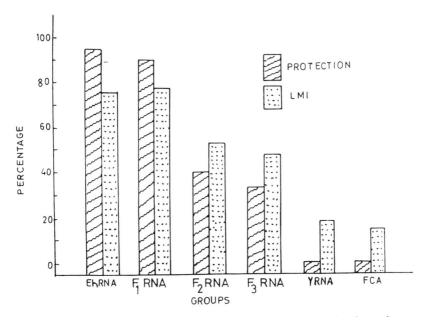

Fig. 22.6. Cellular sensitization in amebal RNA-immunized guinea pigs.

Role of Cell- and Antibody-Dependent Cellular Cytotoxicity

Investigations have revealed the involvement of cells of the immune system to effectively tackle the amebic trophozoites. The human polymorpho-nuclear cell (PMN) kills *E. histolytica* trophozoites of low virulence in vitro in a process independent of complement and specific antiamebic antibodies (38). The trophozoites of virulent HMI isolate of *E. histolytica*, however, has the potential to kill human neutrophils, and neutrophils are able to kill the low-virulence/attenuated trophozoites of *E. histolytica* isolates NIH:200 and NIH:203 (67). The lysis of neutrophils by virulent parasites appears to be an important event in the pathogenesis of the disease. The neutrophilic response constitutes an initial host immune response to invasion by the parasite; and with the cytolysis of neutrophils, hepatocyte destruction has been observed (16). Salata and Ravdin (68) corroborated these in vivo findings by showing destruction of Chang liver monolayers by the trophozoites, which was significantly enhanced in the presence of human neutrophils. The lymphocytes from infected patients or from vaccine-protected animals are also cytotoxic to amebic trophozoites (19,39,119). Highest cytotoxic potentialities of lymphocytes from amebic F-I protein-immunized animals was observed by us (116). The prior stimulation of effector cells with crude amebic extract as well as F-I amebic proteins had an enhancing effect (Fig. 22.7) on the cytotoxicity of effector cells against amebic trophozoites (75). The peritoneal exudate and mesenteric lymph node cells obtained from crude amebic extract of vaccinated gerbil killed trophozoites of *E. histolytica* in vitro (17).

We have observed that not only did the effector cells primed by amebic

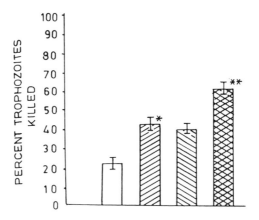

Fig. 22.7. Cytotoxicity of in vivo primed splenic mononuclear lymphocytes (MNC) and peritoneal macrophages toward amebic trophozoites (NIH:200). ▨ Crude amebic extract (CAE) primed MNCs. ☐ F-I amebic protein primed MNCs. ◩ CAE primed macrophages. ▩ F-1 primed macrophages. */**p<0.001. I indicates mean ± SD.

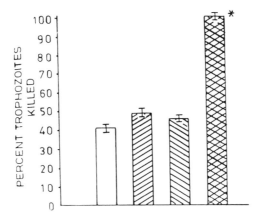

Fig. 22.8. Cytotoxicity of unprimed effector cells in the presence of antiamebic antibodies to amebic trophozoites (NIH:200). ☐ Unprimed MNCs + anti-CAE antibodies. ▨ Unprimed MNC + anti-F-I antibodies. ⊠ Unprimed macrophages + anti-CAE antibodies. ⊠ Unprimed macrophages + anti F-1 antibodies. *$p<0.001$. I indicates mean ± SD.

antigen in vivo kill a significant number of trophozoites, but antiamebic antibodies (i.e., antibodies to crude amebic extract) and anti-F-I antiamebic antibodies also possessed the capacity to enhance cytotoxicity of unprimed effector cells against *E. histolytica* (Fig. 22.8). In addition, the anti F-I antibodies could enhance the cytotoxic potential of in vivo F-I-stimulated effector cells (Fig. 22.9), resulting in cytolysis of almost all trophozoites

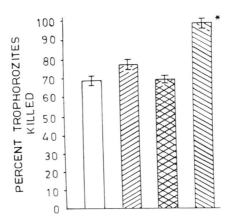

Fig. 22.9. Antibody-dependent cytotoxicity of primed effector cells to amebic trophozoites (NIH:200). ☐ CAE primed MNC + anti-CAE antibodies. F-1 primed MNC + anti-F-1 antibodies. ⊠ CAE primed macrophages + anti-CAE antibodies. ▨ F-1 primed macrophages + F-1 antibodies. I indicates mean ± SD.

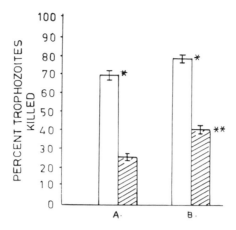

Fig. 22.10. Antibody-mediated cytotoxicity of primed effector cells to trophozoites of attenuated (NIH:200) and virulent (NIH:200 V) sublines of *E. histolytica.* ☐ Subline NIH:200. ▨ Subline NIH:200 V. **A:** CAE primed MNCs + anti-CAE antibodies. **B:** F-1 primed MNCs + anti-F-1 antibodies. */** *p*<0.001. I indicates mean ± SD.

of *E. histolytica* within 3 hours (75,76). It is of interest to note that the cellular cytotoxicity and antibody-dependent cellular cytotoxicity (ADCC) that was effective against attenuated isolates of *E. histolytica* (NIH:200) was significantly (<0.001) less effective against trophozoites of a virulent subline (Fig. 22.10) isolated from *E. histolytica* (NIH:200 V). The monocyte depletion did not alter (Table 22.2) the cytotoxic potential of splenic

Table 22.2. Effect of monocyte depletion and nylon wool fractionation on the cytotoxic potentialities of MNCs against *E. histolytica*.

Effector cells	Anti-amebic serum		
	NGPS	Anti-CAE	Anti-F-1
Unstimulated MNCs	14.33 ± 1.21	41.5 ± 2.07	48.35 ± 2.56
CAE-stimulated MNCs	22.91 ± 3.2	74.33 ± 1.35	ND
CAE-stimulated monocyte-depleted MNCS	25.08 ± 3.35	69.79 ± 2.5	ND
CAE-stimulated NWF lymphocytes	23.5 ± 2.5	73.0 ± 3.0	ND
F-1-stimluated MNCs	43.80 ± 3.5	ND	75.16 ± 4.0
F-1-stimulated monocyte-depleted MNCs	47.16 ± 2.5	ND	78.0 ± 3.0
F-1-stimulated NWF lymphocytes	42.0 ± 3.2	ND	78.0 ± 1.25

Target cell = *E. histolytica* (NIH:200);
NWF = nylon wool fractionated; ND = not done.

mononuclear cells (MNCs). The removal of B lymphocytes from spleen MNC preparations by nylon wool columns also did not alter the cytotoxic activity of splenic MNCs. Because nylon-wool-fractionated lymphocytes retained the cytotoxic potentials (Table 22.2), T cells or null cells appear to play a central role in cytotoxicity against *E. histolytica*. It is unlikely that lymphocytes have natural receptors for *E. histolytica* trophozoites. Therefore the possibility of NK cell activity may be ruled out. Probably a small number of trophozoites are killed nonspecifically by lymphocytes. We have further indicated (75) that Tc and K cells are the main T cell subsets contributing significantly to the induction of cytotoxicity to amebic trophozoites. The antiamebic antibodies to amebic crude extract or F_1 amebic proteins seem to be behaving in an opsonic fashion to induce antibody-dependent cell-mediated cytotoxicity. It is generally agreed that the ADCC is primarily mediated by the IgG class of immunoglobulins (61). However, there are reports implicating IgM as the inducer of ADCC under certain conditions (11,23,121). Secretory IgA has also been shown to mediate antibacterial ADCC (91). We have observed that 2-mercaptoethanol (2-ME) treatment of antiamebic antibodies did not alter the cytotoxicity-inducing capacity (Table 22.3). It was further confirmed by use of purified antiamebic IgM (75). The antiamebic IgM was ineffective in inducing ADCC by lymphocytes or macrophages (Tables 22.3 and 22.4), thus highlighting the nonparticipation of IgM in the ADCC against amebic trophozoites. On the other hand, purified IgG molecules of antiamebic antibodies induce cytotoxicity by splenic lymphocytes and peritoneal macrophages (Tables 22.3 and 22.4). The Fc region of the IgG molecule is required for attachment of effector cells to the antibody-molecule-coated amebic rophozoites, as pretreatment of antiamebic IgG with staphylococcal protein A (which binds specifically to Fc) abolished ADCC by the effector cells (Fig. 22.11).

Although the evidence so generated has provided new insight into the

Table 22.3. Cytotoxicity-inducing capacity of isolated immunoglobulin in association with peritoneal exudate macrophages.

Immunoglobulin	% Trophozoites killed
Anti-CAE	46.33 ± 1.10
Anti-CAE + 2-ME	43.99 ± 1.46
Anti-CAE/IgG	44.77 ± 1.27
Anti-CAE/IgM	13.74 ± 0.54
Anti-F-I	99.08 ± 1.82
Anti-F-I + 2-ME	95.96 ± 2.50
Anti-F-I/IgG	99.62 ± 0.76
Anti-F-I/IgM	24.52 ± 1.55

Target cell = *E. histolytica* (NIH:200).
2-ME = 2-Mercaptoethanol.

Table 22.4. Cytotoxicity-inducing capacity of isolated immunoglobulins in association with nylon-wool-fractionated lymphocytes.

Immunoglobulin	% Trophozoites killed
Anti-CAE	42.94 ± 1.95
Anti-CAE + 2-ME	41.07 ± 5.19
Anti-CAE/IgG	37.94 ± 2.95
Anti-CAE/IgM	11.75 ± 4.92
Anti-F-I	48.35 ± 1.32
Anti-F-I + 2-ME	47.89 ± 1.61
Anti-F-I/IgG	49.24 ± 2.51
Anti-F-I/IgM	13.04 ± 0.85

Target cell = *E. histolytica* (NIH:200).
2-ME = 2-Mercaptoethanol.

immune defense against the amebic parasite, direct evidence for the in vivo role of ADCC is unclear. However, we have now documented that the establishment of amebic infection in guinea pigs is accompanied by suppression of leukocyte migration inhibition (LMI) responses to phytohaemagglutinen (PHA). At day 7 after infection (acute phase of the infection), the response to PHA was lowest. On the other hand, the LMI response to F-I amebic proteins as well as antiamebic anti-F-I antibodies increased gradually with establishment of the amebic infectin (Fig. 22.12). The suppressive ability of effector immune cells to kill the parasite in vitro during the establishment phase paralleled the lower response to PHA (119).

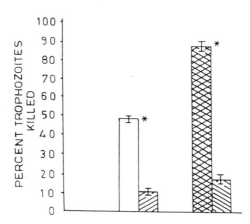

Fig. 22.11. Effect of protein A treatment on the cytotoxicity-inducing potentials of anti-F-I antibodies (IgG) by the effector cells ☐ MNCs + anti-F-1/IgG. ▨ MNCs + protein A-treated anti-F-1/IgG. ▩ Macrophages + anti-F-1/IgG. ◈ Macrophages + protein A-treated anti-F-1/IgG. *$p < 0.001$. I indicates mean ± SD.

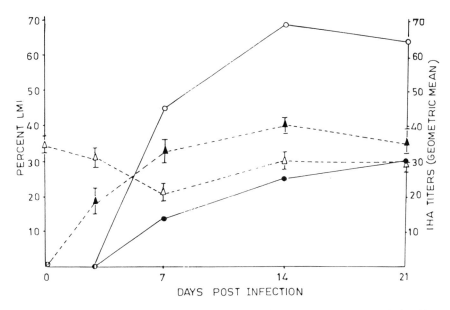

Fig. 22.12. Correlations of specific and non-specific immune responses during the course of cecal amebic infections in guinea pigs. (○—○) anti-Cae antibodies. (•—•) F-1 antibodies. (△—△) LMI with PHA. (▲—▲) LMI with F-1 amebic proteins.

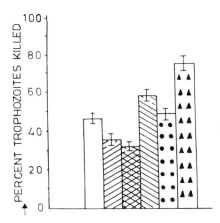

Fig. 22.13. Anti-F-1 anti-amebic antibody-dependent lymphocyte-mediated cytotoxicity to amebic trophozoites during the course of amebic infection in guinea pigs. ☐ Uninfected, ⊘ Day 3 post inoculation (PI). ⊠ Day 7 PI. ⧄ Day 14 PI. •• Day 21 PI. ▲▲ F-1 primed guinea pigs.

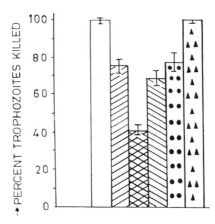

Fig. 22.14. Anti-F-1 antibody-dependent macrophage-mediated cytotoxicity to amebic proteins during the course of amebic infection in guinea pigs. ☐ Uninfected. ▨ Day 3 PI. ▧ Day 7 PI. ◺ Day 14 PI. ⚬• Day 21 PI. ▲▲ F-1 primed macrophages.

The cellular cytotoxicity and ADCC also significantly declined during the first 7 days of infection and increased thereafter, which correlated with the clearing of amebic infection (Figs. 22.13 and 22.14). Our data indicate that the rising anti-F-I antiamebic antibodies and restoration of the cytotoxic response toward amebic trophozoites causes the amebic infection to recede and eventually clear from the host (120). Nevertheless, it would be wrong to assume that the ADCC is the only effector mechanism involved in the protection against *E. histolytica*. In vivo defenses are preceded by a complex network of effector mechanisms that involve various cell types as well as various serum factors. The complexity is even greater when one considers the existence of numerous escape mechanisms directed to the outer membrane of the parasite or its vicinity or to modulation of the lymphoid cell functions.

Acknowledgments. The financial assistance to carry out the part of investigations by the Indian Council of Medical Research, Council of Scientific and Industrial Research, and the Department of Science and Technology (Government of India), New Delhi, is gratefully acknowledged. The scientific assistance provided by Drs. Sawhney, Sharma, Saxena, and Purnima is also gratefully acknowledged.

References

1. Abioye AA, Lewis EA, McFarlane H: Clinical evaluation of serum immunoglobulins in amoebiasis. *Immunology* 1972;23:937–946.

2. Ahmed A, Hag A, Sharma A: Immunology of amoebiasis, in Subrahmanyam D, Radhakrishan V (eds): Bombay, Vakil and Sons, 1983, pp 13–18.
3. Alam M, Ahmed S: Immunogenicity of E. histolytica antigen fraction. *Trans R Soc Trop Med Hyg* 1974;68:370–373.
4. Albach RA, Booden T: In Krier JP (ed): *Amoebae in Parasitic Protozoa* New York, Academic Press, Vol 2. 1978, pp 455–504.
5. Aley SB, Scott WA, Cohn ZA: Plasma membrane of E histolytica. *J Exp Med* 1980;152:391–404.
6. Arroyo-Begovich A, Carabez TA, Ruiz Herrera J: Composicion de la pared cellular di la pared cellular di quistes de Entamoeba invadens. *Arch Invest Med (Mex)* 1978;9(suppl 1):99.
7. Atchley FO, Auernheimer AH, Wasley MA: Precipitate pattern in agar gel with sera from human amoebiasis and E. histolytica antigen. *J Parasitol* 1963;49:313–315.
8. Aust Kettis A, Thortstensson R, Utter G: Antigenicity of E. histolytica strain NIH:200: a survey of clinically relevant antigenic components. *Am J Trop Med Hyg* 1983;32:512–522.
9. Balamuth W, Siddiqui WA: Amoebas and other intestinal protozoa, in Jackson GJ, Herman R, Singer I (eds): *Immunity to Parasite Animals,* Vol 2. New York, Appleton-Century-Crofts, 1978, 439.
10. Biagi FF, Beltran HF, Ortega PA: Remobilization of E. histolytica after exposure to immobilizing antibodies. *Exp Parasitol* 1966;18:87–91.
11. Blair PB, Lane MA, Mar P: Antibody in the sera of tumour bearing mice that mediates spleen cell cytotoxicity towards the autologous tumour. *J Immunol* 1976;116:606–609.
12. Boonpucknavig S, Lynraven CS, Nairn RC: Subcellular localisation of E. histolytica antigen. *Nature* 1967;216:1232–1233.
13. Bos HJ: Fractionation and serological characterization of E. histolytica antigen. *Acta Leiden* 1977;45:105–116.
14. Bos HJ, Leijendekker WJ, Van Den Eizk AA: Entamoeba histolytica cytopathogenicity inducing serum effects on contact dependent and toxin induced lysis of hamster kidney cell monolayers. *Exp Parasitol* 1980;50:343–348.
15. Calderion J, de Lourdes-Munoz M, Acosta HM: Surface redistribution and release of antibody induced caps in Entamoebae. *J Exp Med* 1980;151:184–193.
16. Chadee K, Meerovitch E: The pathogenesis of experimentally induced amoebic liver abscess in the gerbil (Meriones unguiculatus). *Am J Pathol* 1984;117:71–80.
17. Chadee K, Meerovitch E, Moreau F: In vitro and in vivo interaction between trophozoites of E. histolytica and gerbil lymphoid cells. *Infect Immun* 1985;49:828–832.
18. Chang SM, Lin CM, Dusanig DG, Cross JH: Antigenic analysis of two axenized strains of E. histolytica by two dimensional immunoelectrophoresis. *Am J Trop Med Hyg* 1979;28:845–853.
19. Chugh S, Saxena A, Vinayak VK: Interactions between trophozoites of E. histolica and the cells of immune systems. *Aust J Exp Biol Med Sci* 1985;63:1–8.
20. Cole BA, Kent JF: Immunobilization of E. histolytica in vitro by antiserum produced in rabbit. *Proc Soc Exp Biol Med* 1953;83:811–814.

21. Dasgupta A: Immunoglobulin in health and disease. III. Immunoglobulins in the sera of patients with amoebiasis. *Clin Exp Immunol* 1974;16:163–167.

22. De la Torre M, Oritz-Oritz L, De Hoz R, Sepulveda B: Accion de suoro humane immune y de la gammaglobulina antiamoebiana cultivos de E. histolytica. *Arch Invest Med (Mex)* 1973;4(suppl 1):567–570.

23. Dennert G, Lennox ES: Phagocytic cells as effectors in a cell mediated immunity system. *J Immunol* 1973;111:1844–1854.

24. Desimone C, Colli A, Zanzoglu S, et al: Anti Ia reactivity in the sera from subjects with E. histolytica infection. *Trans R Soc Trop Med Hyg* 1984;78:64–68.

25. Diamond LS: Techniques of axenic cultivation of E. histolytica, Schauddin 1903 and E. histolytica like amoebae. *J Parasitol* 1968;54:1047–1056.

26. El-Hennawy M, Abd Raddo H: Hazards of cortisone therapy in hepatic amoebiasis. *J Trop Med Hyg* 1978;81:71–73.

27. Faubert GM, Meerovitch E, McLaughlin J: The presence of liver autoantibodies induced by E. histolytica in the sera from naturally infected humans and immunized rabbits. *Am J Trop Med Hyg* 1978;27:892–896.

28. Ganguly NK, Mahajan RC, Datta DV, et al: Immunoglobulin and complement levels in cases of invasive amoebiasis. *Indian J Med Res* 1978;67:221–226.

29. Ganguly NK, Mahajan RC, Gill NJ, Koshy A: Kinetics of lymphocytes subpopulations and their functions in cases of amoebic liver abscess. *Trans R Soc Trop Med Hyg* 1981;75:807–810.

30. Ghadirian E, Meerovitch E: Vaccination against hepatic amoebiasis in hamsters. *J Parasitol* 1978;64:742–743.

31. Ghadirian E, Meerovitch E: Effect of splenectomy on the size of amoebic liver abscess and metastatic foci in hamsters. *Infect Immun* 1981;31:571–573.

32. Ghadirian E, Meerovitch E: Passive transfer of immunity against hepatic amoebiasis in the hamster by cells. *Parasitol Immunol* 1983;5:369–376.

33. Ghadirian E, Meerovitch E, Hartmann DP: Protection agains amoebic liver abscess in hamsters by means of immunization with amoebic antigen and some of its factors. *Am J Trop Med Hyg* 1980;29:779–784.

34. Ghadirian E, Meerovitch E, Kongshvan PAL: Role of macrophages in host defence against hepatic amoebiasis in hamsters. *Infect Immun* 1983;42:1017–1019.

35. Gitler C, Mogyoros M, Calef E, Rosenberg I: Lethal recognition between E. histolytica and host tissues. *Trans R Soc Med Hyg* 1985;79:581–586.

36. Gold D, Horman LG, Maddison SE, Kagan IG: Immunological studies in hamsters infected with E. histolytica. *J Parasitol* 1978;64:866–873.

37. Goldman M: Evaluation of fluorescent antibody test for amoebiasis using two widely differing amoebae strains as antigen. *Am J Trop Med Hyg* 1966;15:694–700.

38. Guerrant RL, Brush J, Ravdin JI, et al: Interactions between E. histolytica and human polymorphonuclear neutrophils. *J Infect Dis* 1981;143:83–93.

39. Guerrero M, Rois B, Landa L: Interaction between trophozoites of E. histolytica and lymphocytes of patients with invasive amoebiasis, in *Proceedings International Congress on Amoebiasis, Mexico,* 1976, pp 529–539.

40. Healy GR, Visvesvara GR, Kagan IG: Observations on the persistence of antibodies to E. histolytica. *Arch Invest Med (Mex)* 1974;5(suppl 2):495–500.

41. Huldt G, Davis P, Allison A, Schoremmer HU: Interactions between E. histolytica and complement. *Nature* 1978;277:214–216.

42. Jain P, Sawhney S, Vinayak VK: Experimental amoebic infection in guinea pigs immunized with low grade infection. *Trop R Soc Med Hyg* 1980;74:345–350.
43. Kane GJ, Matossian R, Batty I: Fluorochrome labelled antiimmunoglobulin fractions used with stabilized antigen preparations for the assessment of parasitic diseases. *Ann NY Acad Sci* 1971;177:134–145.
44. Kapoor OP: *Amoebic Liver Abscess*. Bombay, S. S. Publication, 1979.
45. Kessel JF, Lewis WP, Pasquel CM, Turer JA: Indirect-haemagglutination and complement fixation tests in amoebiasis. *Am J Trop Med Hyg* 1965;14:540–550.
46. Kretschmer RR, Sepulveda B, Almazan A, Gamboa F: Intradermal reactions to an antigen (histolyticin) obtained from axenically cultivated E histolytica. *Trop Geogr Med* 1972;24:275–281.
47. Krupp IM: Immunoelectrophoretic analysis of several strains of E. histolytica. *Am J Trop Med Hyg* 1966;15:849–854.
48. Krupp IM: Protective immunity to amoebic infection demonstrated in guinea pigs. *Am J Trop Med Hyg* 1974;23:355–360.
49. Lagercrantz R, Hammarstom S, Perlman S, Gutafson BE: Immunological studies in ulcerative colitis. III. Incidence of antibodies to colon antigen in ulcerative colitis and other gastrointestinal diseases. *Clin Exp Immunol* 1966;1:263–276.
50. Landa L, Capin R, Guerrero M: Studies on cellular immunity in invasive amoebiasis, in Sepulveda B, Diamond LS (eds): *Proceedings of International Conference on Amoebiasis*. Mexico City, Institute Mexico del Seguro Social, 1976, pp 661–667.
51. Lundblad G, Huldt G, Elander M, et al: N-Acetyl-glycosaminidase from E. histolytica. *Comp Biochem Physiol [B]* 1981;68:71–76.
52. Maddison SE, Kagan IG, Elsdon-Dew R: Comparison of intradermal and serologic tests in amoebiasis. *Am J Trop Med Hyg* 1968;17:540–547.
53. Mahajan RC, Agarwal SC, Chhuttani PN, Chitkara NL: Coproantibodies in intestinal amoebiasis. *Indian J Med Res* 1972;60:547–550.
54. Neal RA, Robinson GR, Lewis WP, Kessel JF: Comparison of clinical observations on patients infected with E. histolytica with serological titres and the virulence of amoebae to rats. *Trans R Soc Trop Med Hyg* 1968;48:69–75.
55. Nilsson LA, Petchclai B, Elwing H: Application of thin layer immune assay (TIA) for demonstration of antibodies against E. histolytica. *Am J Trop Med Hyg* 1980;29:524–529.
56. Oritz-Oritz L, Sepulveda B, Chevez A: Nuevos estiduo acera de la accion de sueros humanos normales e immues sober el trofozoite de E. histolytica. *Arch Invest Med (Mex)* 1974;5(suppl 2):257–342.
57. Oritz-Oritz L, Zamacona G, Sepulveda B, Capin NR: Cell mediated immunity in patients with amoebic abscess of liver. *Clin Immunol Immunopathol* 1975;4:127–134.
58. Oritz-Oritz L, Capin R, Capin NR, et al: Activation of the alternative pathways of complement by E. histolytica. *Clin Exp Immunol* 1978;34:10–18.
59. Orr TS, Blair AMJN: Potentiated reagin response to egg albumin and con-albumin in Nippostrongylus brasiliensis infected rats. *Life Sci* 1969;8 (part 2):1073–1077.
60. Osisanya JOS, Warhurst DC: Specific antiamoebic immunoglobulins and the

cellulose acetate precipitin test in E. histolytica infection. *Trans R Soc Trop Med Hyg* 1980;74:605–608.

61. Perlmann P, Perlmann H, Biberfeld P: Specific cytotoxic lymphocytes produced by preincubation with antibody complexed target cells. *J Immunol* 1972;108:558–561.

62. Pillai S, Mohimen A: A solid phase sandwich radio immunoassay for E. histolytica proteins and the detection of circulating antigen in amoebiasis. *Gastroenterology* 1982;83:1210–1216.

63. Pinto De Silva P, Martinez-Palomo A, Gonzales-Robles A: Membrane structure and surface coat of E. histolytica: Topochemistry and dynamics of cell surface, cap formation and microexudate. *J Cell Biol* 1975;64:538–550.

64. Prakash O, Tandon BN, Bhalla I, et al: Indirect haemagglutination and amoeba immobilization tests and their evaluation in intestinal extraintestinal amoebiasis. *Am J Trop Med Hyg* 1969;18:670–674.

65. Purnima Chugh S, Nain CK, Vinayak VK: Induction of protective immunity to experimental hepatic amoebic infection in hamsters. *J Hyg Epidemiol Microbiol Immunol* 1987;31:99–105.

66. Ravdin JI, Croft BY, Guerrant RL: Cytopathogenic mechanisms of E. histolytica. *J Exp Med* 1980;152:377–390.

67. Ravdin JI, Murphy CF, Salata RA, et al: The N-acetyl D-galactosamine inhibition lectin of E. histolytica. I. Partial purification and relationships to amoebic in vivo virulence. *J Infect Dis* 1985;141:816–822.

68. Salata RA, Ravdin JI: Lysis of human neutrophils by E. histolytica trophozoites enhanced cytopathogenicity for liver cells. *J Infect Dis* 1986;154:19–26.

69. Salem E, Zaki SA, Moneim WA, et al: Autoantibodies in amoebic colitis. *J Egypt Med Assoc* 1973;56:113–118.

70. Santoro F, Bernal J, Capron A: Complement activation by parasites. *Acta Trop (Basel)* 1979;36:5.

71. Sargeaunt PG, Williams JE: The differentiation of invasive and non-invasive E. histolytica by isoenzyme electrophoresis. *Trans R Soc Trop Med Hyg* 1978;72:519–512.

72. Savant T, Bunnag D, Chogsuchagaissiddhi T, Viriyanond P: Skin test for amoebiasis—an appraisal. *Am J Trop Med Hyg* 1973;22:168–173.

73. Savant T, Viriyanond I, Nimitnogkol N: Blast transformation of lymphocytes in amoebiasis. *Am J Trop Med Hyg* 1973;22:705–710.

74. Sawhney S, Chakravarti RN, Jain P, Vinayak VK: Immunogenicity of axenic E. histolytica and its fractions. *Trans R Soc Trop Med Hyg* 1980;74:26–29.

75. Saxena A, Chugh S, Vinayak VK: Elucidation of cellular population and nature of antiamoebic antibodies in cytotoxicity to E. histolytica (NIH:200). *J Parasitol* 1986;72:434–438.

76. Saxena A, Chugh S, Vinayak VK: Antibody dependent macrophage mediated cytotoxicity against E. histolytica. *J Med Microbiol* 1986;27:17–21.

77. Saxena S, Kaul D, Vinayak VK: Amoebic erythrophagocytosis: significance of membrane cholesterol to phospholipid ratio. *IRCS Med Sci* 1986;14:330–331.

78. Segovia E, Capin R, Landa L: Transformation blastoid de linefocitos estimalodes con antigeno lisosomal en pacients amoebiasis intestinal. *Arch Invest Med (Mex)* 1980;11(suppl 1):225.

79. Sepulveda B: Induccion de immunidad anti amibiasica en primates subhumanos con antigeno lisosomal de E. histolytica. *Arch Invest Med (Mex)* 1980;2(suppl 1):S245–S246.

80. Sepulveda B, Martinez-Palomo A: Immunology of amoebiasis by E. histolytica, in Cohen S and Warren KS (eds): *Immunology of Parasitic Infections.* London, Blackwell, 1982.

81. Sepulveda B, Oritz-Oritz L, Chevaz A, Segura M: Comprobacion de la naturileza immunologica del effecto del suero Y de la gammaglobuline immunes sobre el trofozoite de E. histolytica. *Arch Invest Med (Mex)* 1974;5(suppl 1):343–346.

82. Sepulveda B, Tanimoto-Weki M, Vazquez-Saavedra JA, Landa L: Induccion de immunidad antiamoebiana en al hamster con antigenio obtaindo de cultivos a xenicos de E. histolytica. *Arch Invest Med (Mex)* 1971;2(suppl 1):289–294.

83. Shaalan M, Baker RP: Detection of copro-antibodies in amoebiasis of colon: a preliminary study. *Am J Clin Pathol* 1970;54:615–617.

84. Sharma GL, Naik SR, Vinayak VK: Cell mediated resistance induced by axenic amoebal-RNA protein fraction. *J Hyg Epidemiol Microbiol Immunol* 1984;28:471–480.

85. Sharma GL, Naik SR, Vinayak VK: Immunogenicity of RNA protein fraction isolated from axenic E. histolytica (NIH:200). *Aust J Exp Biol Med Sci* 1984;62:117–125.

86. Sharma P, Das P, Dutta GP: Use of glutaraldehyde treated sheep erythrocytes in indirect haemagglutination test for amoebic coproantibodies. *Indian J Med Res* 1981;74:215–218.

87. Simjee AE, Gathiram V, Coovadia HM, et al: Cell mediated immunity in hepatic amoebiasis. *Trans R Soc Trop Med Hyg* 1985;79:165–168.

88. Stern JJ, Graybill JR, Drutz DJ: Murine amoebiasis: the role of macrophages in host defence. *Amer J Trop Med Hyg* 1984;33:372–380.

89. Stuiver PC, Goud TJLM: Corticosteroid and liver amoebiasis. *Bull Med J* 1978;2:394–395.

90. Swartzwelder JC, Avant WH: Immunity to amoebic infection in dogs. *Amer J Trop Med Hyg* 1952;1:567–571.

91. Tagliabue A, Nencioni L, Villa L, et al: Antibody dependent cell mediated anti-bacterial activity of intestinal lymphocytes and secretory IgA. *Nature* 1983;306:184–186.

92. Tanimoto-Weki M, Calderon P, De la Hoz R, Aguurre Garcia J: Inoculation de trofozoites de E. histolytica in hamsters Eajo la accian di drosas immunosprresorus. *Arch Invest Med (Mex)* 1974;5(suppl 2):441–446.

93. Thomas V, Sinniah B, Lang YB: Assessment of the sensitivity, specificity and reproducibility of the indirect fluorescent antibody techniques in the diagnosis of amoebiasis. *Am J Trop Med Hyg* 1981;30:57–62.

94. Thompson PE, Gradel SK, Schneider CR et al: Preparation and evaluation of standardised amoeba antigen from axenic culture of E. histolytica. *Bull WHO* 1968;30:349–365.

95. Trissl D: Immunology of E. histolytica in human and animals hosts. *Rev Infect Dis* 1982;4:1154–1184.

96. Trissl D, Martinez-Palomo A, Chavez B: Isolation of intact Entamoeba surface coat and caps induced by concanavalin A. *J Cell Biol* 1976;70(part 2):417.

97. Trissl D, Martinez-Palomo A, de la Torre Dela Hoz R, Perez-Suarez E: Sur-

face properties of Entamoeba: increased rates of human erythrocytes phagocytosis in pathogenic strains. *J Exp Med* 1978;148:1137–1145.

98. Trissl D, Martinez-Palomo A, Arguello C, et al: Surface properties related to concanavalin A induced agglutination: a comparative study of several Entamoeba strains. *J Exp Med* 1977;145:652–665.

99. Turner KJ, Feddama L, Quinn EH: Non-specific potentiation of IgE by parasitic infection in man. *Int Arch Allergy Appl Immunol* 1979;58:232–236.

100. Vinayak VK: The specificity of the indirect haemagglutination test in diagnosis of amoebiasis. *Indian J Prev Soc Med* 1975;6:271–276.

101. Vinayak VK: Experimental hepatic amoebiasis: understanding pathogenetic mechanism. *Ann Natl Acad Med Sci* 1980;16:181–197.

102. Vinayak VK, Chugh S: Electrophoretic isoenzymes pattern of isolates of E. histolytica before and after revival of the virulence. *Indian J Med Res* 1985;81:373–377.

103. Vinayak VK, Prakash O, Talwar GP, et al: Significance of indirect haemagglutination test for diagnosis of amoebiasis. *Indian J Med Res* 1974;62:1171–1175.

104. Vinayak VK, Prakash O, Talwar GP, et al: Evaluation of gel diffusion tests for diagnosis of amoebiasis. *Indian J Med Res* 1974;62:1317–1322.

105. Vinayak VK, Mohapatra LN, Tandon BN, Talwar GP: Bentonite flocculation test in amoebiasis. *J Trop Med Hyg* 1974;71:215–219.

106. Vinayak VK, Tandon BN, Talwar GP, Mohapatra LN: Immunoelectrophoresis test in amoebiasis. *Indian J Med Res* 1976;64:661–667.

107. Vinayak VK, Jain P, Sawhney S: Cutaneous histolytica hypersensitivity in apparently healthy subjects, in *Proceedings, First All India Medical Microbiology Association,* 1977, pp 87–88.

108. Vinayak VK, Sawhney S, Sehgal SC, et al: Slide haemagglutination test in amoebiasis. *IRCS Med Sci* 1977;5:258.

109. Vinayak VK, Naik SR, Sawhney S, et al: Studies on the pathogenicity of E. histolytica: virulence of strains of amoeba from symptomatic and asymptomatic cases of amoebiasis. *Indian J Med Res* 1977;66:935–941.

110. Vinayak VK, Chitkara NL, Chhuttani PN: Effect of corticosteroid and irradiation on caecal amoebic infection in rats. *Trans R Soc Trop Med Hyg* 1979;73:266–268.

111. Vinayak VK, Jain P, Bharti G, et al: Cellular and humoral responses in amoebic patients. *Trop Geogr Med* 1980;32:298–302.

112. Vinayak VK, Sawhney S, Jain P, Charkavarti RN: Protective effects of crude and chromatographic fraction of axenic E. histolytica in guinea pigs. *Trans R Soc Trop Med Hyg* 1980;74:483–487.

113. Vinayak VK, Sawhney S, Jain P, et al: Virulence of E. histolytica in rats and its comparisons with the serological responses of the amoebic patients. *Trans R Soc Trop Med Hyg* 1981;75:32–39.

114. Vinayak VK, Sawhney S, Jain P, et al: Immune suppression and experimental amoebiasis in guinea pigs. *Ann Trop Med Parasitol* 1982;76:309–316.

115. Vinayak VK, Sharma GL, Sawhney S, Chugh S: Is amoebic vaccine possible? in Subrahmanyam D, Radhakrishna V (eds): *Recent Advances in Protozoan Diseases.* Bombay, Vakil and Sons, 1983, pp 22–39.

116. Vinayak VK, Sharma GL, Naik SR: Protective immunity induced by ribonucleic acid protein fractions from axenic E histolytica. *Int J Microbiol* 1984;2:9–17.

117. Vinayak VK, Chugh S, Saxena A, Sharma SP: Antibody dependent lymphocytes mediated cytotoxicity in amoebiasis. *Indian J Med Res* 1984;80:421–427.
118. Vinayak VK, Purnima, Nain CK, et al: Specific circulating immune complexes in amoebic liver abscess. *J Clin Microbiol* 1986;23:1088–1090.
119. Vinayak VK, Purnima, Saxena A: Immunoprotective behaviour of plasma membrane associated antigens of axenic E. histolytica. *J Med Microbiol* 1987;65:217–222.
120. Vinayak VK, Saxena A, Malik AK: Alterations of humoral, cell mediated and antibody dependent cell mediated cytotoxic responses during the course of amoebic infection in guinea pigs. Gut, 1987;28:1251–1256.
121. Wahlin B, Perlmann H, Perlmann P: Analysis by a plaque assay of IgG or IgM dependent cytotoxic lymphocytes in human blood. *J Exp Med* 1976;144:1375–1380.
122. World Health Organization: Amoebiasis and its control. *Bull WHO* 1985;63:417–428.
123. Yap EH, Zaman V, Aw SE: The use of cyst antigen in serodiagnosis of amoebiasis. *Bull WHO* 1970;42:533–561.

CHAPTER 23

Molecular Comparisons Among *Entamoeba histolytica* Strains Using DNA and Protein Profiles

Sudha Bhattacharya, Alok Bhattacharya, and Louis S. Diamond

Infection with *Entamoeba histolytica* (amebiasis) is worldwide, extending from the tropics to the subarctic. However, its prevalence is higher, and the disease it produces more serious, in developing countries such as India. Infection with *E. histolytica* occurs following ingestion of the cyst stage of the parasite. The normal habitat of the ameba is the large bowel, and in most infections the relation between host and parasite is believed to be one of commensalism. However, in some infections, for reasons poorly understood, the ameba becomes invasive and enters the colonic mucosa, producing colitis and in the worst cases bloody dysentery. Extraintestinal infection of vital organs such as the liver, lung, and brain occurs through hematogenous spread. Both intestinal and extraintestinal disease are life-threatening. Strains of *E. histolytica* exhibit a broad spectrum of virulence in laboratory animals. At present there are no markers for distinguishing clearly between these strains. None of the indices devised to date, e.g., collagenase activity (3), erythrophagocytosis (10), or isoenzyme patterns (9), provide anything more than the crudest separation of these strains.

We have used restriction enzyme analysis of DNA and antigenic analysis with monoclonal antibodies as means for differentiating among strains. The results of our studies are reported.

Materials and Methods

Organisms and Growth Conditions

All of the *E. histolytica* strains used in this study were originally isolated from patients with active amebic disease. All amebae were maintained axenically in TYI-S-33 medium according to the method of Diamond (2).

DNA Isolation and Analysis

Cells were gently lysed by adding 0.2% sodium dodecyl sulfate (SDS), and DNA was extracted by the standard phenol/chloroform method detailed by Maniatis et al (5). Digestion of DNA with EcoRI, electrophoresis through agarose gels, and cloning of repeated DNA fragments were done essentially as described by Maniatis et al (5).

Generation of Monoclonal Antibodies

Balb/c mice were immunized with either whole cells (10^6 cells/animal) or purified membranes isolated from *E. histolytica* HM-1:IMSS strain. Fusion of the splenic lymphocytes with myeloma cells (P3 × 63 A8.653) was carried out essentially as described by Kearney (4). The screening for secreted antibodies was carried out by an indirect ELISA (BRL kit) using either membrane or soluble cytoplasmic antigens.

Glutaraldehyde fixation of *E. histolytica* cells and monoclonal binding assay were carried out as described elsewhere (1) with minor modifications. The second antibody was labeled with β-galactosidase instead of ^{125}I.

Results

EcoRI Digestion Pattern of Total Genomic DNA from Various Strains of *E. histolytica*

DNA from five strains of *E. histolytica* (HM-1, NIH:200, HK-9, HB:301, and Rahman) was digested with Eco RI and electrophoresed through agarose gels. Prominent, discrete bands were observed in each DNA digest (Fig. 23.1). Although all strains shared a common doublet [0.8 and 0.9 kilobases (kb)], the overall banding pattern of each strain was unique. The closest similarity was between strains HM-1 and HB:301, which differed in only one band. Thus based on EcoRI digestion patterns, it was possible to distinguish among the *E. histolytica* strains.

Cloning of Repeated DNA Fragments and Development of an *E. histolytica*—Specific Probe

The repeated DNA fragments obtained on EcoRI digestion of DNA from *E. histolytica* HM-1 were cloned in the plasmid $_p$TZI8R. Some of these fragments were further subcloned and tested for hybridization with DNA from other *Entamoeba* species (Fig. 23.2). *E. histolytica*-specific DNA probes were thus developed that hybridized with all strains of *E. histolytica* but not with any other *Entamoeba*, including the Laredo strain.

Characterization of Monoclonal Antibodies

Most hybridoma antibodies obtained by us were directed toward the cell surface of *E. histolytica*. This finding was confirmed by ELISA using membrane- and cytoplasm-soluble antigens and glutaraldehyde-fixed whole cells. When these antibodies were used to detect variations of the antigenic

Fig. 23.1. Restriction analysis of genomic DNA. DNA (0.4–1.0µg) was digested with EcoRI, electrophoresed through 0.8% agarose gels, and the DNA bands visualized by ethidium bromide fluorescence. Lanes 1 to 5 are *E. histolytica* strains: 1 = HM-1; 2 = NIH:200; 3 = HK-9 4 = HB:301; and 5 = Rahman. λ *Hind*III molecular weight markers are indicated on the left in Rb.

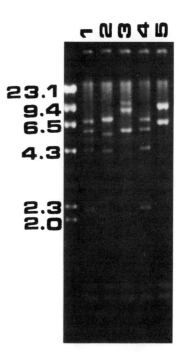

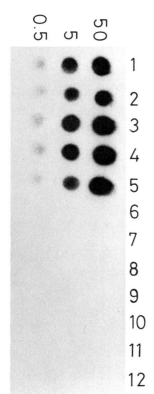

Fig. 23.2. Demonstration of specific hybridization to *E. histolytica*. DNA from various *Entamoeba* species (50, 5, and 0.5 ng) was spotted on a nitrocellulose filter and hybridized with a subfragment of repeated DNA clone from HM-1. Rows 1 to 5 were strains of *E. histolytica:* 1 = HM-1; 2 = NIH:200 3 = HK-9; 4 = HB:301; and 5 = Rahman. Lane 6 = Laredo; lane 7 = Huff; lane 8 = *E. moshkovikii* (FIC); lane 9 = *E. moshkovskii* (CST); lane 10 = *E. invadens* (IP-1); lane 11 = *E. terrapinae* (M); lane 12 = *E. barreti*.

Table 23.1. Reactivity of monoclonal antibodies.

Lines	E. histolytica			E. histolytica-like Laredo	E. moshkovaskii FIC	E. invadens 1652
	HM-1	NIH-200	HB:301			
Membrane Antigens						
Cloned						
3D7	+	−	+	−	−	−
2A6	+	+	+	−	−	−
2F3	+	+	+	−	−	−
3B2	+	+	+	−	−	+(?)
Uncloned						
4D4	+	+	+	−	−	−
4E4	+	+	+	−	−	−
Soluble antigens						
3D1	+	+	ND	−	−	−
1B3	+	+	ND	−	−	−
2D6	+	+	ND	−	−	−

ND = not determined.

determinants among a group of *E. histolytica* strains and other *Entamoeba* species, most of the antibodies were found to behave identically (Table 23.1). All antibodies tested (including a few from uncloned lines) recognized antigens present in different strains of *E. histolytica*. However, none of the antibodies recognized any determinant present in other *Entamoeba* species. This situation seems to hold true for antibodies recognizing both cell surface and cytoplasmic antigens.

Discussion

Probes generated from repeated DNA sequences have been extensively used for the detection and identification of parasites, e.g., *Trypanosoma* (6), *Leishmania* (11), *Schistosoma* (7), and *Brugia* (8). We have demonstrated the feasibility of a similar approach in the identification of *E. histolytica* and in differentiating among various strains. Studies on electrophoretic mobilities of isoenzymes involved in the glycolytic cycle of *Entamoeba* (9) also provide a means to differentiate among *E. histolytica* strains. However, the approach based on restriction enzyme analysis may prove to be of greater analytical value.

The data obtained by us using monoclonal antibodies clearly shows that there is little antigenic variation among *E. histolytica* strains despite the fact that these strains have been obtained from different geographic locations and have varying degrees of virulence in animal models. These data are also in agreement with our results from two-dimensional gel electrophoresis of ^{35}S-methionine-labeled extracts. We did not find any significant difference among the various *E. histolytica* isolates, whereas the *E. histolytica*-like ameba Laredo strain had a completely different pattern (unpublished data).

Summary

All of the five strains of *Entamoeba histolytica* tested revealed the presence of highly repeated DNA, which gave discrete ethidium bromide-stained bands upon digestion with the restriction enzyme EcoRI. The banding pattern was unique to each strain and could thus serve to differentiate the strains from one another. A subclone from one of the repeated DNA clones hybridized preferentially with *E. histolytica* and not with any of the other *Entamoeba* species tested. *E. histolytica* could be further distinguished from other *Entamoeba,* including the Laredo strain, using a panel of monoclonal antibodies raised against the HM-1 strain of *E. histolytica*. None of these antibodies could distinguish *E. histolytica* strains from one another.

References

1. Bhattacharya A, Dorf ME, Springer TS: A shared alloantigenic determinant on Ia antigens encoded by the I-A and I-E subregions: evidence for I region gene duplication. *J Immunol* 1981;127:2488.
2. Diamond LS: Lumen dwelling protozoa; Entamoeba, trichomonads and Giardia, in Jensen JB (ed): *In Vitro Cultivation of Protozoan Parasites.* Boca Raton, CRC Press, 1983, pp 65–109.
3. Gadasi H, Keesler E: Correlation of virulence and collagenolytic activity in Entamoeba histolytica. *Infect Immun* 1983;39:528–531.
4. Kearney JF: Hybridomas and Monoclonal antibodies. In Paul WE (ed): *Fundamental Immunology. New York, Raven Press, 1984, pp 751–766.*
5. Maniatis T, Fritsch EF, Sambrook J: *Molecular Cloning: A Laboratory Manual.* Cold Spring Harbor, New York, Cold Spring Harbor Laboratory, 1982.
6. Massamba NN, Williams RO: Distinction of African trypanosome species using nucleic acid hybridization. *Parasitology* 1984;88:55–65.
7. McCutchan TF, Simpson A, Mullins JA, et al: Differentiation of schistosomes by species, strain and sex by using cloned DNA markers. *Proc Natl Acad Sci USA* 1984;81:889–893.
8. McReynolds LA, Desimone SM, Williams SA: Cloning and comparison of repeated DNA sequences from the human filarial parasite Brugia malayi and the animal parasite Brugia pahangi. *Proc Natl Acad Sci USA* 1986;83:797–801.
9. Sargeaunt PG, Williams JE, Neal RA: A comparative study of Entamoeba histolytica (NIH:200, HK 9, etc.), "E. histolytica-like" and other morphologically identical amoebae using isoenzyme electrophoresis. *Trans R Soc Trop Med Hyg* 1980;74:469–474.
10. Trissl D, Martinez-Palomo A, Dela Torre M, et al: Surface properties of Entamoeba: increased rates of human erythrocyte phagocytosis in pathogenic strains. *J Exp Med* 1978;148:1137–1145.
11. Wirth D, Pratt DM: Rapid identification of Leishmania species by specific hybridization of kinetoplast DNA in cutaneous lesions. *Proc Natl Acad Sci USA* 1982;79:6999–7003.

Part VI
Filariasis

CHAPTER 24

Importance of Antibody Class in Helminth Infections

N.M. Almond and R.M.E. Parkhouse

Antibodies play an important role in the rejection of many parasites. They are essential for immune recognition of antigens incorporated in the surface of large nonphagocytosable helminths. Thus by focusing host defense mechanisms against the intact parasite they may ultimately lead to parasite rejection. The detection of antibodies may also play a role in the diagnosis of parasite infections, particularly when screening for early infections and providing a rational basis for the organization of effective programs for disease control.

Any parasite presents its host with a complex array of antigens, which then elicit a correspondingly complex humoral response. Not all of the antibodies synthesized necessarily play a positive role in the host–parasite relationship. This chapter focuses on the critical role of the antibody heavy-chain class in humoral immunity to parasites. The presentation is broadly divided into two parts. In the first we summarize our knowledge on the properties, functions, and control of expression of immunoglobulin isotypes. In the second we consider how the study of individual class responses to parasite antigens may help us understand the basis of parasite survival and, as well, find practical application in control and diagnosis. Examples are drawn from results of work performed in our laboratory and elsewhere.

Immunoglobulin Heavy-Chain Classes

Five major immunoglobulin (Ig) heavy-chain classes have been identified in mammals: IgM, IgA, IgG, IgE, and IgD. Each is defined by different heavy-chain isotypes, which are structurally and genotypically distinct and which confer their functional properties (Table 24.2). For example, secreted IgM is a polymeric structure consisting of five disulfide-linked subunits, each comprising two μ heavy chains and two light chains (i.e. $\mu_2 L_2$), linked by a joining (J) chain. Its molecular weight is 900 kilodaltons (kDa), and antibodies of this class are believed to be important in the early control of blood bacteremia through their ability to agglutinate particles.

Table 24.1. Variation between human IgG subclasses.

Parameter	IgG$_1$	IgG$_2$	IgG$_3$	IgG$_4$
% total serum Ig in normal patients	65	23	8	4
Complement fixation				
Classical pathway	+ + +	+	+ + + +	±
Alternate	?	?	?	?
Binding				
Monocytes	+ +	−	+ +	−
Granulocytes	?	?	?	?
Blocks anaphylactic sensitization	−	−	−	+
Dominant subclass in responses to	Proteins	Carbohydrates	?	?
pH of elution from staphylococcal protein A	4.3	7.0	4.7	N.D.

N.D. not determined.

Antibodies of the IgA class, on the other hand, are found as monomers or polymers of the basic $\alpha_2 L_2$ immunoglobulin unit. Although present in the serum in modest amounts, antibodies of this class are found in mucosal secretions at much higher concentrations. Subclasses of IgA do exist but have been studied only in higher apes and man. As yet little is known about their relative roles, although in certain disease states IgA antibodies of one or other type may be disproportionately high (40).

Subclasses of IgG ($\gamma_2 L_2$), on the other hand, occur frequently in mammals and often reflect relatively small structural differences. Variations in the pattern of interchain disulfide bonding are commonly found between IgG subclasses. The structural differences account for the distinct biological properties of the various IgG subclasses, such as complement activation and adherence to monocytes, granulocytes, and staphylococcal protein A (Table 24.1). For example, human IgG$_3$ is readily capable of activating complement by the classical pathway, whereas IgG$_2$ is largely ineffective. Consequently, if activation of complement is essential in parasite rejection, it is immediately obvious that the balance of Ig classes and subclasses is a critical factor in determining the outcome of infection. It is worth noting here that in most cases the nomenclature of IgG subclasses in various animal species is based on the relative concentration of each subclass in serum and not by structural or functional homology. This fact must be borne in mind, for example, when attempting to extrapolate to man results obtained in experimental animals.

Control of Immunoglobulin Isotype Expression

Small resting B cells, the putative precursors of all antibody-secreting cells, possess the genes encoding all of the immunoglobulin heavy-chain classes. How the immunoglobulin isotype profile is controlled at the level of gene expression is still not fully understood. Studies have now, however, begun to provide us with an idea of the processes involved.

Accumulated *ad hoc* experience indicates that the balance of immunoglobulin isotypes observed in humoral responses may be reproducibly biased by a number of experimental manipulations. The biochemical structure of the antigen may influence the balance of isotypes elicited. For example, in mice, protein antigens tend to selectively stimulate the production of IgG_1 antibodies, whereas carbohydrate antigens elicit largely IgG_3 antibodies (44). The immunization protocol can also have a marked effect on the isotype profile of antibody responses. Determining factors are the dose of antigen, the route and frequency of administration, and the simultaneous injection of adjuvants. Finally, the balance of antibodies of each heavy-chain isotype is dynamic, and so an important consideration is the interval between immunization and the assay of responses (27). To cite the best known example, IgM antibodies comprise the dominant class of serum antibodies soon after primary immunization, especially when low doses of antigen are used. IgG and IgA antibody responses, on the other hand, develop later, especially when higher antigen doses and multiple immunizations are employed. This reproducible bias of immunoglobulin class balance, associated with varying the regimen of antigen injection, was originally ascribed to differences in "antigen presentation." We now have a greater understanding of what this term may involve in terms of cellular and molecular interactions. Certainly, an important part of immunoglobulin class regulation is mediated by T cells. The requirement for T cells to recognize antigen in association with products of the major histocompatibility gene complex on the surface of antigen-presenting cells thereby provides a role for this phase of the immune response in immunoglobulin class expression.

Indirect evidence for the role which T cells are believed to play in control of the immunoglobulin isotype profile originally came from studies in athymic animals. Congenitally athymic, "nude" mice had lower levels of total serum immunoglobulin than normal mice, with IgA and IgG immunoglobulin classes being selectively depressed (25). This imbalance was not redressed in antibody responses following immunization or infection (7,45).

Coordinated research into parasitology and immunology may contribute to a greater understanding of both subjects. The complex role of T cells in controlling immunoglobulin isotype expression has been particularly illuminated in the case of IgE by the study of experimental parasite infections. Infection with the nematode *Nippostrongylus brasiliensis* selectively enhances IgE synthesis. This observation formed the basis for the identification of distinct T cell populations, which enhance either IgG or IgE responses (21). Further studies extended the concept of soluble factors that regulate, via enhancement or suppression, IgE responses (14,18). Most notable among the T-cell-derived factors are the IgE-binding factors, polypeptide chains of approximately 15 kDa that can promote or suppress

IgE synthesis in cells bearing surface IgE. The positive or negative effect of these factors on the IgE response seems to depend solely on the level of glycosylation of the same simple factor, a process apparently controlled by further distinct T cell populations (14).

The evidence that there is a network of T cells and factors, finely tuned to control the production of potentially damaging IgE antibodies, encouraged speculation as to whether expression of the other heavy-chain classes is similarly regulated. Circumstantial evidence for the existence of such factors was obtained by Rosenberg and Chiller (39). After immunizing rodents with various T-cell-dependent antigens, they observed similar immunoglobulin isotype profiles for both specific antibody and total immunoglobulin responses. From this result they concluded that isotype-specific, antigen-nonspecific T cells were responsible for controlling the immunoglobulin isotypic profile. The development of methods to clone and maintain T cells in culture for prolonged periods has permitted more direct questions to be asked. There is now good evidence that the production of IgG_1 (13,42) and IgA (19,22) antibodies can be influenced by T cells or T-cell-derived soluble factors. A point of interest is that one factor, which enhances B cell differentiation into IgG_1-secreting cells in vitro, has been found to be identical to the lymphokine B cell stimulatory factor 1 (BSF_1) (46). Originally defined as a B lymphocyte growth factor, BSF_1 is also active on mast cells, granulocytes, macrophages, and T cells (8). The ability of one factor to provoke different responses in different cell types is interesting and should be borne in mind when mechanisms of action of these "differentiation" factors are explored. Now that BSF_1 and other growth and differentiation factors have been successfully cloned and expressed (23,28), we can expect a rapid increase in our understanding of this and other related lymphokines. Whether they act through the release of diffusible factors or require cognate cell interaction, T cells may enhance the response of antibodies by acting in two distinct manners. One possibility is that they encourage the "switch" of B cells expressing surface IgM to another, particular class. Alternatively, the T cell may selectively stimulate the proliferation of B cells already expressing the "post-switch" isotype. IgA specific helper T cell clones have been isolated that act in either the former (19) or the latter (22) manner.

Role of Immunoglobulin Isotypes in Parasite Control

The application of hybridoma technology and selective methods for radiolabeling defined parasite components have facilitated the analysis of host–parasite interactions at a molecular level (32). It is now well established that the outcome of many parasitic infections depends on immune recognition of specific parasite antigens followed by focusing of appropriate defense mechanisms. Thus, for example, monoclonal antibodies to parasite surface antigens have been isolated that passively protect rodents against infection with schistosomes (9) and nematodes (1,31). Alternatively, monoclonal antibodies have been used to affinity-purify parasite components for subsequent vaccination studies (43). Empirical "shotgun" tactics, based on monoclonal antibodies and gene cloning, have approached the development of potential vaccines without necessarily understanding the immunobiological basis of protection in natural infections, e.g., malaria. This oversight may be of no consequence, as some antigens may always elicit effective resistance in all hosts. No doubt, however, there are examples where this empirical approach will fail to achieve an effective vaccine. In this situation a thorough understanding of host-parasite interactions is required as a basis for effective vaccine design.

One aspect of antiparasite humoral responses that has been neglected is the relative role played by antibodies of each heavy-chain class and thus the important consequence of the immunoglobulin isotype profile. As indicated in Tables 24.1 and 24.2, antibodies of each isotype mediate a unique range of immune effector functions; antibodies of all classes need not necessarily mediate parasite rejection in vivo. Given a situation where resistance is antibody-mediated, an investigation of the immunoglobulin isotype profile may be useful for identifying mechanisms of rejection. A dramatic study of how the isotype of antibodies directed against a potentially protective antigen may determine the outcome of infection has come from Capron's group. A rat monoclonal antibody of the IgG_{2a} isotype which binds to a 38-kDa surface antigen of schistosomulae of *Schistosoma mansoni* was found to mediate eosinophil adherence in vitro and to protect in vivo (9). A monoclonal antibody of a different immunoglobulin class (IgG_{2c}), but with similar specificity, failed to mediate eosinophil adherence in vitro, and blocked protection in vivo following passive transfer of the IgG_{2a} monoclonal antibody alone (10). This work clearly shows how protection may depend not only on the presence of antibodies of an appropriate specificity but also on the correct balance of antibodies of each heavy-chain class.

We have investigated the importance of antibodies of each heavy-chain class during infection with parasitic nematodes. The parasite systems chosen were (a) the rejection of adult worms of *Trichinella spiralis* from the gastrointestinal tract (2), and (b) the control of circulating microfilariae following transplantation of adult female worms of *Dipetalonema viteae* (3). These systems provide convenient models of infection with two distinct groups of parasitic nematodes: gastrointestinal and filarial. The serological response following infection was compared in strains of mice that differ in their ability to control infection. A fuller account of these experiments and the results are provided in the references cited. Briefly, however, at various times after infection, sera were collected from these mice, and the parasite burden was determined. The concentration of total immunoglobulin and of antiparasite antibodies of each immunoglobulin heavy

Table 24.2. Structure and function of human immunoglobulin classes.

Parameter	IgM	IgA	IgG	IgE	IgD
Mol. wt. (kDa)	900	160 (or 320)	150	200	185
% Serum Ig	6	13	80	0.002	1
Complement fixation					
Classical	+++	−	++[b]	−	−
Alternate	−	+	+	−	−
Binding					
Mast cells	−	−	+[a]	+	−
Granulocytes[a]	−	+	+	?	−
Macrophages[a]	−	+	−	+	−
No. of subclasses	−	2	4	−	−
Function	Produced early after infection; agglutinates well with or without complement	Major Ig class in the secretions of gut and milk	Most abundant Ig within body fluids	Elevated levels during helminth infection; responsible for symptoms of atopic allergy	Present on lymphocytes surface; regulator function?

[a] These cells also have receptors for complement that effectively alter the capacity of the isotope to mediate cytoadherence in vivo.
[b] Variation between subclasses.

chain class were assayed for each serum. Radiolabeled surface or secreted antigens were used to determine the amounts of corresponding antibodies of all heavy-chain isotypes using an immunoglobulin class-specific immuno-coprecipitation assay (2). In addition, antibodies directed against determinants exposed on the surface of living microfilariae of *Dipetalonema viteae* were determined using an immunoglobulin class-specific "surface ELISA" assay (41).

A number of interesting observations arose from this systematic study. The most striking feature was an enormous heterogeneity in the antibody response due to independent variation of each component of the total serological response. It was observed at the following five levels.

1. *Variation in responses to distinct parasite stages.* Selective radiolabeling of *T. spiralis* reveals stage-specific surface and secreted proteins, a feature also seen in other nematodes (11,26,33,36,38). For *T. spiralis* these components are completely non-cross-reacting (29,35,36), and so the host is sequentially immunized against a series of antigenically distinct parasite stages, each at a different time after infection and at different sites. It is not surprising therefore to discover that antibodies that bind to surface components of each stage follow distinct kinetics (2,16,37). For example, the kinetics of IgM antibodies that bind to surface components of the three principal stages of *T. spiralis* (adult worm, and newborn and infective larvae) reflect the exposure of the host to these antigens following infection. Thus antibodies that bind to infective larvae appear first, followed by those recognizing adults and then newborn larvae.

2. *Variations in responses to distinct antigenic compartments of a single parasitic stage.* In many instances proteins that are incorporated into the surface of a parasite possess biochemical properties that are distinct from secretions of the same parasite. In the case of *T. spiralis,* directly radioiodinated surface proteins and metabolically labeled secretions are distinct groups of molecules when analyzed by sodium dodecyl sulfate–polyacrylamide gel electrophoresis (SDS–PAGE) (4,34,36). Antibodies recognizing the labeled surface or secreted components of a single stage appear with different kinetics. For example, IgG antibodies to secretions of infective larvae of *T. spiralis* appear far earlier than IgG antibodies to the surface of the same stage. The distinct kinetics of antibodies formed to components derived from different antigenic compartments of the parasite is an important observation that stresses the independence of each parasite antigen presented to the host.

3. *Variation in responses to individual components from an antigenic compartment of a single stage.* Given that immune recognition of individual parasite stages and their various compartments are independently regulated, how are the several components of one compartment recognized? This question was investigated by qualitative SDS–PAGE analysis of immuno-coprecipitates formed by the various mouse antibody classes

to radiolabeled stage-specific surface and secreted antigens of the parasite. This type of study revealed unique kinetics of antibody production to each of the labeled parasite antigens. Thus antibodies recognizing each surface antigen of *T. spiralis* infective larvae do not appear simultaneously. Jungery and Ogilvie (16), also working with *T. spiralis,* noted that antibodies to the lentil-lectin adherent (glycoprotein) fraction of infective larvae surfaces appeared in the serum of C3H mice as early as 4 days after infection, whereas the lentil-lectin nonadherent (protein) fraction of the surface did not elicit a response until day 30 post infection.

A similar situation of differential recognition of antigens within the same compartment may also occur in filarial infections. Serum taken from C57 Black mice 21 days after the transplantation of adult female *D. viteae* and 14 days after the infection had become patent was found to precipitate a number of Bolton and Hunter reagent-labeled microfilarial surface proteins. A 35-kDa component among them, however, was not recognized by antibodies in the sera of C57 Black mice. The same protein was at the same time highly immunogenic in Balb/c mice.

This observation therefore demonstrates the importance of analyzing responses to each parasite component individually. Each component obviously possesses distinct biochemical and immunogenic properties. The ability to resolve the overall response into its separate components is likely to reveal both "useful" and "counterproductive" responses, which would be otherwise hidden within the total spectrum of the antibodies formed.

4. *Variation in responses to distinct epitopes on a specific antigen.* Antigens can possess a number of distinct antigenic epitopes, and parasite-derived components are no exception. Some epitopes of surface antigens are exposed at the surface of the living parasite, whereas others are obscured in the organization of the intact organism, only to become visible to the immune system after turnover or release from the surface (29). For example, solubilization of surface components of *T. spiralis* has been shown to reveal new distinct epitopes not exposed when the same molecule is incorporated in the nematode cuticle. Most remarkable is the observation that none of the carbohydrate determinants of surface-labeled proteins are exposed on the surface of all three stages of *T. spiralis* (29,30). Immune responses to each epitope, similar to immune responses to different molecules, may also follow distinct kinetics. For example, the response of C57 Black mice to surface labeled solubilized surface components of microfilariae of *D. viteae* includes a marked early humoral response of IgA antibodies (3). This response was measured by an isotype (IgA) specific co-precipitation assay. Interestingly, at the same time, these antisurface IgA molecules failed to bind to the surface of the living worm, as judged by IgA-specific surface immunofluorescence or the surface ELISA test (41). The epitopes recognized by the IgA antibodies from C57 Black mice were therefore unexposed on the surface of intact microfilariae and consequently were probably not a focus for a protective immune response.

The assumption that all the epitopes of a surface molecule are necessarily exposed to the host by an intact parasite is thus erroneous. If mechanisms of protection are being investigated, it is clearly necessary to assay host responses by a number of methods, including, if possible, the living parasite as it is "seen" by the host.

5. *Variations in the responses of individual immunoglobulin classes.* Against this background of independent variation in the kinetics of individual humoral responses to each parasite antigen, one must also superimpose the phenomenon of the balance of immunoglobulin isotypes. The balance of isotypes determines what biological activities may be brought to bear against the invading organism and in this manner may determine the outcome of infection. A systematic dissection of immune humoral responses against parasite antigens at the level of individual heavy-chain classes is therefore an important area of study, particularly when done in the context of longitudinal studies and genetically defined groups (e.g., high and low responders).

Two features were immediately obvious from our investigations in this context: (a) the balance of isotypes is dynamic; and (b) each antibody class can vary independently of the others in response to individual antigens. This independent variation in the balance of heavy-chain classes to each parasite component implies that measurement of total serum immunoglobulin of each isotype does not necessarily reflect the balance of immunoglobulin isotypes synthesized in response to any given parasite antigen. An excellent example was provided by the course of total serum IgG_1 response of mice infected with *T. spiralis* compared with the kinetics of IgG_1 antibodies to surface antigens of either infective larvae or adult worms. Following infection, total serum of IgG_1 rises steadily throughout the first 40 days of infection. IgG_1 antibodies to infective larvae, on the other hand, do not appear until day 25 of infection, after which time their titer rises rapidly. The kinetics of appearance of IgG_1 antiadult surface antigens was again entirely different, with the class of antibody appearing on day 10 post infection, the titer rising rapidly until day 20 and then declining slowly over the subsequent 15 days.

Differences in the balance of immunoglobulin isotypes were not observed solely in the comparison of antibodies against distinct parasite stages. A striking example of independently varying responses of antibodies of one isotype to components of a single parasite stage was seen in IgG_1 and IgG_2 responses to adult *T. spiralis* surface antigens. Surface labeling of this parasite stage restrictively labels only three proteins of 40, 33, and 20 kDa (6). At the peak of the host IgG_1 response only the 20- and 33-kDa components were recognized. At the same time, however, IgG_2 antibodies from the same animals recognized all three components. This finding indicates how exquisite is the level of control exercised in the expression of immunoglobulin isotypes. It is all the more remarkable when, as in this case, all three antigens would have been exposed to the host simultaneously. As these immunoglobulin isotype profiles to closely related antigens are distinct, it suggests a role for cognate cell interaction, rather than diffusible factors in determining the pattern of immunoglobulin isotype expression.

Applications of Immunoglobulin Isotype Studies to the Control of Parasitosis

The systematic dissection of the murine humoral responses following infection with parasitic nematodes is not of academic interest alone. The results provide us with a basis for rational approaches to control via effective vaccine design and accurate serodiagnosis.

Protection

Infection with parasitic nematodes elicits a variety of antibody responses, each potentially different in kinetics or balance of immunoglobulin heavy-chain classes. Not all of these responses, however, contribute to protection. A frequently used experimental approach for detecting potentially protective responses is the comparison of genetically distinct inbred strains of experimental animals. The rationale is to define a resistant strain as a model for a protective host response. For example, the NIH strain of mice are quick to expel adults of *Trichinella spiralis* from their intestines (20), whereas C3H mice reject the worms less effectively (16). A comparison of the patterns of humoral responses made by these two strains of mice, in the most part, failed to distinguish the antibody responses of resistant and susceptible strains. An interesting exception to this rule, however, were the serum IgA responses. Resistant NIH mice mounted a marked IgA antibody response, and, significantly, this reaction was to surface derived antigens of the adult worm and occurred maximally between days 10 and 25 post infection. The peak of this IgA response coincided with the time of complete adult worm rejection. Furthermore, the susceptible C3H mice did not produce such antibodies at this time and did not reject the worms until much later (2). Although a direct demonstration that IgA antibodies mediate protection is still lacking and indeed necessary, the result is highly suggestive, especially as IgA antibodies are associated with mucosal immunity.

Immune Diagnosis

Detection of antibody responses to parasites is an important diagnostic procedure that, at best, uses the exquisite specificity and sensitivity of antigen–antibody reactions to unambiguously identify individual parasite infections. Unfortunately, in many cases the theoretical ability to differ-

entiate closely related parasites has not been realized, largely, one sus-
pects, because of the continued use of poorly defined, frequently cross-
reacting crude parasite extracts. A more rational method of improving
the sensitivity of serodiagnostic tests is through the employment of purified
antigens. In addition to confining the assay to antibodies of restricted antigenic specificity, it may also be advantageous to design assays restricted
at the level of individual immunoglobulin class(es). It may be beneficial
for two reasons: (a) There may be a higher frequency of cross-reactive
antibodies among one isotype compared to another. For example, Weiss
et al (48) found that IgE antibodies present in the sera of a number of
patients with filariasis possessed fewer cross-reactivities than did IgG antibodies present in the same panel of sera. The immunochemical basis for
this was revealed in an immunoblotting study, which showed that the human IgE response in onchocerciasis is restricted and predominantly directed at antigens not recognized by antibodies to other nematode parasites
(5). Thus in this case an assay based on the detection of IgE antibodies
would more accurately diagnose the invading parasite than an assay utilizing the IgG response. (b) The finer dissection of host responses may
identify specific humoral responses that correlate with changes in the
course of disease. Assays capable of predicting the course of future disease
would then possess prognostic as well as diagnostic potential. For example,
Weiss et al (47) and Hussain and Ottesen (12) have both found that IgE,
but not total IgG, responses possess features characteristic of sera taken
from groups of patients presenting distinct symptoms of filarial infection.
Other studies have also shown differential recognition patterns of human
immunoglobulin classes to antigens of *Onchocerca,* and in certain instances
these patterns correlated with different clinical forms of the disease
(5,17,24).

A longitudinal study of serological responses is, of course, the best way
of identifying whether any humoral response may be used to successfully
predict any changes in the course of disease or infection. In the example
given above, detection of IgA antibodies against adults of *T. spiralis* could
be used to identify early rejection of the gastrointestinal stage, a considerably faster procedure than conventional parasitological methods.

A systematic dissection of host immune responses to *D. viteae* also
proved useful in identifying host responses of possible diagnostic potential
(3). Resistant C57 Black mice produced a marked early IgA response to
labeled surface antigens of microfilariae. As mentioned above, these IgA
antibodies do not bind to exposed epitopes on the surface of microfilariae.
Thus although not capable of initiating parasite rejection, the detection
of such antibodies may be used to identify hosts capable of controlling
microfilaremia effectively. In this situation it is not strictly correct to call
this response prognostic, as it does not anticipate the changing course of
infection. However, any simple yet accurate serological assay that obviates
the need for time-consuming parasitological procedures is worth exploring.

In the study of murine responses to infection with *D. viteae* a truly prognostic anti-parasite response was also observed. Although by day 21 post transplantation the levels of circulating microfilariae are still increasing in the more susceptible Balb/c and CBA/N strains, these animals possess a pattern of serum antibody reactivity distinct from the more resistant C57 Black mice. Most obvious was the presence of antibodies precipitating a 35-kDa surface protein of microfilariae only in the less resistant strains. The appearance of antibodies with this specificity precedes the appearance of the high number of circulating microfilariae that eventually develop. A test designed to detect this response could therefore be described as truly prognostic.

Although the above study in mice indicates that prognostic immune responses do exist, they are revealed only by systematic dissection of host immune responses.

Concluding Remarks

The study of the isotype of antibodies elicited in response to specific parasite antigens has been a neglected area in immunoparasitological research. The "universal panacea" of monoclonal antibodies and gene cloning antigens have attracted attention, perhaps at the expense of investigations that attempt to unravel the complexity of host–parasite interactions. Although this review has concentrated on studies of the interaction between the host and nematode parasites, the interplay between mammalian hosts and any metazoan parasite should be expected to express a similar degree of complexity. The most important concept to be remembered is that each immunogenic epitope presented during the course of a parasitic infection is recognized and processed by the immune system, so that the corresponding antibody responses are essentially independent of one another. At first sight this situation presents an unsurmountable problem to the immunoparasitologist to characterize each component response. The application of many technical innovations, including selective radiolabeling, monoclonal antibodies, and gene-cloned antigens should enable this task to be done.

Turning to the importance of the immunoglobulin heavy-chain class of antiparasite antibodies, it should be apparent that the isotype of responses may be important in protection, serodiagnosis, and pathology. With regard to protection, it is worth reemphasizing that it is not the presence of antibodies of a given "protective" class per se that will determine the course of parasitosis. The important factor is the balance of antibodies of those isotypes that mediate parasite rejection with those that "block" this function. Similar considerations apply to pathological features based on antibody-mediated mechanisms. Although so far immunodiagnosis has, at best, assayed for the presence of antibodies of a specific isotype, one may presume that here also it is the balance of the different immunoglobulin classes synthesized that reflect the course of infection most accurately.

In the development of immunodiagnostic tests the emphasis has been on serodiagnostic assays that require minimal effort to obtain antigen samples. It is not necessarily the case for the investigation of protective humoral responses. In this situation, the response of greatest importance is that directed at the parasite surface. Thus it is important to look at local as well as systemic reactions.

Until recently a major problem in performing the type of studies envisaged above is the need for specific antiimmunoglobulin reagents. Now, however, monoclonal antibodies to each immunoglobulin class of a number of relevant mammalian species can be obtained and provide a truly specific and reliable basis for an immunochemical dissection to antiparasite antibody classes.

As our knowledge of immunoglobulin isotypes that are protective in vivo increases, so should our understanding of the basic mechanisms involved in parasite rejection; such understanding may help to define the best way of presenting a potentially protective vaccine so as to elicit the most effective immunity. As mentioned above, the route, dose, and frequency of immunization can alter the nature of the immune response generated. It should therefore be possible to logically develop immunization protocols that maximally raise protective immunity in the host.

Finally, this review has concentrated on the importance of immunoglobulin isotypes in the control of parasitic infection. It is possible, however, to use experimental parasitic infection to help us understand how the balance of isotypes is controlled. Infection with certain parasites can reproducibly elicit particular patterns of response. For example, there is a close correlation between helminthiases and enhanced IgE responses (15). Thus experimental infection may be used to produce the environment necessary for the development of lymphocytes in a given manner. In this way the cellular interactions that control isotype switching of humoral responses may be dissected. The greater our understanding of the immune system, the better is our chance to control infectious organisms in general and parasites in particular.

Acknowledgments. We thank Marlene Bertagne for secretarial assistance. This work was supported in part by grants from the Filariasis component of the UNDP/World Bank/WHO Special Programme for Research and Training in Tropical Diseases and the Commission of the European Communities Research and Development Programme "Science and Technology for Development."

References

1. Aggarwal A, Cuna W, Haque A, et al: Resistance against Brugia malayi microfilariae induced by a monoclonal antibody which promotes killing by macrophages and recognises surface antigens. *Immunology* 1985;54:655–663.

2. Almond NM, Parkhouse RME: Immunoglobulin class specific responses to biochemically defined antigens of Trichinella spiralis. *Parasite Immunol* 1986;8:391–406.

3. Almond NM, Worms M, Harnett W, Parkhouse RME: Variation in class humoral immune responses of different mouse strains to microfilareae of Dipetalonema viteae. *Parasitology* 1987;95:559–568.

4. Almond NM, McLaren DJ, Parkhouse RME: Comparison of the surface and secretions of Trichinella pseudospiralis and T. spiralis. *Parasitology* 1986;93:163–176.

5. Cabrera Z, Cooper MD, Parkhouse RME: Differential recognition patterns of human immunoglobulin classes to antigens of Onchocerca gibsoni. *Trop Med Parasitol* 1986;37:113–116.

6. Clark NWT, Philipp M, Parkhouse RME: Non-covalent interactions result in aggregation of surface antigens of the parasitic nematode Trichinella spiralis. *Biochem J* 1982;206:27–32.

7. Crewther PL, Warner LN: Serum immunoglobulin and antibodies in congenitally athymic (nude) mice. *Aust J Exp Biol Med* 1972;50:625–635.

8. Grabstein K, Eisenman J, Mochizuki D, et al: Purification to homogeneity of B cell stimulating factor. *J Exp Med* 1986;163:1405–1414.

9. Grzych JM, Capron M, Bazin H, Capron A: In vitro and in vivo effector functions of rat IgG$_{2a}$ monoclonal anti-S.mansoni antibodies. *J Immunol* 1982;129:2739–2743.

10. Grzych JM, Capron M, Dissous C, Capron A: Blocking activity of rat monoclonal antibodies in experimental schistosomiasis. *J Immunol* 1984;133:998–1004.

11. Harnett W, Meghji M, Worms MJ, Parkhouse RME: Quantitative and qualitative changes in production of excretions/secretions by Litomosoides carinii during development in the jird (Meriones unguiculatus). *Parasitology* 1986;93:317–331.

12. Hussain R, Ottesen EA: IgE responses in human filariasis. III. Specificities of IgE and IgG antibodies compared by immunoblot analysis. *J Immunol* 1985;135:1415–1420.

13. Isakson PC, Pure E, Vitetta ES, Krammer PH: T-cell derived B-cell differentiation factors: effect on the isotype switch of murine B cells. *J Exp Med* 1982;155:734–748.

14. Ishizaka K: IgE binding factors from rat T lymphocytes, in Pick E (ed): *Lymphokines,* Vol 7. New York, Academic Press, 1983, pp 41–80.

15. Jarrett EEF, Miller HRP: Production and activities of IgE in helminth infections. *Prog Allergy* 1982;31:178–233.

16. Jungery M, Ogilvie BM: Antibody response to stage specific Trichinella spiralis surface antigens. *J Immunol* 1982;129:839–843.

17. Karam M, Weiss N: Seroepidemiological investigation of onchocerciasis in a hyperendemic area of West Africa. *Am J Trop Med Hyg* 1985;34:907–917.

18. Katz DH: Recent studies on the regulation of IgE antibody synthesis in experimental animals and man. *Immunology* 1980;41:1–24.

19. Kawanishi H, Saltzman HE, Strober W: Characteristics and regulatory function of murine Con A induced cloned T cells obtained from Peyer's patches and spleen: mechanism regulating isotype specific immunoglobulin production by Peyer's patch B cells. *J Immunol* 1982;129:475–483.

20. Kennedy MW: Kinetics of establishment and rejection of the enteral phase of a primary infection of Trichinella spiralis in the N.I.H. mouse strain. *Trans R Soc Hyg Trop Med* 1976;70:285.

21. Kishimoto T, Ishizaka K: Regulation of antibody response in vitro. VII. Enhancing soluble factor for IgG and IgE antibody response. *J Immunol* 1973;111:1194–1205.

22. Kiyono H, McGhee JR, Mosteller LM, et al: Murine Peyer's patch T cell clones: characterisation of antigen-specific helper T cells for immunoglobulin A responses. *J Exp Med* 1982;156:1115–1130.

23. Lee F, Yokoto T, Otsuka T, et al: Isolation and characterization of a mouse interleukin cDNA clone that expresses B-cell stimulatory factor 1 activities and T-cell and mast-cell-stimulating activities. *Proc Natl Acad Sci USA* 1986;83:2061–2065.

24. Lucius B, Buttner DW, Kirsten C, Diesfield HJ: A study on antigen recognition by onchocerciasis patients with different clinical forms of the disease. *Parasitology* 1985;92:569–580.

25. Luzzati AL, Jacobson EB: Serum immunoglobulin levels in nude mice. *Eur J Immunol* 1972;2:473–474.

26. Maizels RM, Philipp M, Ogilvie BM: Molecules on the surface of parasitic nematodes as probes of the immune response to infection. *Immunol Rev* 1982;61:111–136.

27. Maurer PH, Callahan HJ: Proteins and polypeptides as antigens. *Methods Enzymol* 1980;70:49–70.

28. Noma Y, Sideras P, Naito T, et al: Cloning of cDNA encoding the murine IgG_1 inducing factor by a novel strategy using the SP6 promoter. *Nature* 1986;319:640–646.

29. Ortega-Pierres G, Chayen A, Clark NWT, Parkhouse RME: The occurrence of antibodies to hidden and exposed determinants of surface antigens of Trichinella spiralis. *Parasitology* 1984;88:359–369.

30. Ortega-Pierres G, Clark NWT, Parkhouse RME: Regional specialisation of the surface of a parasitic nematode. *Parasite Immunol* 1986;8:613–617.

31. Ortega-Pierres G, Mackenzie CD, Parkhouse RME: Protection against Trichinella spiralis induced by a monoclonal antibody that promotes killing of newborn larvae by granulocytes. *Parasite Immunol* 1984;6:275–284.

32. Parkhouse RME (ed): Parasite antigens in disease diagnosis and evasion. *Curr Top Microbiol Immunol* 1985;120.

33. Parkhouse RME, Almond NM: Stage specific antigens of Trichinella spiralis. *Biochem Soc Trans* 1985;13:426–428.

34. Parkhouse RME, Clark NWT: Stage specific secreted and somatic antigens of Trichinella spiralis. *Mol Biol Parasitol* 1983;9:319–327.

35. Parkhouse RME, Philipp M, Ogilvie BM: Characterization of surface antigens of Trichinella spiralis infective larvae. *Parasite Immunol* 1981;3:339–352.

36. Philipp M, Parkhouse RME, Ogilvie BM: Changing proteins on the surface of a parasitic nematode. *Nature* 1980;287:538–540.

37. Philipp M, Taylor PM, Parkhouse RME, Ogilvie BM: Immune responses to stage specific antigens of the parasitic nematode Trichinella spiralis. *J Exp Med* 1981;154:210–215.

38. Philipp M, Worms MJ, McLaren DJ, et al: Surface proteins of a filarial nematode: a major soluble antigen and a host component on the cuticle of Litomosoides carinii. *Parasite Immunol* 1984;6:63–82.

39. Rosenberg Y, Chiller JM: Ability of antigen specific helper cells to effect a class restricted increase in total Ig-secreting cells in spleens after immunisation with antigen. *J Exp Med* 1979;150:517–530.
40. Schluederberg A: Immunoglobulin profiles provide new insight into infectious disease. *Yale J Biol Med* 1982;55:317–320.
41. Schroeder LL: An ELISA to detect antibody specific for surface antigens of parasitic nematodes. *J Immunol Methods* 1985;83:135–139.
42. Sideras P, Bergstedt-Lindquist S, MacDonald HR, Severinson E: Secretion of IgG_1 induction factor by T cell clones and hybridomas. *Eur J Immunol* 1985;15:586–593.
43. Silberstein DS, Despommier DD: Antigens from Trichinella spiralis that induce a protective response in the mouse. *J Immunol* 1984;132:898–904.
44. Slack J, Der Bolian GP, Nahma M, Davie JM: Subclass restriction of murine antibodies. *J Exp Med* 1980;151:853–862.
45. Torrigiani GL: Quantitative estimation of antibody in the immunoglobulin classes of the mouse. II. Thymic dependence of the different classes. *J Immunol* 1972;108:161–164.
46. Vitetta ES, Ohara J, Myers CD, et al: Serological, biochemical and functional identity of B cell stimulating factor 1 and B cell differentiation factor for IgG_1. *J Exp Med* 1985;162:1726–1731.
47. Weiss N, Gualzat M, Wyss T, Betschart B: Detection of IgE-binding Onchocerca volvulus antigen after electrophoretic and immunoenzyme reaction. *Acta Trop (Basel)* 1982;39:373–377.
48. Weiss N, Hussain R, Ottesen EA: IgE antibodies are more species specific than IgG antibodies in human onchocerciasis and lymphatic filariasis. *Immunology* 1982;45:129–137.

CHAPTER 25

Immunodiagnosis of Filariasis

Renu B. Lal

Human lymphatic filariasis, estimated to affect approximately 100 million individuals worldwide, is caused by either lymphatic-dwelling parasites, *Wuchereria bancrofti* and *Brugia malayi,* or a subcutaneous dweller, *Onchocerca volvulus*. These filarial parasites have several important features in common. First, they are all transmitted by biting arthropods (mosquitoes, flies, or midges) and all go through complex life cycles that include slow maturation (often 3 to 12 months) from the infective larval stages carried by the insects to the adult worms (4 to 8 cm in length), which reside either associated with the lymph nodes of the lymphatic system or in the subcutaneous tissue; offspring of these adults, the microfilariae, are 200 to 300 μm in length and either circulate in the blood (*W. bancrofti* and *B. malayi*) or migrate through the skin *(O. volvulus)* while waiting to be ingested by the insect vectors destined to propagate the parasite's life cycle. Second, patent infection is generally not established unless exposure to infective larvae is intense and prolonged. Third, though exposure to infection occurs throughout childhood in endemic regions, most of the pathology of these infections is found in the adult population.

One of the most intriguing aspects of these filarial infections, especially those caused by lymphatic-dwelling parasites, is the broad spectrum of clinical presentation found among individuals in endemic regions (18,24). At one extreme are the many individuals with no clinical manifestations or indications of filarial infection at all, despite clear-cut exposure to the infective larvae. Whether these patients are normal, exposed but not infected, or infected (though clinically asymptomatic) remains to be clarified. A second group, also entirely asymptomatic, is characterized by the presence of microfilariae circulating in the peripheral blood.

Among the symptomatic clinical syndromes, the most common is that characterized by chronic pathology (CP), which includes elephantiasis of the limbs or other regions, hydrocele, and chyluria. The other clinical manifestation is the tropical pulmonary eosinophilia (TPE) syndrome, which is characterized by asthmatic symptoms and later by chronic interstitial lung disease.

The diagnosis of filarial infection in humans depends on either parasitological examination or host antibody responses. As adult worms are essentially never found, parasitological diagnosis is usually made by finding microfilariae in blood or other body fluids. Because *W. bancrofti, B. malayi,* and *B. timori* are nocturnally periodic in most geographic regions, venipuncture must be performed between 2200 and 0200 hours (10 p.m. and 2:00 a.m), when peak parasitemia is reached; however, in many phases of filarial infection, such circulating microfilariae are not detectable (e.g., in prepatency or postpatent, "cryptic" states), so that parasitological diagnosis by conventional means is unfeasible for many infected individuals.

Application of immunologic methods to the diagnosis of lymphatic filariasis has, in general, focused on detection of the host's antibody response to the established parasite (8,19). Serologic assays have, to date, been ineffective in distinguishing between exposure to infective larvae and current or past patent infection. Furthermore, the usefulness of these antibody assays is limited by the problem of extensive cross-reactivity among the antigens of helminthic parasites. Because of these and other limitations in such serological techniques, attention has focused on the detection of parasite antigens in infected patients' blood and other body fluids (14,16).

A number of attempts have been made to detect circulating antigen in lymphatic filarial infections and onchocerciasis. The polyethylene glycol precipitation of immune complexes followed by an ELISA has been utilized to detect circulating filarial antigen in sera from patients with bancroftian filariasis (7,25). Of the various antigen detection systems thus far developed for use in lymphatic filariasis, the two-site immunoassays appear to be the most promising because of ease of performance and the potential for extreme assay sensitivity (15). Using such an assay, the antigenic cross-reactivity of a monoclonal antibody directed against the eggs of the cattle filaria *Onchocerca gibsoni* has been exploited to detect *W. bancrofti* antigen in sera from infected humans (6,10).

Using monoclonal or polyclonal probes, the circulating filarial parasite antigens have been detected in humans infected with *W. bancrofti* (4,10,15,18,28), *B. malayi* (1), and *O. volvulus* (5,20) and in animals infected with *D. immitis* (27), *B. pahangi* (1), and *L. carinii* (3). Most studies have found that parasite antigen is most readily detectable in microfilaremic subjects and is much more difficult to identify in other clinical groups of patients. The inability to detect parasite antigen in these other groups of patients has been attributed to a number of potential problems, including: (a) insufficient sensitivity for detecting the low levels of parasite antigen in patients with either light or amicrofilaremic filarial syndromes; (b) parasite antigen's being complexed with antibody; and (c) insufficient specificity of the antibody probe (either polyclonal or monoclonal) used in these assays.

One of the antigens being detected in these assays has been shown to be a high-molecular-weight [$\sim$ 200 kilodalton (kDa)] glycoprotein (21). The development and characterization of monoclonal antibodies raised against this antigen and the use of these antibodies to detect parasite antigen in the sera of individuals with *W. bancrofti* infection is described. Interestingly, most of the monoclonal antibodies resulting from immunization of mice with this antigen were directed against phosphocholine (PC) epitopes on the parasite antigen. Although this PC determinant is itself not filarial-specific, its abundance on PC-bearing filarial antigens in the circulation makes it a potentially useful target for the immunodiagnosis of infected individuals.

Monoclonal Antibodies to Phosphocholine

Monoclonal antibody made against the 200-kDa antigen (CA_{101}) was shown to be directed against the PC determinant. The anti-PC monoclonal antibodies recognized a broad range of antigens in the *B. malayi* adult worm (BmA) extract, ranging from 200 to 21 kDa, suggesting that a common epitope is present on various antigens. Three other anti-PC monoclonal antibodies [HPC-M_2 and HPC-G_{12} of nonparasite origin (11) and Gib-13 made against extracts of *O. gibsoni* (10)] showed essentially similar banding patterns. The other stages of the parasite, both larvae and microfilariae, were also found to have antigens bearing PC determinants.

Detection of Circulating Parasite Antigen Using Monoclonal Antibody

A monoclonal antibody (CA_{101})-based ELISA was capable of detecting circulating parasite antigen in lymphatic as well as nonlymphatic filarial patients. The antigen was detectable in both periodic (Madras, India) and subperiodic (Cook Island, South Pacific) forms of *W. bancrofti*. The mean antigen levels detected by CA_{101} in the sera of *W. bancrofti*-infected subjects from an endemic area of Madras, India were higher in microfilaremic subjects (geometric mean, 308 ng/ml) compared to the amicrofilaremic patients (CP 27 ng/ml and TPE 33 ng/ml) ($p < 0.004$) (Fig. 25.1). The antigen was present in 50% of patients with CP, 93% of patients with MF, and 56% of patients with TPE. Nineteen North American controls gave values between 0 and 10 ng/ml. This range was defined as negative. Among the nonlymphatic group, the PC antigen was present in 75% of patients infected with *Onchocerca volvulus* (geometric mean 45.3 ng/ml) from an endemic area of Guatemala. Low levels of antigen (24.4 ng/ml) were also detected

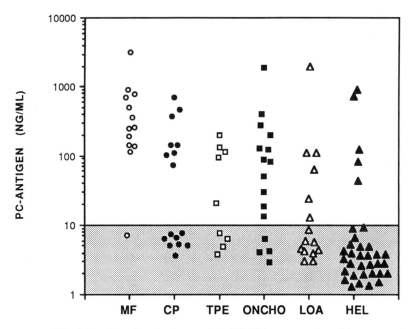

Fig. 25.1. Filarial antigen levels detected in ELISA assay using CA_{101} as probe. Results are expressed as nanogram per milliliter equivalents of filarial antigen interpolated from the reference dose-response curves. Sera from patients with *W. bancrofti* infection were studied: CP = chronic lymphatic pathology. TPE = tropical pulmonary eosinophilia. MF = asymptomatic microfilaremia. Patients infected with *O. volvulus* (ONCHO) and loa loa (LOA) were studied for nonlymphatic forms of filariasis. Among other helminth infections, twelve patients with schistosomiasis, sixteen patients with strongyloidiasis, and two patients each with hookworm infection, trichinosis, and echinococcosis were studied.

in patients infected with Loa loa (Fig. 25.1). Among the nonfilarial helminth-infected group, four patient with strongyloidiasis and one with schistosomiasis was positive.

Phosphocholine as an antigenic determinant has been described in numerous species, e.g., *Streptococcus pneumoniae* (2), *Proteus morganii* (29), *Ascaris suum* (13), *Nippostrongylus brasiliensis* (22), *Haemonchus contortus* (23), and *Toxocara canis* (26). Expression of PC-bearing antigenic determinants has been demonstrated for the filarial parasite of rodents *Dipetalonema viteae* (12), and these determinants were found mainly on certain internal structures (egg, uterine, and intestinal membranes) but not the cuticle. It is possible that these PC antigens are released into the circulation by excretory-secretory processes, in moulting fluids, or as breakdown products that then become detectable by monoclonal antibodies to the PC determinant.

Identification of PC Antigen in Human Sera

Analysis of affinity absorbed target antigens in patients' plasma by immunoblotting and probing with various anti-PC antibodies (Fig. 25.2) demonstrated distinct molecular weight heterogeneity of these circulating, PC-bearing antigens. The presence of 200-, 160-, and 72-kDa circulating antigens has been demonstrated in the limited number of plasma samples studied. The target antigens of Gib-13 have been shown to be approximately 140, 52, 56, and 62 kDa in *W. bancrofti* sera, the latter two molecules also being detected in sera of *O. gibsoni*-infected cattle (9). A serum antigen of 52 kDa and a urine antigen of 67 kDa have also been described for patients with bancroftian filariasis as recognized by Gib-13 (6). The differences in reported findings defining which antigen molecules bear PC determinants could result from differential host responsiveness of individuals from different parts of the world, from antigenic distinction among the parasites themselves, or from technical artifacts in the laboratory es-

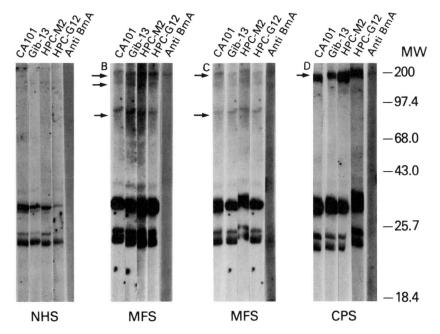

Fig. 25.2. PC antigen determinant of circulating filarial antigens: human serum reacted with anti-BmA Sepharose and probed with different anti-PC MAbs as designated at the top of the figure. **A:** Normal human serum (NHS): antigen < 10 ng/ml. **B:** Microfilaremic serum (MFS): antigen 780 ng/ml. **C:** Microfilaremic serum (MFS): antigen 1200 ng/ml. **D:** Serum from chronic pathology patient (CPS): antigen 700 ng/ml.

timates of molecular weight. Regardless of such reported differences, however, PC is clearly a prominent antigenic determinant of several circulating filarial antigens.

Conclusion

The circulating antigen detected by anti-PC monoclonal antibodies appears operationally to be filaria-specific, as sera from subjects who were infected with nonfilarial helminth parasites only rarely were found to have significant levels of antigen. The antigen was detectable in both periodic and subperiodic forms of lymphatic filariasis and was also present in a nonlymphatic form of filariasis. Thus despite the biological nonspecificity of PC, the detection of PC-containing antigens in the blood may have great practical usefulness. However, more extensive evaluations of the specificity of this assay for other filariae as well as for other types of parasitic (helminthic and protozoal) infections found in filariasis-endemic areas need to be carried out.

References

1. Au AC, Denham DA, Steward MW, et al: Detection of circulating antigens and immune complexes in feline and lymphatic filariasis. *Southeast Asian J Trop Med Hyg* 1981;12:492–498
2. Brundish DE, Baddiley J: Pneumococcal C-substrate, a ribitol teichoic acid containing choline phosphate. *Biochem J* 1968;110:573–577.
3. Dasgupta A, Bala S: Litomosoides carinii—soluble antigen in circulation and immunosuppression in vitro. *Indian J Med Res* 1978;67:30–36.
4. Dasgupta A, Bala S, Dutta SN: Lymphatic filariasis in man: demonstration of circulating antigens in Wuchereria bancrofti infection. *Parasite Immunol* 1984;6:341–348.
5. Des Moutis I, Ouaissi A, Grzych JM, et al: Onchocerca volvulus: detection of circulating antigen by monoclonal antibodies in human onchocerciasis. *Am J Trop Med Hyg* 1983;32:533–542.
6. Dissanayake S, Forsyth KP, Ismail MM, Mitchell GF: Detection of circulating antigen in bancroftian filariasis using a monoclonal antibody. *Am J Trop Med Hyg* 1984;33:1130–1140.
7. Dissanayake S, Galahitiyawa SC, Ismail MM: Immune complexes in Wuchereria bancrofti infection in man. *Bull WHO* 1982;60:919–927.
8. Dissanayake S, Ismail MM: Antibody determination in the diagnosis of Wuchereria bancrofti infections in man. *Bull WHO* 1981;59:753–757.
9. Forsyth KP, Mitchell GF, Copeman DB: Onchocerca gibsoni: increase of circulating egg antigen with chemotherapy in bovines. *Exp Parasitol* 1984;58:41–55.
10. Forsyth KP, Spark R, Kazura J, et al: A monoclonal antibody based immunoradiometric assay for detection of circulating antigens in bancroftian filariasis. *J Immunol* 1985;134:1172–1177.

11. Gearhart PJ, Johnson ND, Douglas R, Hood L: IgG antibodies to phosphorylcholine exhibit more diversity than their IgM counterpart. *Nature* 1981;291:29–34.
12. Gualzate M, Weiss N, Heusser CH: Dipetalonema vitae: phosphorylcholine and non phosphorylcholine antigenic determinants in infective larvae and adult worms. *Exp Parasitol* 1986;61:95–102.
13. Gutman GA, Mitchell GF: Ascaris suum: location of phosphorylcholine in lung larvae. *Exp Parasitol* 1977;43:161–168.
14. Hamilton RG: Application of immunoassay methods in the serodiagnosis of human filariasis. *Rev Infect Dis* 1985;7:837–843.
15. Hamilton RG, Hussain R, Ottesen EA: Immunoradiometric assay for detection of filarial antigens in human serum. *J Immunol* 1984;133:2237–2242.
16. Harinath BC: Immunodiagnosis of bancroftian filariasis—problems and progress. *J Biosci* 1984;6:691–699.
17. Kagan IG: Serodiagnosis of parasitic diseases, Lennette EH, Balros A, Ju Mausler WLH, Tsuant JP (eds): *Manual of Clinical Microbiology,* 3rd ed. Washington, DC, American Society for Microbiology, 1980, pp 724–729.
18. Lal R.B, Paranjape RS, Briles DE et al Circulating parasite antigen(s) in lymphatic filariasis: Use of monoclonal antibodies to phosphocholine for immunodiagnosis. *J. Immunol,* 1987;138:3454–3460.
19. Ottesen EA: Immunopathology of lymphatic filariasis in man. *Springer Semin Immunopathol* 1980;2:373–385.
20. Ouaissi A, Kouemeni LE, Haque A, et al: Detection of circulating antigens in onchocerciasis. *Am J Trop Med Hyg* 1981;30:1211–1218.
21. Paranjape RS, Hussain R, Nutman RB, et al: Identification of circulating parasite antigen in patients with bancroftian filariasis. *Clin Exp Immunol* 1986;63:508–516.
22. Pery P, Luffau G, Charely J, et al: Phosphorylcholine antigens from Nippostrongylus brasiliensis. I. Antiphosphorylcholine antibodies in infected rat and location of phosphorylcholine antigens. *Ann Immunol* 1979;130C:879–888.
23. Pery P, Petit A, Poulain J, Luffau G: Phosphorylcholine bearing components in homogenates of nematodes. *Eur J Immunol* 1974;4:637–639.
24. Piessens WF, Mackenzie CD: Immunology of lymphatic filariasis and onchocerciasis, in Cohen S, Warren KS (eds): *Immunology of Parasite Infections,* Vol 2. Oxford, Blackwell Scientific, 1982, pp 622–653.
25. Prasad GBKS, Reddy MVR, Harinath BC: Detection of filarial antigen in immuno complexes in bancroftian filariasis by ELISA. *Indian J Med Res* 1983;78:780–783.
26. Sugane K, Oshima T: Activation of complement in C-reactive protein positive sera by phosphorylcholine-bearing component isolated from parasite extract. *Parasite Immunol* 1983;5:385–395.
27. Weil GJ, Malone MS, Powers KG, Blair LS: Monoclonal antibodies to parasite antigens found in the serum of Dirofilaria immitis infected dogs. *J Immunol* 1985;134:1185–1191.
28. Weil GJ, Kumar H, Santhanam S, et al: Detection of circulating parasite antigen in bancroftian filariasis by counter immunoelectrophoresis. *Am J Trop Med Hyg* 1986;35:565–570.
29. Williams KR, Claflin JL: Clonotypes of anti-phosphocholine antibodies induced with Proteus morganii (Potter). I. Structural and idiotypic similarities in a diverse repertoire. *J Immunol* 1980;125:2429–2437.

Part VII
Veterinary Vaccines

CHAPTER 26

Development, Production, and Application of Vaccines in Foot-and-Mouth Disease Control in India

B.U. Rao

Foot-and-mouth disease is a crucial animal health hazard, being widespread and endemic in the vast Indian subcontinent, which has a more than 400 million susceptible livestock population. It causes an annual economic loss of many millions of rupees for the country. India may be considered an ideal habitat for the infection to persist, spread, and flourish over centuries with its agroclimatic, socioeconomic conditions. It has mixed farming of different species of livestock, which together provide a conducive ecological milieu for the pathogen. The disease is caused by a single-stranded positive-sense RNA virus of 8 kilobases (kb) length, which itself is a fascinating organism or macromolecule.

Under the existing circumstances and epidemiological situation, the only pragmatic approach for disease control is immunoprophylaxis with supportive zoosanitary and legislative measures. For this purpose, the most important measure is the development and production of a safe, cost-effective vaccine in adequate quantity; then, after its development, it must be made available for judicious application.

There is a vast literature on the subject, but it is not reviewed here. Relevant information is presented, however, with particular reference to vaccines, based on the author's experience, that are considered useful for foot-and-mouth disease control (1).

Problem

The annual economic loss is estimated to be many millions of rupees due to damage to animal health. Such damage reduces production and agricultural operations, including transport, thereby hampering the development of industry and the export market for livestock and their products.

The problem of disease control is complex owing to the interaction of numerous known factors, such as virus variation with many types and subtypes, carrier status of infected or recovered animals, the epidemiological situation with the widespread endemic nature of the infection including periodic disease outbreaks, the varying immune status, and the susceptibility of the population, which consists of several species and age groups. Also to be considered is the emergence of virus mutants, the unrestricted widespread movements and congregation of animals, and the prevailing socioeconomic and farming conditions and practices.

Epidemiology

The epidemiology of foot-and-mouth disease is complex and difficult to understand with its changing patterns of frequency, spread, and the total number of disease outbreaks that occur over a particular period especially in a vast endemic zone such as India, with its different ecological patterns. However, there are three important interacting factors involved: the host population, the virus population, and the environment. The number of foot-and-mouth disease outbreaks reported from various sources were compiled and are depicted in Figure 26.1. The actual number of outbreaks may be higher and difficult to identify.

From the results of processed specimens from disease outbreaks over the period, the prevalence of only four types of foot-and-mouth disease virus (O, A, C, and Asia-1) was confirmed (Fig. 26.2). The virus types of

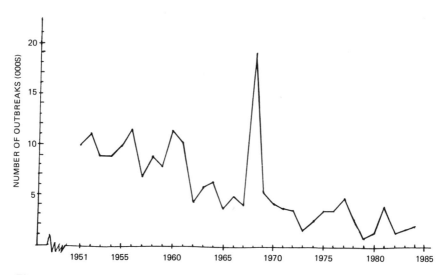

Fig. 26.1. Foot-and-mouth disease in India: 1951–1984: annual number of reported outbreaks. [Data from the *Animal Health Information Bulletin* and Annual Reports of Project Coordinator, AICRP on FMD (ICAR).]

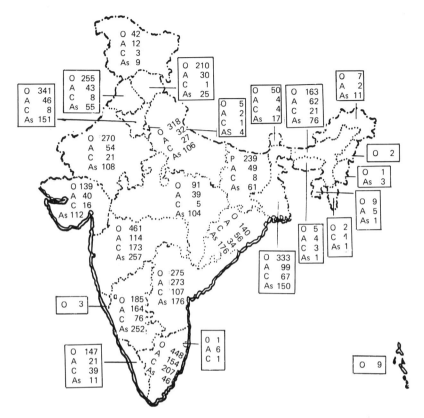

Fig. 26.2. Foot-and-mouth disease: virus types occurrence and distribution: 1971–1985.

South African Territory (SAT) I, II, and III have not yet been encountered in India. The number of foot-and-mouth disease virus types isolated and identified in India during the decade 1976–1985 are shown in Figure 26.3. Virus type O is the most preponderant and widely occurring, followed by types Asia-1 and A with peaks at different periods; type C has shown a low profile.

An All India Coordinated Research Project for epidemiological studies on foot-and-mouth disease was sanctioned by the Indian Council of Agricultural Research and has been in effect since 1971, with a network of Epidemiological Units and Regional Typing Centres backed up by a Central Virus Typing Laboratory and coordinated by a Project Coordinator. The two most important events that occurred have been with the virus type A in 1966, with the introduction of new subtype A_{22} and Asia-1 in 1976–1977 with a mutant strain. Subtype A_{22} was widespread in epidemics and established itself in nearly the entire country.

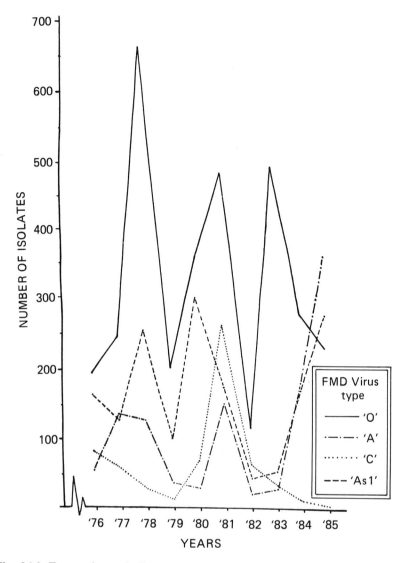

Fig. 26.3. Foot-and-mouth disease: virus types isolated and identified in India: 1976–1985. [Data from Annual Reports of Project Coordinator, AICRP on FMD (ICAR).]

Laboratory Techniques

For the purpose of research, development of vaccines, and diagnosis, various laboratory techniques, e.g., hybridoma, enzyme-linked immunosorbent assay (ELISA), radioimmunoassay, immunoblotting, electrofocusing,

DNA probe, virus crystallization, and hexokinase activity in virus-infected cell cultures, have been standardized and are being used.

In order to develop monoclonal antibodies by the hybridoma technique, cell fusion experiments were carried out using FMD virus types Asia-1 and A_{22} for immunization of BALB/c mice. Suitable protocols were developed for the efficient production of hybridomas (Fig. 26.4). Four clones secreting virus-specific monoclonal antibodies (mAb) to each of the vaccine strains of Asia-1 and A_{22} virus were obtained. The mAbs produced by Asia-1 clones were characterized using ELISA, virus neutralization, and Western blotting techniques. All four clones were positive in ELISA. However, the mAb produced by only one clone was virus-neutralizing in character. With the Western blotting technique, the mAbs produced by this clone were found to recognize the sites on VP1. These mAbs were also used to identify the recombinant DNA product.

Although the mAb produced by the positive clones in cell culture was less, sufficient mAb could be obtained by generating ascites in pristane-treated BALB/c mice. Attempts were also made to step up the growth of the positive clone in cell culture in order to obtain adequate quantities of the mAb.

Various methods of ELISA have been applied in FMD virus research. A double-sandwich micro-ELISA test has been standardized and used to estimate the 146 S viral content in the viral harvest. A sandwich ELISA employing species-specific antibovine immunoglobulin G (IgG)–HRPO (horse radish peroxidase) conjugate has been standardized and applied for detection and estimation of FMD virus antibodies in cattle sera. A competitive ELISA was used to estimate the recombinant DNA product.

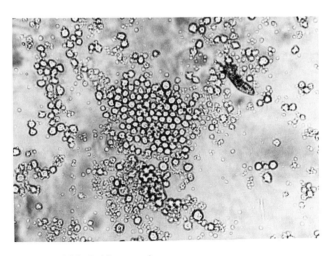

Fig. 26.4. Ten-day-old hybridoma culture.

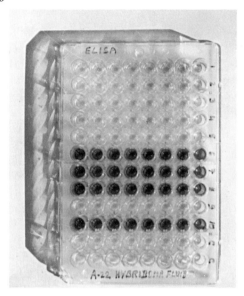

Fig. 26.5. ELISA for detection of monoclonal antibodies in culture fluid.

A highly sensitive sandwich ELISA using FMD virus-coated plates, biotinylated goat antimouse antibody, and streptavidin–peroxidase conjugate was standardized and employed for initial screening of hybridomas (Fig. 26.5). This test was able to detect even weakly positive clones. A modified version of this technique in which the plates were coated with virus type-specific guinea pig immune serum was used to determine the virus type specificity of the mAb.

Western blotting technique in which the viral polypeptides were separated by sodium doderyl sulfate–polyacrylamide electrophoresis (SDS-PAGE) in the presence of urea and then transferred to nitrocellulose filters using a transblot apparatus was standardized. This technique was used to characterize the recombinant DNA product as well as the monoclonal antibodies by revealing them by autoradiography or ELISA. The polypeptides of FMD virus were also separated by electrofocusing. Nucleotide sequencing of the cDNA for major antigen of virus types Asia-1 and O has been initiated.

Attempts were made to develop DNA probes for identification of virus-specific sequences in infected tissues. The cDNA for virus type Asia-1 RNA was used as a probe to detect the FMD virus-specific sequences in infected tissues. Positive signals could be obtained from cell cultures and guinea pig tissues infected with the virus.

To understand the structural configuration of FMD virus, work on virus crystallization was initiated. Crystals of different sizes and morphology were obtained as a result of the preliminary work undertaken with purified virus types O, A, and C. Further studies are under way to confirm the specificity of virus crystals and to determine the x-ray diffraction pattern.

A method to estimate the exact amount of the intact FMD virus particles directly in the cell culture viral harvests using sucrose gradient by ultracentrifugation has been standardized and is being used.

Studies on the hexokinase activity in FMD virus-infected BHK_{21} cells showed a positive correlation between the virus yield and enzyme activity. This enzyme activity could therefore be used as a marker to determine the exact time of virus harvest for vaccine production. In large-scale vaccine production using bioreactors, it is possible to estimate the enzyme activity by developing suitable probes and thereby determine the optimum time of virus harvesting.

Facilities have been developed for application of a radioimmunoassay for FMD virus research. Work on the use of this technique for detection of virus as well as antibodies has been initiated.

Vaccines

Research

The most important single component is cost-effective vaccines, for which there is an urgent need. It is therefore worthwhile to examine the known available alternatives for conducting research and developing immunogens. The methodology for the research and development of vaccines could be chosen broadly from the following.

1. For conventional inactivated virus vaccines, there may be screening and selection of a seed virus representative of a broad antigenic spectrum, including epizootic wild strains, and a suitable donor host system for bulk antigen production with appropriate chemical inactivants and adjuvants.
2. Recombinant DNA technology may be applied for cloning genes of virus strains in a suitable vector in host systems such as *Escherichia coli, Bacillus subtilis,* and virus-like vaccinia either as vehicles and for synthesis of immunogens.
3. Synthetic polypeptides may be obtained by a chemical process.
4. Attenuated stable strains may be obtained by modification of virulent virus via manipulation in heterologous host systems, temperature-sensitive cold mutants, or identification and deletion of virulence genes.
5. The role of nonspecific immunostimulators, regulators, or modulators may be examined singly or in combination with specific immunogens to improve the immunogenic response or to confer protection.

It was first necessary to develop the expertise required to conduct research with recombinant DNA technology as an alternative method for the development of subunit immunogens that are safe and effective. The research program conducted at our institute in collaboration with the Indian Institute of Science, Bangalore, is summarized here.

Virus type Asia-1 vaccine strain (63/72 Mukteswar isolate) was selected as a model for research. The viral RNA was isolated, purified, and fractionated into A^+ and A^- RNA fractions. Complementary DNA for A^+ RNA was synthesized and directly cloned in the expression-vector-carrying promoter region of P-galactosidase gene (2). The clones were screened for the production of major antigen (VP1) using the double-sandwich enzyme-linked immunosorbent assay (ELISA) (Fig. 26.6).

The positive clones were further screened for maximum production of the VP1 protein. The clone producing VP1 in large quantities was selected. The presence of VP1 sequences was tested for by partial restriction mapping and immunoprecipitation experiments (Fig. 26.7). The extract prepared from the clone was injected into guinea pigs, and the specific immune response was confirmed. Preliminary immune response studies in cattle are in progress. Other studies indicated that the expressed protein isolated by 40 to 60% ammonium sulfate precipitation competed with the whole virus in ELISA tests.

Cloning of the cDNA for virus type A_5 viral RNA was initiated. Viral RNA was prepared by the following published procedure. The complementary DNA for the viral RNA was prepared, as was the recombinant

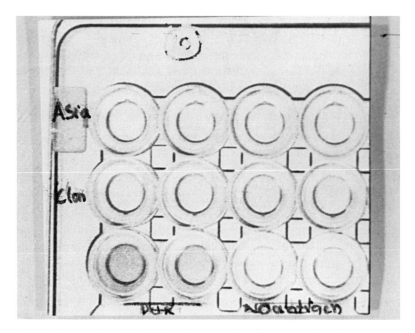

Fig. 26.6. Competitive ELISA for the detection of expressed antigen.

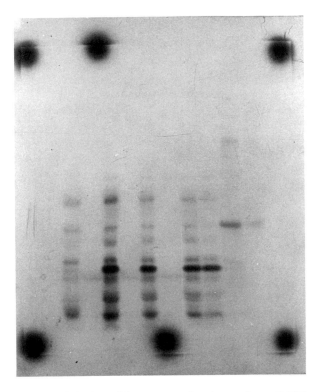

Fig. 26.7. Immunoprecipitation of the expressed antigen using anti-VP1 antibodies.

DNA with PBR 322 plasmid. Preliminary work with virus type O is also in progress.

Development

During the past three decades, there has been a complete transformation in the production technology for inactivated FMD virus vaccines. In brief, the donor host system used for propagation of the virus has changed from the tongue epithelium of live animals, to explants of the same from slaughtered cattle, to primary cell cultures, to continuous cell lines. The cells have been grown on stationary glass surfaces, roller bottles, microcarriers, and at present fermenter vessels of different capacities.

The development efforts were initially concentrated on the technical know-how of handling biosystems such as cell lines, screening and selection of FMD virus strains, and handling of fermenters. The required monolayer and suspension cell lines of BHK_{21} Cl_{13} were obtained, grown, and maintained in suitable medium prepared with indigenously available chemicals. The cells were tested for susceptibility to various FMD virus types. The most susceptible cells for each type of virus were selected. Cattle-tongue-passaged vaccine strains of virus types O, A, C, and Asia-1 were adapted to BHK_{21} cells. The pilot scale vaccine production with field trials were undertaken, and the vaccine was found potent.

The slaughterhouse cattle serum was treated with polyethylene glycol (PEG) to remove the antibodies and used for media preparation. It was tested for microbial load, toxicity, and ability to support cell growth.

Because the vaccines adjuvanted with aluminum hydroxide gel are not potent in pigs, vaccines adjuvanted with either mineral oil or DEAE-dextran have been developed especially for use in these animals. A beginning has been made in the development of such oil-adjuvanted vaccines, and they are being evaluated in sheep and cattle.

Because the cost of handling, preservation, packing, and transport of conventional virus-adsorbed aluminum hydroxide gel vaccines is high, an attempt has been made to develop freeze-dried, inactivated virus vaccines. Freeze-dried vaccines of virus types A, C, and Asia-1 adjuvanted with "Quil-A" alone were found to be potent when tested in animals.

Ultrafiltration using membranes with a suitable molecular weight cutoff value was found to be an effective method of virus concentration, with maximum recovery of virus antigen. An alternate, simpler method for virus concentration by PEG treatment and filtration is also being tried that has helped in the development of potent vaccines with reduced volume of dose.

The microcarrier cell culture technique was standardized in either roller bottles or stirred culture vessels of 5 L capacity, with an increase in the yield of cells and virus antigen for producing highly potent vaccines. This method was useful for obtaining a high concentration of virus antigen for either diagnostic purpose or the production of vaccines for pigs.

Limited attempts have been made to develop live attenuated strains by propagation of selected virulent strains of O, A, C, and Asia-1 in a heterologous host system such as mice and testing for their immunogenicity. The results have been encouraging. The application of live attenuated modified strains has not been explored owing to a lack of stability and the varying virulence of the strains to different species and age groups of animals by the conventional method of developing of such strains. However, it is expected that the live attenuated virus vaccines will be comparatively cost-effective, especially in endemic zones and for use in relatively resistant, large indigenous population.

The approach of nonspecific immunostimulation using *M. phlei* has given encouraging results and requires further investigation.

Production

At present, large-scale cell culture technology using BHK_{21} cells in continuous suspension systems in fermenter vessels each of 300 L capacity

with aluminum hydroxide gel and saponin as adjuvants and either formaldehyde or acetylethylenimine as inactivant is in vogue for the production of conventional inactivated vaccines.

At present four institutions involved in vaccine manufacture have col-
laborated internationally in terms of technical know-how and instrumen-
tation. The present manufacturing capacity in India is about 45 million
quadruvalent vaccine doses per annum. The production technology at one
of the institutions is depicted in Figure 26.8.

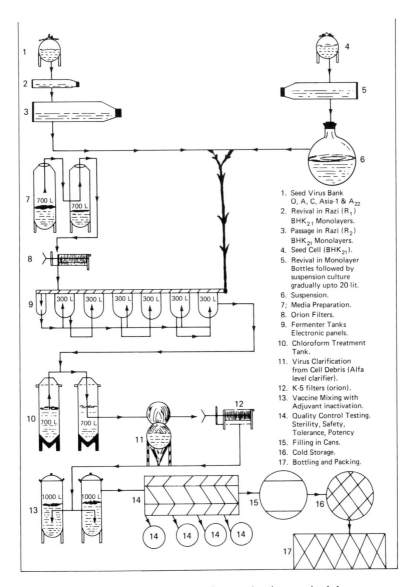

1. Seed Virus Bank
   O, A, C, Asia-1 & A_{22}
2. Revival in Razi (R_1)
   BHK_{21} Monolayers.
3. Passage in Razi (R_2)
   BHK_{21} Monolayers.
4. Seed Cell (BHK_{21}).
5. Revival in Monolayer
   Bottles followed by
   suspension culture
   gradually upto 20 lit.
6. Suspension.
7; Media Preparation.
8. Orion Filters.
9. Fermenter Tanks
   Electronic panels.
10. Chloroform Treatment
    Tank.
11. Virus Clarification
    from Cell Debris (Alfa
    level clarifier).
12. K-5 filters (orion).
13. Vaccine Mixing with
    Adjuvant inactivation.
14. Quality Control Testing.
    Sterility, Safety,
    Tolerance, Potency
15. Filling in Cans.
16. Cold Storage.
17. Bottling and Packing.

Fig. 26.8. Foot-and-mouth disease vaccine production methodology.

These institutes have built up an infrastructure fulfilling P2/P3 containment facilities with biosafety and disease security systems using controlled environmental conditions: air filtration through Pre and Hepa filters, affluent treatment of laboratory washings, restricted entry with change of clothing and bath system, incineration to ash of infected materials, and supportive services of an uninterrupted, continuous supply of electricity, water as per specifications, and communication.

The policy and programs of the state and central government departments largely determine the investments to be made on the various aspects, including requirements and production of vaccines. The production of vaccine is geared to demand, which at present is less than the manufacturing capability. However, it is expected to increase with the programs now being developed for disease control.

The present vaccines contain selected representative, appropriate strains of virus types O, A, C, and Asia-1. A battery of in vitro and in vivo laboratory tests for sterility, safety, and potency are being used for quality control. In brief, samples at different stages of vaccine processing are screened: the serum and environment for bacterial load; cell culture for contamination, especially with bacterial and mycoplasma; karyotyping of cells with master and working seed cell cultures and for virus strains; estimation of complement-fixing units; infectivity titers; plaque characteristics; 146S content and protein. Moreover, biological tests for safety in cattle and potency in guinea pigs and cattle are conducted routinely. The quality control of vaccine at present is the primary responsibility of the manufacturer. However, there is a need to develop and strictly enforce the regulations under the Drugs and Cosmetics Act to ensure a quality, effective product.

The sale price is about rupees 3.50 per quadruvalent dose of inactivated virus vaccine. This cost may be subsidized by the state and/or the federal agency in order to support specific programs for disease control. The half-life of the vaccine at 4 to 8°C has been observed to preserve its potency up to not less than a period of 24 months.

Packaging is done by placing the vaccine vials with labeled circular instructions in thermocol boxes with ice packs. The packs are then enclosed in corrugated cartons.

Application of Vaccines for Disease Control

Foot-and-mouth disease control is difficult because of the various interacting factors and the prevailing conditions in India. However, there is an urgent need, felt by all concerned, to plan the strategy of disease control and implement the program. A National Advisory Committee for this purpose has been constituted by the Government of India.

In brief, the policy/program adopted by a number of countries may be stated as follows: For free zones, there is prevention of contact with infection, strict quarantine of livestock and their products, stamping out (by slaughter) of all suspected or infected animals, and proper disposal of contaminated products and surveillance. For semi-free zones, the protocol mostly relies on a combination of strict quarantine, zoosanitary measures, slaughter of infected animals, and wherever necessary a compulsory regular immunization program with inactivated vaccines and surveillance.

For endemic zones, epidemological studies are conducted, and valuable animal populations are vaccinated; there may also be an immunization program in selected areas with inactivated vaccines coupled with supportive zoosanitary measures and legislative support.

Because India is an endemic zone with widespread infection and the prevailing epidemiological situation exists, it is difficult to prevent contact with the pathogen in a large susceptible population. In brief, the disease control program with limited resources has been as follows.

Vaccination of valuable, productive, susceptible livestock in organized farms, as well as breeding stock and draught animals, is being widely practiced. A beginning has been made to extend this protocol to a mass immunization program in milkshed areas in a population or zone.

Normally for this purpose the quadruvalent potent vaccine containing virus types O, A, C, and Asia-1 is being used. In addition, during limited sporadic disease outbreaks—either to prevent spread of infection or to develop an immune belt or zone—a homologous monovalent vaccine (after identifying the virus type) is being used to avoid disease outbreaks. There is also surveillance for exotic virus strains and monitoring of disease outbreaks with spread and emergence of mutants and the antigenic relation to vaccine strain as supports for the vaccination and immunization program.

Other pilot projects have been developed and implemented for a systematic regular compulsory free vaccination program for the entire population in selected areas in four districts and a naturally bound hilly area in the southern region of India. The goal is to develop and maintain disease-free zones.

Supporting zoosanitary measures with quarantine and legislative regulatory measures within the area or concerning the entry of animals from outside this zone have been instituted.

This program is to be gradually extended, based on the experience gained and the resources available, to more selected areas in the near future, with a disease control program for the entire country.

Future Program

The future program that is envisaged, planned, and to be implemented briefly is as follows.

1. Investment of larger resources of personnel, material, and money to understand and evolve a strategy to find a solution to the crucial animal health problem.
2. Large-scale cell propagation with continuous suspension cell cultures or perfusion cultures in fermentor vessels, e.g., those of 1500 L capacity and automatic process control.
3. Production of bulk virus antigen, preserved and reconstituted for vaccines.
4. Training and transfer of technical know-how; making available plant and machinery technology to more manufacturers.
5. An infrastructrure facility fulfilling P_3 conditions with controlled environmental conditions, containment of disease security, ensuring an uninterrupted supply of inputs, and supportive services for institutions conducting research or vaccine manufacture and quality control.
6. A continuous surveillance, monitoring, and forecasting system for epidemiological trends, with emphasis on the emergence and spread of virus mutants, their antigenic relation to vaccine strains, exotic viruses, timely dissemination of information, and development of disease models helpful for disease control.
7. Research programs for the development and evaluation of alternate methodologies for the production of cost-effective immunogens by conventional inactivated or attenuated virus vaccines, recombinant DNA technology, synthetic peptides, nonspecific immunostimulation, or a combination of the same.
8. Application of laboratory techniques for research, diagnostics, and mass screening of sera by microtechniques; use of monoclonal antibodies, DNA probes, etc.
9. For disease control, planning of the strategy and its implementation with an organizational setup, time schedule, periodical evaluation, and progress review by developing disease-free zones and extending them to eventually include the entire subcontinent.

Acknowledgments. I wish to record the support and encouragement given by the Director of the Institute. My sincere thanks are due to my colleagues Drs. G. Butchaiah, V.V.S. Suryanarayana, and D. Rama Rao for their assistance in preparing the manuscript and to Mr. E.R. Nair and Mrs. Jayalakshmi for secretarial help.

References

1. Rao BU: *Foot-and-Mouth Disease, Indian Year Book of Veterinary & Animal Sciences.* Hissar, Haryana Agricultural University, 1980, pp 45–55.
2. Suryanarayana VVS, Rao BU, Padayatty JD: Cloning and expression of the cDNA for the major antigen of foot-and-mouth disease virus type Asia-1 63/72. *Curr Sci* 1985;54:1044–1048.

CHAPTER 27

Cell Culture Vaccine Against Bovine Tropical Theileriosis

R.D. Sharma

Bovine tropical theileriosis caused by *Theileria annulata* is widely prevalent in tropical and subtropical countries of northern Africa, southern Europe, and Asia (25). Exotic cattle and their cross-breds and young indigenous calves are highly susceptible, whereas adult Zebu cattle are considerably resistant to the disease (15). The disease has assumed paramount importance in India with the intensification of cross-breeding programs under the livestock improvement policy aimed at enhancing milk production in the country.

The disease is transmitted transstadially principally by ticks of *Hyalomma* sp. Occurrence of the disease is seasonal, i.e., during the summer and rainy months (April to October). The pertinent signs are lymphadenopathy, fever, anemia, weakness, recumbency, and the intriguing feature of continuous feeding and rumination until the advanced stages of the disease. Unusual signs observed are hemoglobinuria, cutaneous eruptions, protrusive eyeballs, and cerebral involvement. Diagnosis is confirmed by demonstration of macroschizonts in lymph node biopsy smears and piroplasms in erythrocytes. Punched necrotic ulcers in the abomasum is the pathognomonic postmortem lesion.

Various chemotherapeutic agents, e.g., berenil, halofuginone lactate, and parvaquone, have been tried with varying degrees of success. Buparvaquone (BW720C) from Cooper's Animal Health Division, U.K., has produced encouraging results (10,20). Keeping in view the prevalence of the disease, the limited chemotherapy, and the impracticable control of vector ticks under field conditions, the need for developing suitable immunoprophylactic measures has been greatly felt in India and various other countries. Research has been undertaken on various methods of immunization, which have been tried from time to time with varying degrees of success.

1. Quantum of infection method
2. Irradiation method
3. Infection and treatment method
4. Nonspecific immunization
5. Cell culture vaccine

Methods 1 to 4 have been extensively used in the past and have their own merits and demirts (3,8,14). Currently, cell culture vaccine is being tried in various countries with encouraging results.

The development of a suitable cell culture vaccine has necessitated satisfactory techniques for in vitro cultivation of *T. annulata*. Lymphoid cells infected with *T. annulata* were cultivated in monolayer (24), in association with BHK cells (7), and in suspension culture (6). Malmquist et al (9) established *T. parva* in cell culture and propagated it without the feeder layer in medium containing Eagle's minimum essential medium with 20% fetal calf serum. *T. annulata* has been successfully grown in medium RPMI-1640 supplemented with 20% fetal bovine serum (16). Hashemi-Fesharki and Shad-Del (4) used three strains of *T. annulata* schizonts grown in suspension as a vaccine, and they observed immunity for up to one year. Pipano (13) reported full protection from 2×10^6 cells of *T. annulata* macroschizonts grown in vitro when cattle were challenged with homologous strains. Pipano (14) reported that periods of cultivation ranging from several months to more than three years were needed to totally attenuate various *T. annulata* isolates. Ozkoc and Pipano (12) attenuated Ankara strain of *T. annulata* after 250 to 300 in vitro passages and used it as a vaccine. Stepanova et al (22) immunized about 30,000 cattle (in the field) with cell culture vaccines of five strains of *T. annulata*, with satisfactory results. Gill et al (2) immunized fifteen cross-bred calves with cell cultures grown at the fifteenth passage. Subramanian et al (23) immunized calves with cultures of infected lymphocytes at passage 42 with IVRI strain of *T. annulata*, with encouraging results. Similarly Singh (21) observed encouraging results with in vitro grown schizonts.

The prohibitive cost of fetal bovine serum is a major constraint in the in vitro studies of *T. annulata* and mass vaccine production. It is more so in countries such as India, where it involves cumbersome and lengthy import procedures. Brown (1) used 5 to 20% normal calf serum in medium for in vitro propagation of *T. annulata*. Hooshmand-Rad (5) substituted bovine serum with equine serum and ovine serum in modified Eagle's medium. Sharma and Nichani (17) isolated *T. annulata* macroschizonts in cell culture using precolostral and normal bovine serum. Nichani et al (11) substituted fetal bovine serum in medium RPMI-1640 (Gibco) with precolostral, normal calf, sheep, goat, and horse sera for successful in vitro cultivation of *T. annulata* (Hisar), thus making in vitro culture most cost-effective and least foreign-dependent, thereby greatly reducing the cost of production of a cell culture vaccine. Sharma and Shukla (18) used precolostral calf serum for prolonged in vitro propagation of macroschizonts and showed encouraging results in immunization trials with this culture. Sharma et al (19) have found that lymphoblastoid cell cultures infected with macroschizonts of *T. annulata* (Hisar) showed a mild thermal and no parasitological reaction when inoculated with 10^6 and 10^8 cells at the 100th passage in calves 3 to 5 months of age. The calves showed maximum indirect fluorescent antibody titers and withstood lethal, homologous tick challenge.

Cell culture vaccine is safe for all breeds and ages of cattle. This vaccine can protect cattle from severe tick challenge. The dose of such a vaccine can easily be monitored. The vaccine has few side reactions and can safely be used in pregnant and high-yielding cattle (14).

More experimentation is needed on transportation, storage, and standardization of cell-culture-grown schizont vaccine for field application. Identification and characterization of cell surface antigens of various strains at various stages of the parasite's life cycle must be done to determine the common immunogens so as to study cross-protection. Application of molecular biology techniques to the isolation of relevant parasite DNA and production of the antigen or its component in vitro for testing its immunogenicity must be carried out to produce a specific and safe immune response. The nature of the antigenic changes induced on the surface of macroschizont-infected cells need to be determined for the production of inactivated macroschizont-derived vaccines.

References

1. Brown CGD: Propagation of Theileria, in Maramorosch K, Hirumi H (eds.): Practical Tissue Culture Application. New York, Academic Press, 1979, p 223.
2. Gill BS, Bhattacharyulu Y, Kaur D: Vaccination against bovine tropical theileriosis (Theileria annulata). Nature 1976;264:355–356.
3. Gill BS, Bhattacharyulu Y: Bovine theileriosis in India, in Henson JB, Campbell M (eds.): Theileriosis. Ottawa, I.D.R.C., 1977, p 8.
4. Hashemi-Fesharki R, Shad-Del F: Vaccination of calves and milking cows with different strains of Theileria annulata. Am J Vet Res 1973;34:1465–1467.
5. Hooshmand-Rad P: The growth of Theileria annulata infected cells in suspension culture. Trop Anim Health Prod 1975;7:23–27.
6. Hooshmand-Rad P, Hashemi-Fesharki R: The effect of virulence on cultivation of Theileria annulata strains in lymphoid cells which have been cultured in suspension. Arch Inst Razi 1968;20:85–89.
7. Hulliger L: Cultivation of three species of Theileria in lympoid cells in vitro. J. Protozool 1965;12:649–655.
8. Irvin AD, Gill BS; Immunisation against theileriosis: appraisal and future perspectives, in Proceedings International Conference on the Advances in the Control of Theileriosis. Nairobi, 9–13 February 1981.
9. Malmquist WA, Nyindo MBA, Brown CGD: East coast fever: cultivation in vitro of bovine spleen cell lines infected and transformed by Theileria parva. Trop Anim Health Prod 1970;2:130–145.
10. McHardy N: Antitheilerial activity of BW720C (buparvaquone): a comparison with parvaquone. Res Vet Sci 1985;39:29–33.
11. Nichani AK, Sharma RD, Sarup S, Shukla PC: In vitro cultivation of Theileria annulata (Hisar) by using different homologous and heterologous sera, in Pro-

ceedings Second Asian Congres of Parasitology. C.D.R.I. Lucknow, 13–16 February 1986.

12. Ozkoc U, Pipano E: Trials with cell culture vaccine against theileriosis in Turkey, in Proceedings International Conference on the Advances in the Control of Theileriosis. Nairobi, 9–13 February 1981.

13. Pipano E: Basic principles of Theileria annulata control, in Henson JB, Campbell M (eds.): Theileriosis. Ottawa, I.D.R.C., 1977, p 55

14. Pipano E: Schizont and tick stages in immunisation against Theileria annulata infection, in Proceedings International Conference on the Advances in the Control of Theileriosis, Nairobi, 9–13 February 1981.

15. Sharma RD: Some epidemiological observations on tropical theileriosis in India, in Gautam OP, Sharma RD, Dhar S (eds.): Haemoprotozoan diseases of Domestic Animals. Hisar, Haryana Agricultural University, 1980.

16. Sharma RD, Brown CGD: Isolation, in vitro establishment and cryopreservation of lymphoblastoid cell cultures infected with Theileria annulata, in Gautam OP, Sharma RD, Dhar S (eds.): Haemoprotozoan Diseases of Domestic Animals. Hisar, Haryana Agricultural University, 1980.

17. Sharma RD, Nichani AK: Evaluation of bovine sera for in vitro propagation of Theileria annulata (Hisar) and assessment of its growth characteristics, in Proceedings, National Symposium on Recent Advances on Control and Prevention of Diseases of Dairy Animals. Ranchi, Birsa Agricultural University, 7–9 October 1985.

18. Sharma RD, Shukla PC: Immunisation trials with lymphoblastoid cell cultures infected with T. annulata (Hisar-strain), in Proceedings, Sixth National Congress of Prasitology. Pantnagar, University of Agriculture and Technology, 19–21 March 1985.

19. Sharma RD, Shukla PC, Nichani AK, Rakha NK, Sarup S: Immunization trials against bovine tropical theileriosis with a cell culture vaccine. Presented at the World Veterinary Congress, Montreal, 16–21 August 1987.

20. Sharma RD, Talukdar JN, Rakha NK, Nichani AK: Chemotherapeutic trials against bovine tropical theileriosis. Presented at 12th WAAVP Conference, Montreal, 12–15 August 1987.

21. Singh DK: Tropical theileriosis in India. Presented at the Workshop on orientation and coordination of Research on Tropical Theileriosis. Edinburgh, 22–23 September 1986.

22. Stepanova NI, Zablotskii VT, Rasulov KI: Vaccine prophylaxis of bovine theileriosis. Trudy Vsasoyu-Znogo Inst Eksperimental Vet 1982;56:10–18.

23. Subramanian G, Ray D, Naithani RC: In vitro culture and attenuation of macroschizonts of Theileria annulata (Dschunkowsky and Luhs, 1904) and in vivo use as a vaccine. Indian J Anim Sci 1986;56:174–182.

24. Tsur I, Adler S: Cultivation of Theileria annulata schizonts in monolayer tissue cultures. Refuah Vet 1962;19:225–234.

25. Uilenberg G: Theilerial species of domestic livestock, in Proceedings International Conference on the Advances in the Control of Theileriosis. Nairobi, 9–13 February 1981.

Part VIII
Leprosy

CHAPTER 28

Present Approaches to Immunotherapy and Immunoprophylaxis for Leprosy

G.P. Talwar, R. Mukherjee, S.A. Zaheer,
A.K. Sharma, H.K. Kar, R.S. Misra,
and A. Mukherjee

Development of a vaccine against leprosy is not an easy undertaking. The causative organism, *Mycobacterium leprae,* even though discovered more than 100 years back, has not been cultivated so far in vitro in a cell-free medium. No experimental animal model exactly simulating human lepromatous leprosy, the form of leprosy against which a vaccine is ideally required, is available. The latent period of the disease is several years. The difficulty of evaluating a potential immunoprophylactic agent is compounded by the fact that most humans are naturally resistant to the disease and do not require a vaccine. In highly endemic areas, where everyone is expected to be exposed to infection, fewer than 0.5 to 1.0% develop the disease (28). Not withstanding these limitations, attempts have been made to develop vaccines. Seven vaccines have been proposed (Table 28.1), and at present four candidate vaccines are in clinical trials.

BCG

The earliest attempts of immunization against leprosy were made with bacillus Calmette-Guérin (BCG). Three control trials were carried out during the early 1960s. In Karimui, Papua, New Guinea, immunization with BCG conferred 46% protection in the general population over a 6-year period of observation. In the age group 5 to 14 years, the protection was a shade better, i.e., 56% (20). The highest reported efficacy of BCG vaccination against leprosy was in Uganda. Contacts of leprosy patients in the age group up to 15 years were immunized and observed over a period of 6 years; protection of the order of 80% was obtained (2). It may be mentioned, however, that the form of leprosy that developed in Uganda is mainly of the tuberculoid type. In Burma, on the other hand, BCG vaccination did not give significant protection in the age group 5 to 14 years. However, vaccination in the under-4-years age group in the general population had 40% protection over a 9-year period (1). Incidently, in Burma, at the first follow-up of 4 years, 6 to 15% more cases of leprosy

Table 28.1. Candidate vaccines under development.

Name and type	Research group
Category I (based on *M. leprae)*	
Killed M. leprae	WHO IMMLEP Task Force (6)
Killed *M. leprae* + BCG	Convit et al (4)
Acetoacetylated *M. leprae*	Talwar (26)
Category II (based on cultivable mycobacteria)	
BCG	Already in use
BCG + *Mycobacterium vaccae*	Stanford et al (25)
Killed ICRC bacillus	Deo et al (5)
Killed *Mycobacterium w*	Talwar (26)

were recorded in the vaccinated group. The protective effect was noted only after a delay of several years. It is possible that these patients were incubating the disease and vaccination led to an adjuvant effect of expression. BCG could also cause this effect by stimulating the wrong clone of antigenic cells.

In India, a large trial of 260,000 subjects was carried out in Chingelput district, Tamil Nadu. Subjects vaccinated with BCG were followed for 15 years to determine the effect of immunization on tuberculosis and leprosy. BCG was without beneficial effect on tuberculosis during the first 5 years. However, at follow-up for the subsequent 10 years, 25 to 30% protection was noted against both tuberculosis and leprosy (14,27).

Vaccines Based on *M. leprae*

Killed *M. leprae*

The killed *M. leprae* vaccine, on which the WHO/IMMLEP Task Force has invested much effort, is based on the following observations. Mice immunized with killed *M. leprae* resist successfully footpad challenge with live *M. leprae* (24). Killed *M. leprae* in saline without any additional adjuvant induces delayed hypersensitivity to *M. leprae* antigens in guinea pigs (16). From these observations and on the usual logic of past experience with communicable diseases, killed *M. leprae* should serve as a vaccine. However, clinical observations in patients show that the lepromatous leprosy (LL) patients who are loaded with adequate antigenic stimulus from live and killed *M. leprae* do not regress. The self-limiting character of the disease is encountered only in tuberculoid leprosy (TT) patients. Killed *M. leprae* have so far not been used in clinical trials; and the clinical acumen on the disease would warrant its utility only in those patients whose immune status would lead to TT type of leprosy. It is known that repeated administration of lepromin (killed *M. leprae* preparations) does not lead to conversion or improvement in LL patients. Lepromin can be considered a minivaccination with killed *M. leprae* antigens.

The WHO Task Force has conducted a dose-response tolerance study in Norway on healthy subjects immunized with killed *M. leprae* vaccine (6,11). Lepromin conversion was observed in 95% of subjects. However, vaccine trials based solely on killed *M. leprae* have not been undertaken. WHO/IMMLEP has started a clinical trial in Malawi with a combination of live BCG and killed *M. leprae* in the general population to evaluate its immunoprophylactic potential (7).

Killed *M. leprae* with BCG

BCG is no doubt an excellent adjuvant and stimulates the immune system in a general way. It also shares several antigens with *M. leprae*. The lack of consistent protection with BCG led to the idea of its use in conjunction with killed *M. leprae*. This combination has been used in Venezuela by Convit et al (4). A large number of patients were given these vaccines in an immunotherapeutic modality. Multiple injections, numbering up to eight, were given at intervals of 6 to 12 weeks. In many cases immunization with this mixture led to faster clinical recovery even though the patients continued to receive chemotherapy.

Conversion to a positive reaction to leprosin was observed in 38% of BB-BL (active) patients.* 259 treated Among the inactive category of patients, the conversion was 63%. A much higher conversion was noted in the "indeterminate" group of patients (4).

Acetoacetylated *M. leprae*

Realizing that a key defect in LL patients is their inability to respond to some key antigens of *M. leprae*, we prepared hapten derivatives of *M. leprae* with the idea of breaking the tolerance to these antigens (9). Acetoacetylation of varying degree was carried out. Acetoacetylated *M. leprae* was fully protective in mice against challenge with live *M. leprae* (Fig. 28.1). Leukocytes from LL patients could elicit lymphokines as assayed by leukocyte migration inhibition assay, and they were anergic in this respect to native *M. leprae* (10) (Fig. 28.2).

Vaccines Employing Cultivable Mycobacteria

Many years back we carried out investigations to determine if a cultivable mycobacterium can fulfill the demands of an antileprosy vaccine (26).

---

*BB-BL denote histopathological grading of the lesion in the patient on Ridley Joppling scale.

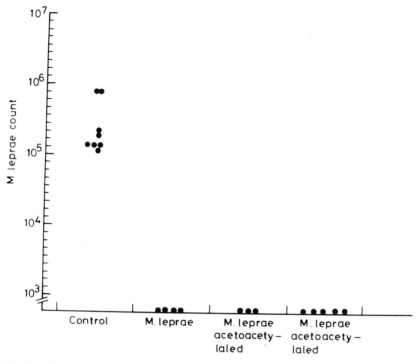

Fig. 28.1. Protection to live *M. leprae* challenge in footpads in female BALB/c mice immunized with irradiated *M. leprae* and acetoacetylated derivatives of *M. leprae*.

Studies were carried out on 15 standard strains of mycobacteria (including BCG) and some atypical isolates. The bacilli were coded and screened with leukocytes from a panel of six patients with tuberculoid leprosy (TT). This strategy was based on the belief that TT patients have an immune status that can restrict the infection successfully and eliminate the mycobacteria. In view of previous findings indicating the primary role of cell-mediated immunity (CMI) in resistance to the disease, tests of CMI were carried out, including the antigen-driven blast transformation and the leukocyte migration inhibition assay. Lepromin (killed *M. leprae* antigens) was used in each case as a reference antigen. Those bacilli that evoked a response similar to that produced by *M. leprae* lepromin were selected for further studies (19). Five mycobacterial strains were selected from the 15 strains based on this criteria. Their immunogenic potency was evaluated by measuring the swelling of draining lymph nodes, as Shepard (24) had correlated this response with immunogenicity of mycobacteria in mice. By this test *Mycobacterium w* gave the best results. It was most effective in causing the enlargement of lymph nodes with an almost 100% positivity index (18).

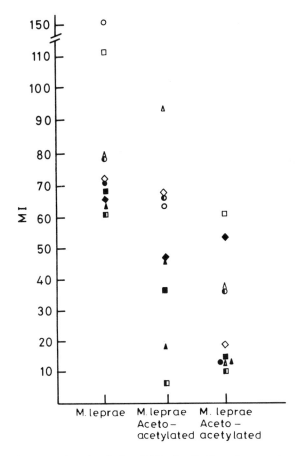

Fig. 28.2. Leukocyte migration index (MI) of cells from lepromatous leprosy patients with *M. leprae* and two acetoacetylated derivatives of *M. leprae*. Migration index of leukocytes from a given patient tested with different antigens is represented by a common symbol. [Data from Fotedar et al (10).]

Studies were also carried out in guinea pigs, in which the ability of the selected mycobacteria to evoke delayed-type hypersensitivity (DTH) on challenge with *M. leprae* was determined. *M. w* and ICRC bacillus gave DTH responses closer to those of *M. leprae* (17). Similar studies carried out in mice also demonstrated that *M. w* produced better DTH than ICRC bacillus (8).

Finally, a verdict had to be obtained in patients. Lepromin-like preparations were made from the four mycobacterial strains selected on the basis of LTT, LMIT, i.e., *M. w*, ICRC bacillus, *M. vaccae*, *M. gordonae*, *M. phlei*, and *M. leprae*. They were given to six clinical centers located in different parts of India so as to take into account the effects of environmental factors. The clinicians were asked to test the preparations in

TT as well as LL patients (Fig. 28.3). Our aim was to pick up one or more strains of bacilli that produced in TT patients a DTH skin reaction similar to that obtained with *M. leprae* lepromin. There was a further requirement that the bacillus should also be able to evoke a response in LL patients, in whom *M. leprae* antigens normally do not evoke a response. What we wanted was a bacillus that shared antigens with *M. leprae* but that also had additional antigens capable of producing immunological responsiveness in LL patients (26).

Mitsuda-type antigens were prepared from *M. leprae, M. vaccae, M. phlei, M. gordonae, M. w,* and ICRC bacillus; they were to be studied in a double-blind manner at six centers in India: the All India Institute of Medical Sciences, New Delhi (13); Victoria Hospital, Andhra Pradesh (15); Maulana Azad Medical College, New Delhi (23); and Polambacum, Tamil Nadu (21). Patients at these centers were carefully selected according to the Ridley-Jopling scale and belonged to the LL and TT categories.

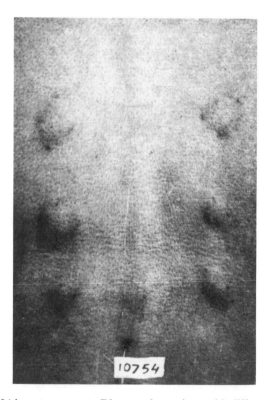

Fig. 28.3. The 24-hour response to Dharmendra antigen with different mycobacteria in a borerline tuberculoid patient. At left, downward: *M. phlei, M. leprae,* ICRC bacillus. At right, downward: *M. vaccae, M. w, M. leprae.* [From Girdhar and Desikan (12), with permission.]

The overall results of this study showed that the *M. w* and ICRC bacilli gave the best correlative DTH reaction with *M. leprae* in the tuberculoid spectrum of the disease. In addition, unlike *M. leprae, M. w* and ICRC also induced a DTH reaction in the lepromatous patients.

ICRC Bacillus

The Indian Cancer Research Centre (ICRC) strain of *Mycobacterium* was initially cultivated in 1958 from human lepromatous nodules using conditioned medium of human spinal ganglia cultures. It belongs to the *M. avium–intracellulare* group of organisms. For many years the work of the Institute centered around the hypothesis that the bacillus was a cultivable form of *M. leprae*. The work of our group, which started in 1974, on cultivable mycobacteria revealed the properties of this bacillus as a potential candidate vaccine.

Currently it is under clinical trial for both immunoprophylaxis and immunotherapy. Vaccine has been found to bring about persistent lepromin conversion in 53% of lepromatous leprosy patients (5). Conversion was high (95%) in lepromin-negative healthy contacts of patients in endemic areas. Phase III clinical trial is in progress in the general population in an endemic area in Maharashtra. Immunization is carried out on volunteers in the general population. The effect of vaccination on the incidence of the disease is to be determined in the population immunized.

Mycobacterium w

The background of selection of *M. w* as a candidate leprosy vaccine is described above. Toxicological studies on the bacillus were carried out as per the guidelines of the Indian Council of Medical Research in two species of animals. *M. w* was adjudged safe and devoid of toxicity. In other experiments conducted with live *M. w*, administration of large doses of the bacillus by various routes did not cause any mortality of mice and guinea pigs, confirming the nonpathogenic nature of the bacillus. *M. w* is a fast-growing bacillus. It resembles mycobacteria belonging to Runyon's group IV on the basis of its metabolic and growth properties (22). It is, however, not completely indentical to the presently listed mycobacteria in this group and may be a variant or a new strain.

Phase I clinical studies with this bacillus were carried out at the School of Tropical Medicine (3). Thirty-two well documented patients with LL who were bacteriologically negative after several years of chemotherapeutic treatment were the subjects of study. The vaccine was given as a single intradermal dose of 5×10^7 autoclaved bacillus in 0.1 ml saline. Of the 32 subjects who from previous records and preimmunization testing were consistently negative to lepromin, 20 became distinctly Mitsuda-positive at 4 weeks after vaccination. No side effects of vaccination were noted except for a single patient, who had a severe local reaction at the vaccination site with ulceration. This patient was diagnosed as suffering from pulmonary tuberculosis. No significant systemic side effects were seen in the others. The lepromin conversion was not transient but was stable for several months. A retest was done after 7 to 11 months, when these patients formed lymphokines in the presence of *M. leprae*, measurable in an LMIT assay. Histopathologic study showed typical granuloma formation with activated macrophages surrounded by a cuff of lymphocytes (3).

With this background study demonstrating *M. w* as a promising potential antileprosy vaccine, phase II/III clinical trials were begun on this vaccine after obtaining the necessary clearance from the Drug Controller of India. Trials are being conducted in two major hospitals in Delhi—Safdarjung and Ram Manohar Lohia Hospitals—with the active cooperation of the Institute of Pathology, Indian Council of Medical Research. The trial aims to evaluate the benefits of this potential antileprosy vaccine when used as an immunotherapeutic agent in patients with multibacillary disease. Evaluation is based on clinical, bacteriological, immunological, and histopathological parameters.

Lepromin-negative BB, BL, and LL patients were included in the trial. Half of them were given the vaccine, and the other half received only the placebo. All patients were maintained on a standard regimen of multidrug therapy. So far 90 leprosy patients (21 BB, 18 BL, and 51 LL) have been inducted into the trial. The vaccine is given at 3-month intervals. The first dose of 1×10^8 killed bacilli is administered intradermally. Subsequent doses of the vaccine are 5×10^7 killed bacilli. A maximum of eight doses of vaccine is employed.

After the first dose of the vaccine 75% patients in the BB category became lepromin-positive, 81% after the second dose, and 100% after the third dose. In BL-type patients the conversion rates were 43%, 60%, and 82% after the first, second, and third doses of the vaccine. In polar LL patients the conversion rates were 8.4%, 8.4, 15.6%, and 74%, respectively. Lepromin conversion was well correlated with clinical improvement (Fig. 28.4) and rapid bacterial clearance. Histopathologically, a significant number of patients demonstrated upgrading.

Concluding Comments

Research continues on understanding the immunological deficit in leprosy, although three candidate vaccines have been developed based on different rationales and are currently undergoing clinical trials. Immunotherapeutic trials are in progress with BCG + *M. leprae* in Venezuela and irradiated ICRC bacilli and killed *M. w* in India. Data on the former are more extensive, having been recorded over a 13-year period. The vaccine has

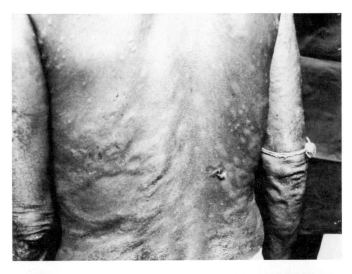

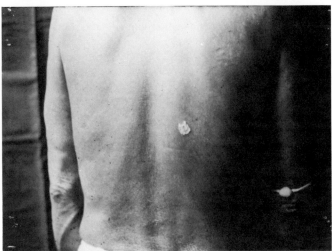

Fig. 28.4. Clinical improvement seen in a borderline lepromatous leprosy patient (R.P.S.) following combined chemotherapy and immunotherapy with *M. w.* Top At the start of the trial. Bottom: After 4 months of therapy with two doses of the *M. w.* vaccine. A significant decrease in the infiltration of the lesions is noticeable.

proved its utility; however, despite repeated vaccine doses and high doses of BCG, a significant number of LL patients remain negative to lepromin.

Immunotherapeutic trials with *M. w* have been started more recently. After 1 year of observing 91 subjects, the results have been encouraging, with notable clinical improvement and bacterial clearance. In BB, BL, and LL patients, on whom these observations are based, repeated injections of *M. w* have led to high lepromin conversion. Further follow-up and extension of these trials are required to confirm the efficacy of this vaccine.

References

1. Bechelli LM, Gallego Garbojosa P, Engler V, et al: BCG vaccination of children against leprosy: preliminary findings of the WHO-controlled trial in Burma. *Bull WHO* 1970;42:235–281.
2. Brown JAK, Stone MM, Sutherland I: Trial of BCG vaccination against leprosy in Uganda. *Lepr Rev* 1969;40:3–7.
3. Chaudhuri S, Fotedar A, Talwar GP: Lepromin conversion in repeatedly lepromin negative BL/LL patients after immunization with autoclaved Mycobacterium w. *Int J Lepr* 1983;51:159–168.
4. Convit J, Aranzazu M, Zuniga M, et al: Immunotherapy and immunoprophylaxis of leprosy. *Lepr Rev* 1983;(special)475–605.
5. Deo MG, Bapat CV, Chullawalla RG, Bhatki WS: Potential anti leprosy vaccine from killed ICRC bacilli: clinicopathological study. *Indian J Med Res* 1981;74:164–177.
6. Expanded programme on immunization global status report. *WHO Weekly Epidemiol Rec* 1985;60:261–263.
7. Fine PEM, Ponnighaus JM, Maine N: The relationship between delayed type hypersensitivity and protective immunity induced by mycobacterial vaccines in man. *Lepr Rev* 1986;57(suppl 2):275–283.
8. Fotedar A, Mehra NK, Mustafa AS, Talwar GP: Local reactions to intradermal instillation of Mycobacterium w and ICRC bacilli in mice. *Lepr India* 1978;50:520–533.
9. Fotedar A: Immunology of leprosy: experimental induction of immunity to leprosy. PhD thesis, All India Institute of Medical Sciences, New Delhi, 1982.
10. Fotedar A, Mustafa AS, Narang BS, Talwar GP: Improved leukocyte migration inhibition response of leukocytes from lepromatous leprosy patients with hapten modified M. leprae. *Clin Exp Immunol* 1982;49:317–324.
11. Gill HK, Mustafa AS, Godal T: Induction of delayed hypersensitivity in human volunteers immunised with a candidate leprosy vaccine consisting of Mycobacterium leprae. *Bull WHO* 1986;64:121–126.
12. Girdhar BK, Desikan KV: Results of skin tests with five different Mycobacteria. *Lepr India* 1978;50:555–559.
13. Govil DC, Bhutani LK: Delayed hypersensitivity skin reactions to lepromin and antigens prepared from four other Mycobacteria. *Lepr India* 1978;50:550–554.
14. Gupte MD: Working papers for joint ICMR/WHO meeting on leprosy vaccine trials, Madras, India, 1986.
15. Hogerzeil LM, Prabhudass N: Delayed hypersensitivity skin reactions to lepromins prepared from M. leprae and selected cultivable Mycobacteria. *Lepr India* 1978;50:560–565.
16. Mehra V, Bloom BR: The induction of cell-mediated immunity to human M. leprae in the guinea pig, in Talwar GP (ed): *Progress in Immunology of Leprosy* London, Heinemann, 1983, pp 123–130.

17. Mustafa AS, Talwar GP: Delayed hypersensitivity skin reaction to homologous and heterologous antigens in guinea pigs immunized with M. leprae and four selected cultivable mycobacterial strains. *Lepr India* 1978;50:509–519.
18. Mustafa AS, Talwar GP: Enlargement of draining lymph nodes in mice by four selected cultivable strains of Mycobacteria. *Lepr India* 1978;50:534–538.
19. Mustafa AS, Talwar GP: Five cultivatable mycobacterial strains giving blast transformation and leukocyte migration inhibition of leukocytes analogous to Mycobacterium leprae. *Lepr India* 1978;50:498–508.
20. Russel DA, Scott GC, Wigley SC: BCG and prophylaxis—the Karimui trial. *Int J Lepr* 1968;36:618.
21. Sahib HSM, Vellut C: Some observations on skin reactions induced by lepromin and four other mycobacterial antigens. *Lepr India* 1978;50:579–587.
22. Saxena VK, Singh US, Singh AK: Bacteriological study of a rapidly growing strain of Mycobacterium. *Lepr India* 1978;50:588.
23. Sharma RC, Singh R: Comparative study of skin reactions in leprosy patients to M. leprae–lepromin and to antigens from cultivable saprophytic mycobacteria. *Lepr India* 1978;50:572–578.
24. Shepard CC: Vaccination of mice against M. leprae. *Int J Lepr* 1976;44:222–622.
25. Stanford JL, Shield MJ, Rook GAW: How environmental mycobacteria may predetermine the protective efficacy of BCG: hypothesis I. *Tubercle* 1981;62:55–62.
26. Talwar GP: Towards development of a vaccine against leprosy. *Lepr India* 1978;50:442–497.
27. Tripathy SP: The case for BCG. *Ann Natl Acad Med Sci (India)* 1983;19:11–12.
28. Wardekar RV: Some aspects of epidemiology of leprosy. *Indian J Med Res* 1972;60:1733–1743.

CHAPTER 29

Armadillo-Derived Killed *M. leprae:* Candidate Vaccine Against Leprosy

Abu Salim Mustafa

Leprosy is a chronic mycobacterial disease caused by *Mycobacterium leprae*. Worldwide, there are about 10 million to 15 million people suffering from this disease. It is not only the number of patients but also the disabilities inflicted on them that have made leprosy among the priority diseases to be controlled and eventually eradicated. Man is the only known natural host of *M. leprae,* and it is from man to man that the disease seems to be carried. For the successful control of leprosy, two approaches have been advocated.

1. Block the transmission of *M. leprae* from patients to healthy people by early detection and effective treatment of the infectious patients.
2. Render the people at the risk of infection inhospitable territory for *M. leprae* growth by either chemoprophylaxis or immunoprophylaxis.

The control programs running at centers throughout the world have utilized the approach of the early detection of infectious cases and making them noninfectious by effective chemotherapy. However, these programs have not been successful in reducing the leprosy incidence to any appreciable extent. Most of the patients try to hide their disease because of the social stigma attached to leprosy. *M. leprae* by itself is nontoxic and well tolerated by the body. Therefore most of the patients do not come for treatment until the disease is full-blown, and they may have infected several others before they are diagnosed and effectively treated. An additional problem is the patient infected with DDS-resistant strains of *M. leprae*. Multiple drug therapy for such patients is effective, but expensive. Chemoprophylaxis on a mass scale is impracticable owing to the large number of people at risk and the high cost. Immunoprophylaxis may be the only feasible alternative for controlling leprosy. The vaccines being evaluated for this purpose are based on: (a) *M. leprae* antigens; (b) cultivable mycobacteria (sharing with *M. leprae* the antigens involved in cell-mediated immunity functions) either alone, i.e., *Mycobacterium w* (29) and ICRC bacillus (2), or in combination with *M. leprae*, i.e., live BCG + killed *M. leprae* (1). The vaccines based on cultivable mycobacteria may be used for immunoprophylactic as well as immunotherapeutic purposes. An *M. leprae* vaccine can have a role only in immunoprophylaxis, but it may be the best for this purpose as it has all the antigens relevant for protection. The *M. leprae* vaccine may consist of either killed *M. leprae* or viable *M. leprae* attenuated for its pathogenicity. However, because of an inability to grow *M. leprae* in vitro, the organism cannot be attenuated for its pathogenicity. Killed *M. leprae* will also be required in large quantities in order to be sufficient for mass vaccination.

In 1960 the mouse footpad model for the cultivation of *M. leprae* was discovered by Shepard (23); but because of the limited growth of bacteria in this system, human leproma biopsy specimens continued to be the only source of *M. leprae* for experimental work. The human source had its own limits; and therefore until 1971 it was not conceivable to obtain the quantities of bacteria sufficient for mass vaccination. In 1971 Kirchheimers and Storrs showed unabated growth of *M. leprae* in an experimentally infected nine-banded armadillo. The yield of *M. leprae* in many armadillo tissues was 100 to 1000 times greater than in human leproma tissues (11,12). The Immunology of Leprosy (IMMLEP) Programme of the UNDP/World Bank/WHO Special Programme for Research and Training in Tropical Diseases, soon after its inception in 1974 realized the enormous potentials of the discovery made by Kirchheimers and Storrs, and it quickly established an armadillo-derived *M. leprae* bank under the supervision of Dr. R.J.W. Rees at the National Institute of Medical Research, Mill Hill, London. Since then, *M. leprae* from this bank has been supplied to qualified investigators throughout the world.

One of the main objectives of IMMLEP has been to develop a vaccine against leprosy. As large quantities of *M. leprae* were available from armadillos, the development of a vaccine based on killed *M. leprae* was foreseeable. However, because bacilli were to be purified from infected armadillo tissues, the purification procedure required that a number of criteria be fulfilled: (a) maximum yield of intact bacilli; (b) nontoxicity of purified bacteria; (c) minimum contamination with armadillo tissue, and (d) no loss of important antigens. All these criteria to a large extent were fulfilled in the protocol 1/79 developed by Dr. Philip Draper (30). The purified bacilli were tested by a number of investigators collaborating with the IMMLEP Programme for the efficacy of *M. leprae* vaccine in the animal models.

Animal Vaccination Studies

Mice, guinea pigs, and armadillos were the animals used to determine the efficacy of *M. leprae* vaccine. Mice were injected with *M. leprae* in aqueous suspensions either intradermally or in the footpads. Guinea pigs and armadillos were immunized intradermally. The delayed-type hypersensitivity (DTH) skin reactions at the immunization sites and to recall *M. leprae* antigens at distant sites were recorded. Immunization with intact *M. leprae* induced a positive DTH skin response in all the animals tested (13,21,24,26). Physical disruption of the organisms resulted in loss of their immunogenicity as evidenced by failure to induce a positive DTH skin response (27). *M. leprae* killed by autoclaving showed increased immunogenicity, whereas the immunogenicity of BCG was greatly reduced by autoclaving (25). In mice, induction of DTH to *M. leprae* antigens correlated well with protection (24). Immunization with live as well as with killed *M. leprae* protected the animals against a viable *M. leprae* challenge in the footpads (25,27). The protection against a viable challenge and the positive DTH skin responses lasted for at least a year after vaccination of mice with killed *M. leprae* (24). In guinea pigs also the positive DTH skin response persisted for at least 1 year (13). The vaccination of mice with killed *M. leprae* into either footpads or thigh caused life-long enlargement of the draining lymph nodes, a parameter that correlated with protection against *M. leprae* infection (24,25). Passive transfer of splenocytes from killed *M. leprae*-vaccinated mice to naive recipients induced cell-mediated immunity (CMI) responses to *M. leprae* antigens and protected them against viable *M. leprae* challenge but not against *L. monocytogenes* challenge (9). This experiment showed the specificity of protective immunity induced after the transfer of immune splenocytes.

Induction of CMI Responses to *M. leprae* in Humans

After establishment of the good performance of killed *M. leprae* vaccine in animal models, the next step was to assess its efficacy in humans. However, due to the long incubation period of the disease, the field trial to assess the efficacy of an antileprosy vaccine in an endemic population requires more than 10 years of follow-up. The low incidence of the disease is another problem. Even in the highly endemic areas, prevalence is about 0.5 per 100 people. Only 20% of these patients develop lepromatous disease, and they are probably the best indicators of vaccine efficacy. One may have to vaccinate more than 1000 people to obtain a reduction in one detectable lepromatous patient over a decade. Therefore before embarking on a large-scale clinical trial, it was considered necessary to assess the efficacy of killed *M. leprae* in a small population in a nonendemic country using appropriate indicators. One such indicator is the induction of *M. leprae*-specific CMI responses, which correlate well with the resistance against lepromatous disease. It has been clearly observed in patients themselves. The paucibacillary tuberculoid patients have localized disease with skin lesions heavily infiltrated by mononuclear cells. These patients have a low level of circulating antibodies but show a high CMI response to *M. leprae* antigens as evidenced by positive DTH skin responses, in vitro transformation of their lymphocytes, and the production of leukocyte migration inhibition factors (8). The multibacillary lepromatous patients have generalized disease with high levels of circulating antibodies but deficient CMI responses to *M. leprae* antigens (8). More evidence for a correlation between CMI and protection against lepromatous leprosy originates from the studies of Dharmendra and Chatterjee (3). After 20 years of follow-up, these authors found eight lepromatous patients among 16 healthy contacts who were repeatedly lepromin skin-test-negative, but none among 524 skin-test-positive contacts of leprosy patients.

The studies in contacts and the observations in the patients strongly indicate a role of CMI in protection against leprosy, especially the lepromatous type. Therefore as a first step toward testing an antileprosy vaccine in humans, vaccination studies with killed *M. leprae* were conducted in BCG-prevaccinated healthy subjects in order to determine (a) the optimal dose of killed *M. leprae* required to induce *M. leprae*-specific CMI responses without causing unacceptable side effects, and (b) the modulatory effects of *M. leprae* vaccination on preexisting sensitization to BCG/PPD.

The study design has been described in detail elsewhere (5). Before starting the trial, ethical clearance was obtained from appropriate authorities. Thirty-nine volunteers vaccinated with BCG during childhood were divided into five groups, with seven to nine subjects in each group. The volunteers in groups 1 to 4 were vaccinated with four different doses of killed *M. leprae,* i.e., 1.5×10^7, 5×10^7, 1.5×10^8, and 5×10^8 bacilli. The fifth group of unvaccinated subjects served as controls. The local reactions at vaccine sites were assessed at various time intervals up to 1 year. The volunteers were followed for other side effects, e.g., fever, loss of appetite, and chronic enlargement of draining lymph nodes.

The local reactions at the vaccination site were the only vaccine-related side effects. Induration at the vaccine site was maximum at 3 weeks of immunization and was clearly dose-related. At the two highest doses of the vaccine, ulceration and scar formation were also observed. However, the scars at 1 year did not look much different than a BCG scar (5). The skin reactions at the vaccination site are equivalent to the late Mitsuda reaction in response to integral lepromin, which is considered to be the indication of an induced CMI response to *M. leprae* antigens. The strong local reactions at the vaccine site in the volunteers vaccinated with 1.5×10^8 and 5×10^8 killed *M. leprae* suggest the induction of a strong CMI response to *M. leprae* antigens.

DTH Skin Responses to PPD and *M. leprae* Soluble Antigens

All the volunteers were skin-tested before and at 3 months after vaccination, with coded samples of purified protein derivative (PPD) of *M. tuberculosis* and *M. leprae* soluble antigens (MLSAs). Prior to vaccination with killed *M. leprae,* 31 of the 37 BCG-vaccinated subjects showed a positive DTH skin response to PPD, but only 11 of the 37 subjects showed a positive DTH skin response to MLSAs (5). Pönnighaus and Fine (22) have also reported a low percentage of MLSA responders (14 to 36%) at 9 months of BCG vaccination. It may mean that in most of the subjects studied the T cell-activating antigens of BCG that give long-lasting sensitization do not cross-react with *M. leprae.* At 3 months after vaccination with killed *M. leprae,* all of the subjects vaccinated with 1.5×10^8 and 5×10^8 bacilli showed a strong positive skin test response to MLSAs. The responses to PPD were neither positively nor negatively modulated (5). Smelt et al (28) could also induce positive DTH responses to soluble *M. leprae* skin test antigens, but not to PPD, following immunization of skin test-negative healthy subjects with 2×10^8 killed armadillo-derived *M. leprae.* No better sensitization was obtained by a mixture of live BCG and killed *M. leprae.*

Lymphocyte Transformation Test Response

Peripheral blood mononuclear cells (PBMCs) were obtained from the blood of volunteers prior to vaccination and at 1, 3, 6, and 12 months after vaccination. PBMCs were used to assess the in vitro lymphocyte transformation in response to *M. leprae* and other antigens. The vaccination of group 3 and group 4 subjects with 1.5×10^8 and 5.0×10^8 killed *M. leprae* resulted into significantly enhanced in vitro proliferation of their PBMCs in response to *M. leprae* antigens (7). The enhanced lymphocyte transformation test (LTT) responses peaked at 3 months and persisted up to 1 year of the observation period. The proliferation of PBMCs in response to PPD, tetanus toxoid, diphtheria toxoid, and streptokinase-streptodornase was also slightly enhanced after vaccination; but compared to the prevaccination response, the differences were not statistically significant (7).

Antibody Responses

Serum for analyzing antibody levels was separated from the peripheral blood of volunteers before and at various times after vaccination with killed *M. leprae.* In the sera of group 3 and group 4 volunteers collected at 6 months and 1 year after vaccination with killed *M. leprae,* increased antibody titers were observed against total *M. leprae* sonicate and *M. leprae*-specific epitope on 36kDa antigen (6). The antibodies to phenolic glycolipid-1 (PGL-1) were not detected in these sera (5), probably because of the absence of PGL-1 from vaccine preparations. It may be advantageous, as PGL-1 has been shown to activate suppressor cells in lepromatous patients (14).

M. leprae-Specific T Cell Clones

The characteristics of *M. leprae*-activated T cells were studied by establishing human T cell clones from the PBMCs of group 3 and group 4 individuals. Forty-two T cell clones against soluble *M. leprae* antigens were raised from six subjects. Evidence for *M. leprae*-specific sensitization of the volunteers after vaccination with killed *M. leprae* was obtained from the reactivity pattern of the T cell clones in response to *M. leprae*, BCG, and other mycobacteria. In proliferative assays, 11 clones were specific to *M. leprae* and did not respond to BCG. Other T cell clones responded to *M. leprae* as well as to BCG (16). *M. leprae*-specific T cell clones produced biologically important lymphokines, such as interleukin 2 and γ-interferon, in response to *M. leprae* but not to BCG (17). All of the clones had CD4$^+$, the helper phenotype. However, they also exhibited antigen-dependent cytotoxicity and killed antigen-presenting cells such as monocytes/macrophages. *M. leprae*-specific T cell clones killed *M. leprae* pulsed targets but not BCG/PPD pulsed targets (17).

Helper and cytotoxic T cells usually belong to two different lineages. Helper T cells are CD4$^+$, proliferate in response to antigen, produce interleukin 2 and other lymphokines, and are devoid of cytotoxic activity. The cytotoxic T cells are CD8$^+$, are poor in producing interleukin 2, and require exogenous interleukin 2 for their proliferation and the production of other lymphokines, e.g., γ-interferon (10). However, it is interesting that *M. leprae*-induced human T cell clones had the characteristics of both helper and cytotoxic T cells. BCG-induced T cell clones from BCG-vaccinated healthy subjects have also been shown to have similar characteristics (18,19).

The T cells with helper as well as cytotoxic activity may have both beneficial and harmful effects in vivo. They might help in the clearance of *M. leprae* infection by producing lymphokines, which increase microbicidal capacity of infected macrophages. However, by killing infected macrophages and other cells of macrophage lineage, e.g., Schwann cells, cytotoxic T cells may enhance dissemination of the disease and exacerbation of the nerve damage. The cytotoxic activity of T cells may also be beneficial for the host under the circumstances, where the infected macrophages have somehow become incompetent to kill the ingested bacilli. The bacilli from lysed macrophages can be taken up by fresh and more competent monocytes, thereby decreasing the load of bacilli per cell and increasing the chance of bacterial clearance.

Identification of Recombinant Mycobacterial Antigens by T Cell Clones

An advancement has been the cloning and expression of *M. leprae* genome in foreign hosts. Five major *M. leprae* proteins of molecular weight 65,

36, 28, 18, and 12 kilodaltons (kDa) have been identified with the use of mouse monoclonal antibodies (31). In order to identify T-cell-activating antigens/epitopes, *M. leprae*-induced T cell clones from killed *M. leprae*-vaccinated subjects were tested against the above five recombinant proteins. Five T cell clones from two subjects responded to an epitope on the 18-kDa *M. leprae* protein. This epitope seems to be specific to *M. leprae,* as the T cell clones did not respond to other mycobacteria except with a weak response to *M. schrofulaceum* (16). Another cross-reactive T cell clone responded to an epitope on the 65-kDa proteins of *M. leprae* and *M. tuberculosis* (17). The reactivity pattern of the 65-kDa responding T cell clone suggested that the epitope recognized by this clone is highly shared among different species of mycobacteria. Similar strategies have been followed to identify T-cell-activating epitopes on *M. tuberculosis* recombinant proteins using T cell clones from tuberculosis patients (20).

All of the mouse monoclonal antibodies reacting to the protein antigens of *M. leprae* lysate recognized the epitopes restricted to the above five recombinant proteins of *M. leprae* (4). In our studies only 6 of the 22 T cell clones tested responded to two of these proteins. The reasons for such low reactivity of T cell clones to recombinant antigens are not clear. Most *M. leprae* epitopes/antigens recognized by T cells may be different than the epitopes/antigens recognized by B cells. Alternatively, mouse B cells may be limited with respect to their repertoire recognized by *M. leprae* antigens. To identify other T-cell-activating epitopes/antigens, strategies should be developed to test the recombinant *M. leprae* library against helper T cell clones as probes. However, the recombinant library will have its own limitations if the epitopes recognized by *M. leprae* T cell clones are nonprotein in nature.

The T cell clones can also be used to map the T cell epitopes on the recombinant antigens by making minilibraries from individual recombinants, as reported by Mehra et al (15) for determining the antibody epitopes. The interesting epitopes could easily be synthesized in vitro, and the ones that are specific to *M. leprae* could be evaluated for their diagnostic potential in skin tests. The epitopes recognized by helper T cells may potentially be manipulated for the development of new-generation vaccines.

Summary

An immunoprophylactic antileprosy vaccine is urgently required for control and eventual eradication of leprosy. The studies directed toward the development of a vaccine based on *M. leprae* antigens have been feasible owing to the availability of large quantities of antigenically intact *M. leprae* from experimentally infected armadillo tissues. Immunization with armadillo-derived, killed *M. leprae* induced a long-lasting DTH skin response in animal models and protection against viable challenge in the mouse footpad. In human volunteers, at 1.5×10^8 and 5×10^8 bacilli per person, killed *M. leprae* were able to induce a positive DTH skin response and significantly enhanced lymphocyte transformation to *M. leprae* antigens. *M. leprae*-specific, $CD4^+$ multifunctional T cell clones were raised from such volunteers. These T cell clones have been used to identify recombinant antigens/epitopes of mycobacteria, which in future may be exploited to develop specific diagnostic tests and new-generation vaccines.

References

1. Convit J, Aranazu N, Ulrich M, et al: Investigations related to the development of a leprosy vaccine. *Int J Lepr* 1983;51:531–539.
2. Deo MG, Bapat CV, Bhalerao V, et al: Antileprosy potentials of ICRC vaccine: a study in patients and healthy volunteers. *Int J Lepr* 1983;51:540–549.
3. Dharmendra, Chatterjee KR: Prognostic value of the lepromin test in contacts of leprosy cases. *Lepr India* 1955;27:149–152.
4. Engers HD, Bloom BR, Godal T: Monoclonal antibodies against mycobacterial antigens. *Immunol Today* 1985;6:345–348.
5. Gill HK, Mustafa AS, Godal T: Induction of delayed-type hypersensitivity in human volunteers immunized with a candidate leprosy vaccine consisting of killed Mycobacterium leprae. *Bull WHO* 1986;64:121–126.
6. Gill HK, Mustafa AS, Ivanyi J, et al: Humoral immune responses to M. leprae in human volunteers vaccinated with killed, armadillo derived M. leprae. *Lepr Rev* 1986;57(suppl 2):293–300.
7. Gill HK, Mustafa AS, Godal T: In vitro proliferation of lymphocytes from human volunteers vaccinated with armadillo-derived, killed M. leprae. *Int J Lepr* 1987;55:30–35.
8. Godal T: Immunological aspects of leprosy—present status. *Prog Allergy* 1978;25:211–242.
9. Graham L Jr, Navalkar RG: Evaluation of Mycobacterium leprae immunogenicity via adoptive transfer studies. *Infect Immun* 1984;43:79–83.
10. Kaufmann SHE, Chiplunkar S, Flesch I, de Libero G: Possible role of helper and cytolytic T cells in mycobacterial infections. *Lepr Rev* 1986;57(suppl 2):101–111.
11. Kirchheimer WF, Storrs EE: Attempts to establish the armadillo (Dasypus novemcinctus Linn.) as a model for the study of leprosy. I. Report of lepromatoid leprosy in an experimentally infected armadillo. *Int J Lepr* 1971;39:693–702.
12. Kirchheimer WF, Storrs EE, Binford CH: Attempts to establish the armadillo (Dasypus novemcinctus Lynn.) as a model for the study of leprosy. II. Histologic and bacteriologic post-mortem findings in lepromatoid leprosy in the armadillo. *Int J Lepr* 1972;40:229–242.
13. Mehra V, Bloom BR: Induction of cell mediated immunity to Mycobacterium leprae in mice. *Infect Immun* 1979;23:787–994.
14. Mehra V, Brennan PJ, Rada E, et al: Lymphocyte suppression in leprosy induced by unique M. leprae glycolipid. *Nature* 1984;308:194–196.
15. Mehra V, Sweester D, Young RA: Efficient mapping of protein antigenic determinants. *Proc Natl Acad Sci USA* 1986;83:7013–7017.

16. Mustafa AS, Gill HK, Nerland A, et al: Human T-cell clones recognize a major M. leprae protein antigen expressed in E. coli. *Nature* 1986;319:63–66.
17. Mustafa AS, Oftung F, Gill HK, Natvig I: Characteristics of human T cell clones from BCG and killed M. leprae vaccinated subjects and tuberculosis patients: recognition of recombinant mycobacterial antigens. *Lepr Rev* 1986;57(supply 2):123–130.
18. Mustafa AS, Kvalheim G, Degre M, Godal T: Mycobacterium bovis BCG induced human T cell clones from BCG vaccinated healthy subjects: antigen specificity and lymphokine production. *Infect Immun* 1986;53:491–497.
19. Mustafa AS, Godal T: BCG induced CD4$^+$ cytotoxic T cells from BCG vaccinated healthy subjects: relation between cytotoxicity and suppression in vitro. *Clin Exp Immunol* 1987;69:255–262.
20. Oftung F, Mustafa AS, Husson R, et al: Human T cell clones recognize two abundant Mycobacterium tuberculosis protein antigens expressed in Escherichia coli. *J Immunol* 1987;138:927–931.
21. Patel PJ, Lefford MJ: Induction of cell-mediated immunity to Mycobacterium leprae in mice. *Infect Immun* 1979;19:87–93.
22. Pönnighaus JM, Fine PEM: The Karonga prevention trial—which BCG? *Lepr Rev* 1986;57(suppl 2):285–292.
23. Shepard CC: The experimental disease that follows the injection of human leprosy bacilli into foot-pads of mice. *J Exp Med* 1960;112:445–454.
24. Shepard CC: Animal vaccination studies with Mycobacterium leprae. *Int J Lepr* 1984;51:519–523.
25. Shepard CC, Walker LL, van Landingham R: Heat stability of Mycobacterium leprae immunogenicity. *Infect Immun* 1978;22:87–93.
26. Shepard Draper P, Rees RJW, Lowe C: Effect of purification steps on the immunogenicity of Mycobacterium leprae. *Br J Exp Pathol* 1980;61:375–379.
27. Shepard CC, Minagawa F, van Landingham R, Walker LL: Foot pad enlargement as a measure of induced immunity to Mycobacterium leprae. *Int J Lepr* 1980;48:371–381.
28. Smelt AHM, Rees RJW, Liew FY: Induction of delayed-type hypersensitivity to Mycobacterium leprae in healthy individuals. *Clin Exp Immunol* 1981;44:501–506.
29. Talwar GP, Fotedar A: Two candidate anti-leprosy vaccines—current status of their development. *Int J Lepr* 1983;51:550–552.
30. World Health Organization: Report of the Fifth Meeting of the Scientific Working Group on the Immunology of Leprosy. Protocol 1/79. TDR/SWG/IMMLEP (5) 80.3, 1980.
31. Young RA, Mehra V, Sweester D, et al: Genes for the major protein antigens of the leprosy parasite Mycobacterium leprae. *Nature* 1985;316:450–452.

CHAPTER 30

Mycobacterium w: Candidate Vaccine Against Leprosy with Antigens Cross-Reactive with Three Major Protein Antigens of *Mycobacterium leprae*

Abu Salim Mustafa

Leprosy is a chronic mycobacterial infection caused by *Mycobacterium leprae*. There are about 15 million people in the world suffering from this disabilitating disease. Although *M. leprae* was among the first organisms discovered to be associated with a human disease, it has not yet been cultivated on synthetic media in vitro. The growth of *M. leprae* in nine-banded armadillos for the first time provided enough bacilli for immunological and biochemical studies. Vaccination with purified killed *M. leprae* obtained from infected armadillo tissues induces long-lasting cell mediated immunity (CMI) in mice (37), guinea pigs (20), and humans (6,7). Immunization of mice with killed *M. leprae* protects them against viable challenge in the footpads (38).

These studies along with the nontoxic and nonpathogenic nature of killed *M. leprae* make it an ideal candidate for an antileprosy vaccine. However, it has a few potential limitations. Armadillos are not bred in captivity, and the supply of wild armadillos is limited; thus armadillo-derived killed *M. leprae* vaccine may run into the problem of short supply. The cost of killed *M. leprae* vaccine is high [approximately $5 (US) per dose], which may not be economically affordable by developing countries, where leprosy is endemic. Killed *M. leprae* vaccine given alone can be of use in immunoprophylaxis but not in immunotherapy of lepromatous leprosy patients who are anergic to *M. leprae* antigens, unless it is given in combination with other immunostimulants, e.g., killed *M. leprae* + BCG (4). Therefore studies have focused on developing an antileprosy vaccine based on cultivable mycobacteria, which are available in unlimited supply and can be produced inexpensively.

There is no faithful short-term experimental system available to test the efficacy of an antileprosy vaccine. Humans are the only reliable model of this disease. Evaluation of a vaccine in humans, because of the low incidence and long incubation period, takes at least 10 years of follow-up of a large vaccinated population. However, a correlation between CMI to *M. leprae* antigens and resistance against the disease is well established. Therefore the first criteria for suitability of a cultivatable mycobacterium as a candidate antileprosy vaccine should be its cross-reactivity with *M. leprae* in the antigens involved in CMI functions. In order to be used for immunotherapy of lepromatous leprosy patients, the selected mycobacteria should also have antigens that do not cross-react with *M. leprae*. The helper effects provided by *M. leprae* noncross-reactive antigens may facilitate the restoration of *M. leprae* reactive CMI in these patients. As briefly discussed in the next paragraph, *Mycobacterium w*, a cultivable mycobacterium was found to fulfill these criteria.

Mycobacterium w

The peripheral blood mononuclear cells (PBMCs) from tuberculoid leprosy patients proliferated in response to *Mycobacterium w* in a manner analogous to *M. leprae* (24). Immunization of guinea pigs with *M. w* and *M. leprae* induced a homologous and cross-reactive delayed-type hypersensitivity (DTH) skin response (25). Like killed *M. leprae*, injection of mice with *M. w* caused chronic enlargement of draining lymph nodes (26) and long-lasting delayed infiltration of mononuclear cells in the mouse footpad (5). Lepromin-like preparations prepared from *M. leprae* and *M. w* gave a comparable DTH skin response in tuberculoid leprosy patients (9,10,15,27,34,36). These studies showed that *M. w* is a potent immunogen that has antigens with considerable cross-reactivity with *M. leprae* antigens in CMI functions. Stronger DTH responses to homologous antigens in guinea pigs immunized with *M. w* (25) and positive DTH responses to *M. w* antigens in lepromatous patients (9,10,15,27,34,36) suggested that *M. w* also has antigens that are not present in *M. leprae*. Thus it could be an ideal candidate for an antileprosy vaccine and may serve the purpose of immunoprophylaxis and immunotherapy. Although evaluation of immunoprophylactic effects of *M. w* will require many years of follow-up of a large vaccinated population, its immunotherapeutic potentials were studied in repeatedly lepromin-negative lepromatous leprosy patients. Sixty percent of these patients converted to lepromin positivity after vaccination with *M. w* and showed improvements in their bacteriological and clinical status (3). These in vivo effects of *M. w* may have been due to the effect of helper factors, e.g., interleukin 2 and other lymphokines produced in response to *M. leprae* non-cross-reactive antigens, to which the cells of lepromatous patients respond. The cells capable of responding to *M. leprae* antigens in lepromatous patients probably existed but were funtctionally incompetent owing to suppressor (21) or other (11,12) mechanisms. These cells may have been activated and amplified by helper factors in such a way that they gained the ability to respond to *M. leprae* antigens on a subsequent challenge (13). Alternatively, virgin cells capable of responding to *M. leprae* antigens may have been recruited and amplified in the presence of the cascade of lymphokines. Whatever the mechanisms may be, it is possible that both *M. leprae* cross-reactive and non-cross-reactive antigens of *M. w* were involved. However, the identity of such antigens was unknown.

The recombinant genomic DNA libraries of *M. leprae* and *M. tuberculosis* DNA have been constructed in the phage vector λgt11. The major protein antigens of *M. leprae* and *M. tuberculosis* were identified when recombinant phage were expressed in *Escherichia coli,* and the expressed products were screened by *M. leprae* and *M. tuberculosis* reactive monoclonal antibodies (44,45). The establishment of techniques to test the reactivity of T cells against the recombinant antigens (28) has made it possible to identify some of the antigens of *M. w,* which either cross-react or do not cross-react with *M. leprae.* The establishment of such T cell lines and clones and their identification of the recombinant antigens is described in the following sections.

Response of T Cell Lines and Clones to *M. leprae*, BCG, and *M. w*

The T cell lines were established against *M. leprae* and *M. w* from the PBMCs of BCG and killed *M. leprae* vaccinated subjects (6) according to standard protocols (22). For comparison, T cell lines were also raised against BCG from these subjects (Table 30.1). Each cell line was tested for its reactivity to the antigens of *M. leprae, M. w,* and BCG. All of the seven T cell lines raised against *M. leprae* proliferated to *M. leprae* and *M. w,* and six of these cell lines responded to BCG (Table 31.1). The five T cell lines raised against *M. w* responded to *M. leprae, M. w,* and BCG. Although all of the eight T cell lines raised against BCG responded to BCG, only three and two of them, respectively, responded to *M. leprae* and *M. w* (Table 30.1). These results suggest that in the subjects tested the dominant T-cell-activating antigens of *M. leprae* and *M. w* cross-react with each other and with BCG, but the dominant T cell antigens of BCG may not cross-react with *M. leprae* and *M. w.* This conclusion is supported from our other studies, where most (70%) of the *M. leprae*-induced T cell clones responded to BCG, but only a small fraction (<10%) of the BCG-

Table 30.1. Proliferation of T cell lines raised from BCG and killed *M. leprae* vaccinated subjects to *M. leprae,* BCG, and *M.w.*

T cell lines raised against	Proliferative response[a] of T cell lines to		
	M. leprae	*M. w*	BCG
M. leprae	7/7	7/7	6/7
M. w	5/5	5/5	5/5
BCG	3/8	2/8	8/8

[a]Positive/tested.

induced T cell clones from BCG-vaccinated people responded to *M. leprae* (29,30).

The positive response of *M. leprae*-induced T cell lines to *M. w* and vice versa suggests that one or more antigens of these mycobacteria are cross-reactive. The question of cross-reactivity for single specificities was addressed using CD4$^+$ human T cell clones established against *M. leprae* from BCG- and killed *M. leprae*- vaccinated subjects (22). In proliferative assays, one of the three *M. leprae*-specific and five of the 12 cross-reactive T cell clones responded to *M. w* (Table 30.2). It confirms that *M. leprae* and *M. w* have antigens that are cross-reactive in defined specificities but are not identical in their antigenic makeup. Proliferation to *M. w* of one or more cross-reactive T cell clones from a single subject would have been the reason for responsiveness of all of the *M. leprae*-induced T cell lines to *M. w*. The positive response of all of the *M. leprae*-induced cross-reactive T cell clones to BCG (Table 30.2) indicates that *M. w* lacks some of the cross-reactive antigens of *M. leprae* shared by BCG. A unique feature of *M. w* is its cross-reactivity with *M. leprae* for an epitope not shared by BCG or any other cultivable mycobacteria tested (22). As is discussed in the next section, this epitope resides on an immunologically important protein.

The T cell clones that proliferated to the antigens of *M. leprae*, BCG, and *M. w* were cytotoxic for macrophages pulsed with these mycobacteria. In the presence of a given antigen, correlation was found between proliferative and cytotoxic activities of the T cell clones. The T cell clones that were *M. leprae*-specific in proliferation exhibited *M. leprae* specificity in the cytotoxicity assay, and the T cell clones that in addition to *M. leprae* proliferated to *M. w* or BCG were cytotoxic for macrophages pulsed with the respective mycobacteria (22). There was a quantitative difference in the number of cloned T cells required for cytotoxicity and proliferation or lymphokine production. The optimal cytotoxicity required ten times more cells than optimal proliferation or lymphokine production (23).

Table 30.2. Proliferative response of *M. leprae*-induced T cell clones in response to *M. leprae*, BCG, and *M. w*.

Donors	Proliferation of T cell clones[a] in response to the antigens of		
	M. leprae	BCG	*M. w*
M. leprae specific T cell clones			
ATT	1/1	0/1	1/1
NHH	2/2	0/2	0/2
Cross-reactive T cell clones			
ATT	7/7	7/7	3/7
JM	3/3	3/3	2/3
AF	2/2	2/2	1/2

[a]Positive/tested.

What could be the in vivo implications of these in vitro findings? *M. leprae* is an obligate intracellular parasite of macrophages, a cell type that is endowed with bactericidal activity. It is also involved in antigen presentation to T cells for their antigen-induced proliferation and lymphokine production. The lymphokines so produced enhance the bactericidal activity of macrophages. If macophages somehow become incompetent to kill the ingested mycobacteria, as in lepromatous leprosy patients, they may be killed by cytotoxic activity of otherwise helper T cells. The mycobacteria released from such incompetent macrophages could be taken up by more competent monocytes, which will have better killing machinery than mycobacteria-laden macrophages. The cytotoxic potential of helper T cells may thus provide additional mecanisms to clear infections. Kaufmann et al (17) have demonstrated the inhibitory activity of CD8[+] cytotoxic T cell clones on the proliferation of mycobacteria in the macrophages in vitro. The cross-reactivity of *M. leprae* and *M. w* antigens in the cytotoxic functions mediated by CD4[+] cells may be advantageous for the induction of protection against leprosy in the subjects vaccinated with *M. w*.

Identification of Recombinant Antigens Stimulating T Cell Lines and Clones

The above described T cell lines and clones were used as tools to identify the recombinant antigens that are either shared or are different in *M. leprae*, BCG, and *M. w*. Among the 20 T cell lines established from BCG- and killed *M. leprae*-vaccinated subjects, seven T cell lines responded to one or more of the recombinant antigens tested (22) (Table 30.3). All of the seven T cell lines responded to the 65-kilodalton (kDa) antigen of *M. tuberculosis* (Table 30.3). Two each of these cell lines were raised against *M. leprae* and BCG, and three cell lines were established against *M. w* antigens. The T cell clone JMA7, which responded to *M. leprae*, BCG, and *M. w* antigens, proliferated to the 65-kDa antigens of *M. leprae* and *M. tuberculosis* (30) (Table 30.4). The 65-kDa antigen is the most extensively studied mycobacterial antigen. This antigen is highly cross-reactive among mycobacterial species in antibody and T cell reactivities (2,19,30). It may also have antibody and T cell specific epitopes (2,33). The amino acid sequence analysis of 65-kDa *M. leprae* and *M. tuberculosis* antigens revealed more than 95% homology (39). Immunization studies demonstrate that the 65-kDa antigen induced CMI responses against whole bacilli in experimental animals (8,18,40). In limiting dilution assays, 20% of the *M. tuberculosis* responding T cells react to the 65-kDa antigen (18). The 65-kDa antigen has an epitope that may be involved in the protection against an autoimmune diseases, adjuvant arthritis in rats (41).

Mycobacterium leprae and *M. w* induced T cell lines from the donor NHH responded to the 18-kDa *M. leprae* antigen (Table 30.3). *M. leprae-*

Table 30.3. Identification of recombinant antigens recognized by T cell lines.

T cell lines raised against	Response of T cell lines to recombinant antigens								
	M. leprae						M. tuberculosis		
	65 kDa	36 kDa	28 kDa	18 kDa	12 kDa	13B3	65 kDa	19 kDa	14 kDa
Donor AF									
M. leprae	NT	–	–	–	–	–	–	–	–
BCG	NT	–	–	–	–	–	–	–	–
M. w	NT	–	–	–	–	–	+	–	–
Donor JM									
M. leprae	NT	–	–	–	–	–	+	–	–
BCG	NT	NT	NT	–	NT	–	+	–	–
M. w	NT	–	–	–	–	–	+	–	–
Donor NHH									
M. leprae	NT	–	–	+	–	+	+	–	–
BCG	NT	–	–	–	–	+	+	+	+
M. w	NT	–	+	+	–	–	+	+	–

Table 30.4. Identification of recombinant antigens recognized by CD4$^+$ T cell clones.

T cell clones	Mycobacteria			Recombinant antigens of M. leprae						Recombinant antigens of M. tuberculosis		
	M. leprae	BCG	M. w	65 kDa	36 kDa	28 kDa	18 kDa	12 kDa	13B3	65 kDa	19 kDa	14 kDa
ATT6/10F	+	–	+	–	–	–	+	–	–	–	–	–
JM A7	+	+	+	+	–	–	–	–	–	+	–	–
ATT 2/3D	+	+	–	–	–	–	–	–	+	–	–	–

specific T cell clone ATT 6/10F, which responded to *M. w* but not to BCG and other mycobacteria tested (22), proliferated in the presence of 18-kDa *M. leprae* antigen (Table 30.4). Earlier studies have shown that 18-kDa *M. leprae* antigen is *M. leprae*-specific (28). However, this otherwise *M. leprae*-specific antigen is cross-reactive with *M. w*. The T cell lines raised against BCG and *M. w* from the donor N.H.H. responded to the *M. tuberculosis* 19-kDa antigen (Table 30.3). This antigen has been shown to have a T cell epitope specific to the mycobacteria of the tuberculosis complex (33). Reactivity of the *M. w*-induced T cell line from the donor N.H.H. (Table 30.3) shows that 19-kDa *M. tuberculosis* antigen is shared by *M. w*. The 18-kDa *M. leprae* and the 19-KDa *M. tuberculosis* antigens are the proteins of almost identical molecular weight but are totally different in nature, as suggested by the specific reactivities of the antibodies and T cells (16,28,30,33). DNA encoding these proteins do not hybridize (16), and their amino acid sequences lack identity (43). The responses of T cell lines and clones show that *M. w* has epitopes that are cross-reactive with both of these proteins.

The *M. w*-induced NHH T cell line responded to a third *M. leprae* antigen of 28-kDa (Table 30.3). This antigen was identified by antibody probes but was not demonstrated to have T cell epitopes. Response of one of the *M. w*-induced T cell lines for the first time has shown the presence of a T cell epitope on this antigen and suggests that 28-kDa antigen is yet another *M. leprae* antigen shared by *M. w*.

A recombinant antigen, isolated using T cell clones as primary probes and tentatively named 13B3, was found to be specific to *M. leprae* and to the mycobacteria of tuberculosis complex (31). Screening the T cell lines and T cell clones revealed that *M. leprae*- and BCG-induced T cell lines from the donor NHH responded to this antigen (22), (Table 30.3). The T cell clone ATT2/3D that responded to 13B3 antigen proliferated to *M. leprae* and BCG but not to *M. w* (Table 30.4). None of the *M. w*-induced cell lines responded to 13B3 antigen. Thus 13B3 antigen has an epitope shared by *M. leprae* and BCG but not by *M. w*.

In conclusion, the results discussed above demonstrate that *M. w* has antigens that cross-react with at least three major protein antigens of *M. leprae*. However, *M. w* has additional identifiable antigens that are not shared by *M. leprae*. The *M. leprae* cross-reactive as well as non-cross-reactive antigens of *M. w* may play a role in the induction of protective immunity to *M. leprae* infection in the subjects vaccinated with *M. w*.

18-kDa and 65-kDa *M. leprae* antigens: Stress Proteins

To get an insight into the nature of antigens shared between *M. leprae* and *M. w,* the amino acid sequences of the major protein antigens of *M. leprae* and *M. tuberculosis* proteins were searched for homology with other proteins in the data bank. The 18-kDa *M. leprae* protein and the 65-kDa and 70-kDa *M. leprae* and *M. tuberculosis* proteins were found to have significant homology with heat shock proteins (43). The 65-kDa *M. leprae* and *M. tuberculosis* proteins are 60% identical in their amino acid sequences with the 60-kDa *E. coli* major heat shock protein Gro EL (43). The 18- kDa *M. leprae* protein is 31% identical in 127 overlapping amino acid sequences with the low-molecular-weight soybean heat shock proteins (32). Heat shock proteins are stress proteins: Their synthesis is selectively increased under stress conditions. Macrophages provide a hostile environment for mycobacteria. It is conceivable that under these stressful conditions the stress proteins such as 18-kDa and 65-kDa antigens are produced in abundance and become the major targets of the immune response. Stress proteins in a variety of infectious agents, i.e., *Coxiella* (42), *Plasmodium* (1), *Schistosoma* (14), filaria (35), and mycobacteria (43) have been identified as targets of the immune response. If stress proteins are the major targets of an immune response, *M. w* may be an ideal candidate for an antileprosy vaccine, as it shares the antigenic epitopes on the two stress proteins of *M. leprae*.

References

1. Bianco AE, Favaloro JM, Burkot TR, et al: A repetitive antigen of *Plasmodium falciparum* that is homologous to heat shock protein 70 of *Drosophila melanogaster*. *Proc Natl Acad Sci USA* 1986;83:8713.
2. Buchanan TM, Nomaguchi H, Anderson DC, et al: Characterization of antibody-reactive epitopes on the 65 kilodalton protein of *Mycobacterium leprae*. *Infect immun* 1987;55:1000.
3. Chaudhuri S, Fotedar A, Talwar GP: Lepromin conversion in repeatedly lepromin negative BL/LL patients after immunization with autoclaved *Mycobacterium w*. *Int J Lepr* 1983;51:159.
4. Convit J, Aranazu M, Ulrich M, et al: Immunotherapy with a mixture of *Mycobacterium leprae* and BCG in different forms of leprosy and in Mitsuda negative contacts. *Int J Lepr* 1982;50:415.
5. Fotedar A, Mehra NK, Mustafa AS, Talwar GP: Local reactions to intradermal instillation of *Mycobacterium w* and ICRC bacilli in mice. *Lepr India* 1978;50:520.
6. Gill HK, Mustafa AS, Godal T: Induction of delayed type hypersensitivity in human volunteers immunized with a candidate leprosy vaccine consisting of killed *Mycobacterium leprae*. *Bull WHO* 1986;64:121.
7. Gill HK, Mustafa AS, Godal T: *In vitro* proliferation of lymphocytes from human volunteers vaccinated with armadillo derived, killed *M. leprae*. *Int J Lepr* 1987;55:30.
8. Gillis TP, Job CK: Purification of the 65 KD protein from *Mycobacterium gordonae* and use in skin test response to *Mycobacterium leprae*. *Int J Lepr* 1987;55:54.
9. Girdhar BK, Desikan KV: Results of skin tests with five different mycobacteria. *Lepr India* 1978;50:555.

10. Govil DC, Bhutani LK: Delayed hypersensitivity skin reactions to lepromin and antigens from four other mycobacteria. *Lepr India* 1978;50:550.

11. Haregewoin A, Godal T, Mustafa AS, et al: T-cell conditioned media reverse T-cell unresponsiveness in lepromatous leprosy. *Nature* 1983;303:342.

12. Haregewoin A, Mustafa AS, Helle I, et al: Reversal by interleukin-2 of the T-cell unresponsiveness of lepromatous leprosy to *Mycobacterium leprae*. *Immunol Rev* 1984;80:77.

13. Haregewoin A, Longley J, Bjune G, et al: The role of interleukin-2 (IL-2) in the specific unresponsiveness of lepromatous leprosy to *Mycobacterium leprae:* studies *in vitro* and *in vivo*. *Immunol lett* 1986;11:249.

14. Hedstrom R, Culpepper J, Harrison RA, et al: A major immunogen in *Schistosoma mansoni* infections is homologous to the heat shock protein HSP70. *J Exp Med* 1987;165:1430.

15. Hogerzeil LM, Prabhudass N: Delayed hypersensitivity skin reactions to lepromins prepared from *M. leprae* and selected cultivable mycobacteria. Investigations at the Victoria Hospital, Dichpalli. *Lepr India* 1978;50:560.

16. Husson R, Young RA: Genes for the major antigens of *Mycobacterium tuberculosis:* the etiologic agents of tuberculosis and leprosy share an immunodominant antigen. *Proc Natl Acad Sci USA* 1987;84:1379.

17. Kaufmann SHE, Chiplunkar S, Flesch I, De Libero G: Possible role of helper and cytolytic T cells in mycobacterial infections. *Lepr Rev* 1986;57(suppl 2):101.

18. Kaufmann SHE, Vath U, Thole JER, et al: Enumeration of T cells reactive with *Mycobacterium tuberculosis* organisms and specific for the recombinant mycobacterial 64-kDa protein. *Eur J Immunol* 1987;17:351.

19. Lamb JR, Ivanyi J, Rees ADM, et al: Mapping of T cell epitopes using recombinant antigens and synthetic peptides. *EMBO J* 1987;6:1245.

20. Mehra V, Bloom BR: Induction of cell mediated immunity to *Mycobacterium leprae* in guinea pigs. *Infect Immun* 1979;23:787.

21. Mehra V, Convit J, Rubinstein A, Bloom BR: Activated suppressor cells in leprosy. *J Immunol* 1982;129:1946.

22. Mustafa AS: Identification of T cell activating antigens shared between three candidated antileprosy vaccines, killed *M. leprae, Mycobacterium bovis* BCG and *Mycobacterium w. Int J Lepr* (in press, 1988).

23. Mustafa AS, Godal T: BCG induced CD4$^+$ cytotoxic T cells from BCG vaccinated healthy subjects: relation between cytotoxicity and suppression in vitro. *Clin Exp Immunol* 1987;69:255.

24. Mustafa AS, Talwar GP: Five cultivable mycobacterial strains giving blast transformation and leukocyte migration inhibition of leukocytes analogous to *Mycobacterium leprae*. *Lepr India* 1978;50:498.

25. Mustafa AS, Talwar GP: Delayed hypersensitivity skin reactions to homologous and heterologous antigens in guineapigs immunized with *M. leprae* and four selected cultivable mycobacterial strains. *Lepr India* 1978;50:509.

26. Mustafa AS, Talwar GP: Enlargement of draining lymphnodes in mice injected with four selected cultivable mycobacterial strains. *Lepr India* 1978;50:534.

27. Mustafa AS, Talwar GP: Early and late reactions in tuberculoid and lepromatous leprosy patients with lepromins from *Mycobacterium leprae* and five selected cultivable mycobacteria. *Lepr India* 1978;50:566.

28. Mustafa AS, Gill HK, Nerland A, et al: Human T cell clones recognize a major *M. leprae* protein expressed in *E. coli*. *Nature* 1986;319:63.

29. Mustafa AS, Kvalheim G, Degre M, Godal T: *Mycobacterium bovis* BCG induced human T-cell clones from BCG vaccinated healthy subjects: antigen specificity and lymphkine production. *Infect Immun* 1986;53:491.
30. Mustafa AS, Oftung F, Gill HK, Natvig I: Characteristics of human T-cell clones from BCG and killed *M. leprae* vaccinated subjects and tuberculosis patients: recognition of recombinant mycobacterial antigens. *Lepr Rev* 1986;57(suppl 2):123.
31. Mustafa AS, Oftung F, Deggerdal A, et al: Gene isolation using human T lymphocyte probes. Isolation of a gene that expresses a T cell antigen specific for *Mycobacterium bovis* BCG and pathogenic mycobacteria. *J Immunol* (in press, 1988).
32. Nerland AH, Mustafa AS, Sweetser D, et al: A protein antigen of *Mycobacterium leprae* is related to a family of small heat shock proteins. *J Bacteriol* (in press, 1988).
33. Oftung F, Mustafa AS, Husson R, et al: Human T cell clones recognize two abundant *M. tuberculosis* proteins expressed in *E. coli*. *J Immunol* 1987;138:927.
34. Sahib HS, Vellut C: Some observations on skin reactions induced by lepromin and four other mycobacterial antigens. *Lepr India* 1978;50:579.
35. Selkirk ME, Rutherford PJ, Danham DA, et al: Cloned antigen genes of *Brugia filarial* parasite. Biochemistry Society Symposium (in press, 1988).
36. Sharma RC, Singh R: Comparative study of skin reactions in leprosy patients to *M. leprae*-lepromin and to antigens from cultivable saprophytic mycobacteria. *Lepr India* 1987;50:572.
37. Shephard CC, Minagawa F, Landingham RV, Walker LL: Footpad enlargement as a measure of induced immunity to *Mycobacterium leprae*. *Int J Lepr* 1980;48:371.
38. Shepard CC, Walker LL, Landingham RV: Heat stability of *Mycobacterium leprae* immunogenicity. *Infect Immun* 1978;22:87.
39. Shinnick TM, Sweetser D, Thole J, et al: The etiologic agents of leprosy and tuberculosis share an immunoreactive protein antigen with the vaccine strain *Mycobacterium bovis* BCG. *Infect Immun* 1987;55:1932.
40. Thole JER, Keulen WJ, Kolk AHJ, et al: Characterization, sequence determination, and immunogenicity of a 64-kilodalton protein of *Mycobacterium bovis* BCG expressed in *Escherichia coli* K-12. *Infect Immun* 1987;55:1466.
41. Van Eden W, Thole JER, Van der Zee R, et al: Cloning of the mycobacterial epitope recognized by T lymphocytes in adjuvant arthritis. *Nature* 1988;331:171.
42. Wodkin MH, Williams JC: A heat shock operon in *Coxiella burnetii* produces a major antigen homologous to a protein both in Mycobacteria and *E. coli*. *J Bacteriol* (in press, 1988).
43. Young D, Lathigra R, Hendrix R, et al: Stress proteins are immune targets in leprosy and tuberculosis. *Proc Natl Acad Sci USA* (in press, 1988).
44. Young RA, Bloom BR, Grosskinsky CM, et al: Dissection of *Mycobacterium tuberculosis* antigens using recombinant DNA. *Proc Natl Acad Sci USA* 1985;82:2583.
45. Young RA, Mehra V, Sweetser D, et al: Genes for the major protein antigens of the leprosy parasite *Mycobacterium leprae*. *Nature* 1985;316:450.

CHAPTER 31

Molecular Approaches to Developing a Vaccine for Leprosy

Vijay Mehra, Robert L. Modlin, Thomas H. Rea,
William R. Jacobs, Scott B. Snapper,
Jacinto Convit, and Barry R. Bloom

Leprosy, a chronic infectious disease afflicting 10 million to 15 million people, is caused by the obligate intracellular parasite *Mycobacterium leprae*. Although *M. leprae* was the first identified bacterial pathogen of man, it remains one of the few human pathogens that cannot yet be grown in culture. The inability to grow leprosy bacillus in culture has severely limited the understanding of the bacillus and the disease.

Leprosy is a spectral disease that presents a diversity of clinical manifestations (2). At one pole of the spectrum, tuberculoid leprosy patients have a few localized lesions with few discernible organisms and a high level of cell-mediated immunity (CMI) that ultimately kills and clears the bacilli, although often with concomitant damage to the nerves. In contrast, lepromatous patients exhibit a selective immunological unresponsiveness to antigens of *M. leprae* and have numerous skin lesions containing extraordinary high numbers of acid-fast bacilli, e.g., 10^{10}/g of tissue. Antibodies to *M. leprae* are found throughout the spectrum; the highest levels occur in the lepromatous disease, indicating that they are unlikely to play a major role in protection. There is a striking inverse correlation between the level of CMI to antigens of *M. leprae* and the growth of bacilli in the tissues.

Specific Unresponsiveness in Lepromatous Leprosy

Most patients with lepromatous leprosy are able to exhibit CMI to common recall antigens, such as PPD, candidin and SK-SD. They respond perfectly well to the related organisms, e.g., BCG and *Mycobacterium tuberculosis,* that have most of the known protein and glycoprotein antigens cross-reactive with *M. leprae*. Thus there is a selective and specific cell-mediated unresponsiveness in vivo and in vitro to antigens of *M. leprae*.

Immunologically the most challenging question in lepromatous leprosy concerns the mechanisms that govern the specific unresponsiveness of patients to *M. leprae* antigens. We suggested the hypothesis that there might be one or a small number of unique antigens or determinants associated with *M. leprae* capable of inducing active suppression of the responses of potentially reactive helper T cells (3). We developed a simple assay to test the ability of *M. leprae* antigens to induce suppression of a proliferative response of peripheral blood lymphocytes to a mitogen, concanavalin A (Con A), used at suboptimal doses. In the more than 200 patients studied, suppression was observed in 84% of lepromatous and borderline patients but not in tuberculoid patients, lepromin-positive contacts, or normal donors. The in vitro suppression was found to be mediated by both adherent and nonadherent subsets of the peripheral mononuclear cell population (18).

It is our impression that macrophage-induced in vitro suppression of mitogen responses is related to the extent of disease and bacillary load and is found less frequently in patients whose disease is detected at an early stage. Moreover, this type of suppression is not antigen-specific, and it fails to explain the selective immunological unresponsiveness seen in lepromatous patients.

Consequently, we sought to characterize the nonadherent suppressor cell population. The studies conducted with T-suppressor and T-helper subsets, isolated on FACS (florescence-activated cell sorter) using monoclonal OKT8 antibodies, demonstrated that the suppressor activity was contained in the 30% CD8$^+$ subset of peripheral T cells (19). In addition, 50% of the CD8$^+$ cells expressed Fc receptor and HLA-DR (Ia) antigens (17). Furthermore, using the phenolic glycolipid unique to *M. leprae*, discovered by Brennan's group (13), it was observed that it induced suppression of mitogen responses in lepromatous patients as well as whole *M. leprae* did. A significant portion of the suppression induced by the phenolic glycolipid could be eliminated by removing the 3'-methyl group of the terminal dimethyl glucose, whereas removal of mycolic acid side chains had no effect on suppression (16). Finally, our recent observations indicate that chemically synthesized disaccharide conjugated to bovine serum albumin is fully capable of inducing the suppressor response in vitro.

A critical test of the suppressor T cell hypothesis was whether cells actually infiltrating lesions of lepromatous patients could be shown to manifest antigen-induced suppressive activity similar to that observed from the cells of peripheral blood of such patients. We have recently developed the procedure for isolating T cell subsets directly from biopsies of lesions and establish them as short-term lines and then as clones (24). The CD8 lines established from lepromatous and tuberculoid lesions were tested for in vitro lepromin-induced suppressor activity. About half the CD8 lines from lepromatous lesions, but none from tuberculoid lesions, had suppressor activity (23). When individual clones obtained from these CD8 lines were tested for suppressor activity, their ability to suppress the response of *M. leprae* specific CD4 helper T cell clones to lepromin was restricted by the major histocompatibility complex (MHC) class II antigens

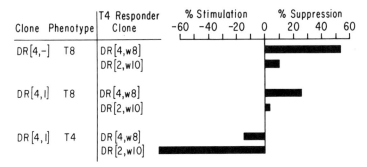

Fig. 31.1. MHC class II-restricted suppression of the lepromin response of CD4 clones by CD8 clones from lepromatous leprosy skin lesions. Antigen-reactive CD4 cloned cells were cultured with irradiated autologous PBMC, with and without lepromin and with and without CD8 clones. ³H-Thymidine incorporation was measured at 72 hours. [From Modlin et al (23), with permission.]

(Fig. 31.1). We believe that these studies provide strong evidence for functional suppressor cells within lesions of a human disease, and that the CD8 suppressor cells are MHC class II restricted.

Although some previous efforts to detect antigen-specific suppression by lymphocytes from lepromatous patients have not been successful (27,31), it is gratifying to note that de Vries and his co-workers (8,28) as well as Sasazuki (personal communication) have not only demonstrated the suppression of lepromin responses by CD8 cells from lepromatous patients but have also observed it to be MHC restricted.

Furthermore, using CD8 clones obtained directly from skin lesions of lepromatous patients and antibodies against framework determinants on human TCR α, β, and γ polypeptides, with Brenner we examined the nature of the antigen receptor of human Ts cells (22). These Ts clones were noted to rearrange TCR β genes, express messenger RNA for α and β chains of the TCR, and express CD3-associated TCR α,β structures on their cell surface (Fig. 31.2); but they do not express the γ chain, indicating that antigen recognition by at least some human CD8⁺ suppressor cells is likely to be mediated by TCR α,β heterodimers.

We fully appreciate that there may be other types of suppressor cells, e.g., suppressor inducers and anti-idiotypic suppressors, in addition to the antigen-specific probable effector-suppressor cells described here. Second, it is possible that mechanisms other than suppression may be involved in the failure of patients with leprosy to fully respond to *M. leprae,* e.g., antigen down-regulation of the CD4 T helper cell. Third, suppressor cells have been found in the peripheral blood and lesions of borderline patients, indicating that there is a quantitative interaction between antigen-responsive helper cells and antigen-specific suppressor cells that is likely to be involved in the final clinical manifestations of the disease.

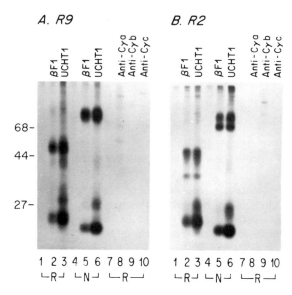

Fig. 31.2. Immunoprecipitation of T cell receptors from CD8 Ts clones. Surface
^{125}I-labeled R9 (**A**) and R2 (**B**) were detergent-solublized and immunoprecipitated
with anti-CD3 (UCHT1) or anti-TCR (BF1) antibodies and analyzed by SDS-PAGE
under reducing (R) and nonreducing (N) conditions. Lanes 7 to 10 show immunoprecipitation with normal rabbit serum (NRS) and rabbit antibodies to TCR γ
chain peptides. R9 and R2 cells expressed two CD3-associated species of 40 to
50kDa under reducing conditions, whereas under nonreducing conditions all of
the CD3-associated species were disulfide-linked. To determine whether these
species represented the protein products of TCR α or β genes, the monoclonal
antibody BF1, which has framework reactivity against TCR β subunit and is capable
of immunoprecipitating the disulfide-linked TCR α–β complex, was used. This
antibody immunoprecipitated polypeptides (40 to 50kDa reduced, 80–90kDa non-
reduced) that appeared identical to those co-immunoprecipitated with anti-CD3
mAb. Thus CD8^{+} suppressor clones express BF1-reactive, heterodimeric CD3-
associated subunits characteristic of TCR α–β complexes. [Adapted from Modlin
et al (22).]

In no system are the mechanisms by which suppressor cells act to block
specific antigen responses are known. It is unclear whether they are involved in depressing interleukin 2 (IL-2) or gamma interferon (IFN-γ)
production, down-regulating or killing T-helper cells, or killing or blocking
the function of antigen-presenting cells.

Vaccine Strategies

The first major breakthrough in overcoming the unresponsiveness in patients with lepromatous leprosy was made by Convit and his colleagues
in Venezuela, who showed that a mixture of live BCG-plus killed purified

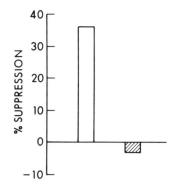

Fig. 31.3. Lepromin-induced suppression of Con A response of peripheral blood lymphocytes in lepromatous leprosy patients tested before (□) and subsequent to (▨) immunotherapy with BCG plus *M. leprae*. [Adapted from Mehra et al (17).]

M. leprae was capable of converting a proportion of skin-test-negative lepromatous patients to lepromin skin test positivity (6). Thus far, several hundred lepromatous leprosy patients have been vaccinated, and the results show immunologic conversion to positive skin test reactivity, clearance of bacilli from the skin, histopathological upgrading, and clinical improvement in approximately 75% of borderline lepromatous patients and 57% of the polar lepromatous patients. These results in patients who have been immunologically unresponsive for long periods of time are rather dramatic.

We have had an opportunity to examine in vitro suppressor T cell activity and activation markers in ten patients prior and subsequent to immunotherapy (17). The results indicated that in 10 of 10 patients the in vitro suppressor activity and expression of Ia antigens was decreased to normal levels following successful immunotherapy (Fig. 31.3).

It is important to unravel the immunological mechanism by which it is accomplished, as BCG alone offers variable protection against leprosy. Analysis to date of four major trials of BCG against leprosy indicate that BCG has some but not sufficient protective efficacy everywhere it has been tested (Table 31.1).

Because it is possible to overcome unresponsiveness in patients with active lepromatous leprosy using the combined vaccine, it should be possible to immunize and protect normal individuals in endemic areas with such a vaccine as well. There are currently two major controlled preventive vaccine trails under way against leprosy sponsored by the IMMLEP committee of WHO to evaluate the effectiveness of killed *M. leprae* and BCG versus BCG alone: one in 68,000 contacts of lepromatous patients in Venezuela, and the second in a total population of 125,000 people in Malawi. In the population in Malawi, case-controlled and cohort data indicate that BCG vaccination alone in this population engenders at least 50% protection against leprosy (11), and the possibility that the combined vaccine might be even more effective will be tested.

Table 31.1. Vaccine trials of BCG against leprosy.

Area	No. of subjects (age at vaccination)	Year started	Follow-up (years)	% Protection
Burma	28,220 (0–14)	1964	9	20
South India (Chingleput)	210,337 (all ages)	1968	5–10	30
New Guinea	5,000 (all ages)	1962	10	44
Uganda	16,150 (childhood contacts)	1960	8	80

Adapted from Fine (11).

However, even if the armadillo-derived vaccine should be effective, there is a real question whether its cost and the limited supply would permit it to be widely used. In addition, there is a certain percentage of lepromatous leprosy patients who do not respond to this vaccine. For this reason, it is necessary to search for other approaches: (a) a cultivable mycobacterium that has antigens cross-reactive with *M. leprae* as well as additional antigens capable of evoking CMI to *M. leprae* in lepromatous leprosy patients; (b) potentially protective antigens in *M. leprae*.

In India, two cultivable mycobacterial strains have been reported to be effective at inducing CMI in lepromatous patients. An atypical mycobacterium, discovered by Talwar and his colleagues (5) and found to be promising as an antileprosy vaccine, is discussed extensively elsewhere in this volume. The other strain of cultivable mycobacterium also found to be a promising candidate for vaccine was isolated at The Indian Cancer Research Centre (ICRC) from the nodules of a lepromatous leprosy patient. It belongs to the *M. avium-intracellulare* group of mycobacteria (7). Encouraging results have been obtained with both these bacilli, and we look forward to the results of clinical trials comparing these two and the combined BCG + *M. leprae* vaccine.

Search for a Unique Antigen

The advances in monoclonal antibodies, recombinant DNA technology, and the development of specific T cell clones have provided tools for identifying and producing protective antigens. With the availability of large amounts of armadillo-derived *M. leprae* it became possible to raise monoclonal antibodies to *M. leprae*, and that permitted identification of unique and cross-reactive epitopes on major protein antigens of *M. leprae* (10). These monoclonal antibodies were used to probe the recombinant DNA expression library of *M. leprae* prepared by Young et al (32) to identify and isolate genes encoding protein antigens. A recombinant DNA expression library of genomic DNA of *M. leprae* sheared to 1- to 7-kilobase (kb) pieces mechanically was constructed in λgt11 phage vector, which is capable of driving the expression of foreign DNA with *Escherichia coli* transcription and translation signals. The foreign protein is expressed as a β-galactosidase fusion protein (32). The *M. leprae* DNA library contained 2.5×10^6 recombinant phage. Because the mycobacterial genome is approximately 2×10^6 basepairs in length the λgt11 recombinant phage contains mycobacterial DNA fragments whose end points occur at nearly every basepair throughout the genome. This ensures that all coding sequences are inserted in the correct transcriptional orientation and translational frame to be expressed as a fusion protein with the β-galactosidase encoded in λgt11. The library was screened with monoclonal antibodies directed against *M. leprae* antigens to isolate the genes encoding the five major protein antigens of *M. leprae*, i.e., 65-, 36-, 28-, 18-, and 12-kDa antigens.

The fact that seven monoclonal antibodies (mAbs) recognized the 65-kDa antigen, and that the individual anti-65-kDa mAbs reacted with some but not all recombinant clones isolated using pooled anti-65 kDa antibodies suggested that there were multiple epitopes on this antigen, as was also shown by competition inhibition studies (12). It has been possible to map and identify 14 epitopes on the 65-kDa antigen by serological methods (4). Using the recombinant DNA expression strategy and anti-65-kDa mAbs as probes, we have mapped six epitopes on 65-kDa antigen, obtained the DNA sequences of the regions encoding them, and deduced the amino acid sequences from DNA sequences for all of them (20).

The recombinant antigens/synthetic peptides have been used to screen T cell clones in an attempt to identify those antigens required for induction of T helper/killer cell responses that are likely to be important in protection. The first strategy was to screen T cell clones from individuals sensitized to *M. leprae* against five recombinant antigens identified by monoclonal antibodies. Several laboratories have identified T cell clones that respond to the 65-, 36-, and 18-kDa antigens (9,25,29). However, they represent only a small proportion of the T cell clones responding to intact *M. leprae*. On the other hand, recent studies indicate that a large proportion of T cell clones reactive with intact *M. leprae* are stimulated by highly purified cell walls (21). Clearly, new methods are needed to facilitate identification of antigens recognized by T cells for which antibodies are presently not available.

Lamb and Young (15) have developed an important procedure by which *M. leprae* antigens are separated by sodium dodecyl sulfate–polyacrylamide gel electrophoresis (SDS-PAGE) and transferred onto nitrocellulose filters. The nitrocellulose is cut into small strips and used to test the reactivity of T cells/clones. Using the modification of this procedure where the nitrocellulose is converted into fine antigen-bearing particles and then used to test the T cells (1), new antigens have been detected. The next step is to produce polyclonal or monoclonal antibodies to the proteins bound to nitrocellulose and use these antibodies to screen the recombinant

Table 31.2. Advantages of recombinant BCG vaccine vehicle.

1. BCG is currently the most widely used vaccine in the world: It has been used in more than 2.5 billion people since 1948.
2. BCG has a low degree of toxicity and a mortality rate of about 60 per billion, about 100-fold less than that of the smallpox vaccine.
3. BCG is the only vaccine recommended to be given at birth.
4. BCG requires only a single immunization, which engenders cell-mediated immunity to tuberculoproteins for 5 to 50 years.
5. BCG is the most effective known adjuvant for induction of cell-mediated immunity in animals and man.
6. BCG costs $0.55 per dose.

DNA library to isolate and characterize the genes for these antigens. Another strategy developed to screen for antigens for which antibodies are not available involves screening the recombinant DNA library directly with T cell clones; it is based on a sib selection strategy used in early studies of bacterial genetics (26,30). It has the disadvantage of requiring large numbers of cloned as well as feeder cells.

Once the antigens useful for protection are identified, our aim is to make a recombinant mycobacterial vaccine expressing the antigens important for engendering the right kind of immunity required for protection against leprosy.

Development of a Mycobacterial Multivaccine Vehicle

There are a number of compelling reasons that suggest that the live, attenuated BCG vaccine used against tuberculosis would represent an important approach to developing an ideal recombinant vaccine. It would offer certain unique advantages in the development of a multivaccine vehicle (Table 31.2). If BCG could be appropriately engineered, it might serve to immunize against protective antigens not only of mycobacteria but also for those of a variety of other pathogens to which T cell immunity is critical for protection.

There are three major hurdles in developing such a recombinant mycobacterial vaccine. The first is to develop methods for effectively introducing foreign DNA into mycobacteria through its waxy coat. The second is to develop means for the stable maintenance of these foreign genes in BCG. Finally, it will be necessary to develop means for ensuring expression of the foreign antigens. Jacobs et al (14) have optimized conditions for producing spheroplasts of *Mycobacterium smegmatis* and the subsequent introduction of foreign DNA into them. It was done using mycobacteriophages, and transfection efficiencies of 10^4 to 10^5 pfu/µl DNA were routinely obtained.

Furthermore, they have constructed a novel vector termed "shuttle phasmid," which has the capability of exchanging DNA between *E. coli* and mycobacteria, thereby permitting manipulation of mycobacterial DNA in *E. coli* and the transfer of foreign DNA into *M. smegmatis* by transfection. Recombinant phage molecules isolated from *M. smegmatis* can then be efficiently transferred to BCG vaccine strains by infection (14).

The shuttle phasmid consists of mycobacterial phage TM4, into which is introduced an *E. coli* cosmid pHC79 that contains an origin of replication and a selectable marker in *E. coli*. The first shuttle phasmid developed is shown in Figure 31.4. Currently, efforts are being made to develop conditions for stable expression of cloned genes in mycobacteria.

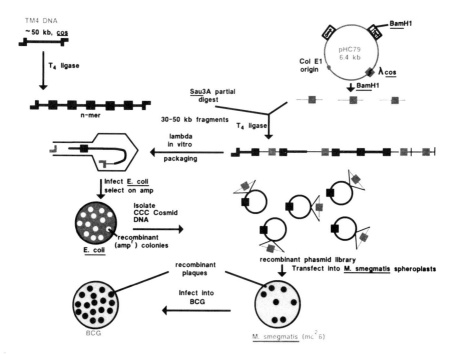

Fig. 31.4. Construction of shuttle phasmid. Mycobacteriophage TM4 is a temperate phage that was found capable of replicating in *M. smegmatis,* BCG, and *M. tuberculosis.* This phage also has a double-stranded DNA genome of 50 kilobases and possesses cohesive ends. The vector consists of a mycobacterial phage (TM4), into which is introduced an *E. coli* cosmid (pHC79) that contains an origin of replication and a selectable marker in *E. coli.* The phage DNA is ligated, partially digested with *Sau* 3A, and ligated to the *E. coli* cosmid pHC79 that has been cleaved with *Bam* H1. After packaging in vitro into lambda phage, the phasmid can then be transduced and grown in large amounts as covalently closed circular DNA in *E. coli.* This phasmid DNA is then transfected into *M. smegmatis.* The recombinant phasmid replicates as phage in *M. smegmatis,* and these recombinant phasmids can be used to transfer DNA into BCG vaccine strains with high efficiency by infection. [From Jacobs et al (14), with permission.]

It will be necessary to define transcriptional and translational signals in BCG to allow foreign genes to be expressed at high levels. Our goal is to develop BCG into a multivaccine vehicle by stably expressing in recombinant BCG a set of genes encoding protective antigens of specific pathogens for which CMI may be important for protection, e.g., *M. leprae, M. tuberculosis, Leishmania,* and *Schistosoma.*

Acknowledgments. We wish to express our gratitude to our collaborators, Drs. Patrick Brennan and Shirley Hunter (Colorado State University), R. Young (Whitehead Institute), and N. Aranzazu and M. Pinardi (Institute de Biomedicine, Venezuela). We also wish to acknowledge Margareta Tuckman for technical assistance and Angela Stockhausen, Rosalia Cawley, and Marie Rizzo for secretarial assistance.

The work was supported by NIH grants AI 07118, 02111, and 10702, 23545; by the Irvington House Institute for Medical Research; and by the UNDP/World Bank/WHO Special Programme for Research and Training in Tropical Diseases (IMMLEP) and Special Programme for Vaccine Development (IMMTUB).

References

1. Abou-Zeid C, Filley E, Steele J, Rook GAW: A simple new method for using antigens separated by polyacrylamide gel electrophoresis to stimulate lymphocytes in vitro after converting bands cut from Western blots into antigen-bearing particles. *J Immunol Methods* 1987;98:5–10.
2. Bloom BR, Godal T: Selective primary helath care: strategies for control of disease in the developing world. V. Leprosy. *Rev Infect Dis* 1983;5:765–780.
3. Bloom BR, Mehra V: Immunological unresponsiveness in leprosy. *Immunol Rev* 1984;80:5–28.
4. Buchanan TM, Nomaguchi H, Anderson DC, et al: Characterization of antibody-reactive epitopes on the 65-kilodalton protein of Mycobacterium leprae. *Infect Immun* 1987;55:1000–1003.
5. Chaudhuri S, Fotedar AR, Talwar GP: Lepromin conversion in repeatedly lepromin negative BL/LL patients after immunization with autoclaved Mycobacterium w. *Int J Lepr* 1983;51:159–168.
6. Convit J, Aranzazu N, Ulrich M, et al: Immunotherapy with a mixture of Mycobacterium leprae BCG in different forms of leprosy and in Mitsuda-negative contacts. *Int J Lepr* 1982;50:415–424.
7. Deo MG, Bapat CV, Chullawalla RG, Bhatki WS: Potential anti leprosy vaccine from killed ICRC bacilli-clinicopathological study. *Indian J Med Res* 1981;74:164–177.
8. DeVries RRP, Ottenhoff THM, Shuguang L, Young RA: HLA class II restricted helper and suppressor clones reactive with Mycobacterium leprae. *Lepr Rev* 1986;57(suppl. 2):113–121.

9. Emmrich F, Thole J, van Embden J, Kaufmann SHE: A recombinant 64 kilodalton protein of Mycobacterium bovis BCG specifically stimulates human T4 clones reactive to mycobacterial antigens. *J Exp Med* 1986;163:1024–1029.

10. Engers HD, Abe M, Bloom BR, et al: Workshop: results of a World Health Organization-sponsored workshop on monoclonal antibodies to Mycobacterium leprae. *Infect Immun* 1985;48:603–605.

11. Fine PEM: The role of BCG in the control of leprosy. *Ethiop Med J* 1985;23:179–191.

12. Gillis TP, Buchanan TM: Production and partial characterization of monoclonal antibodies to Mycobacterium leprae. *Infect Immun* 1982;37:172–178.

13. Hunter SW, Fugiwara T, Brennan PJ: Structure and antigenicity of the major specific glycolipid antigen of Mycobacterium leprae. *J Biol Chem* 1982;257:15072–15078.

14. Jacobs WR, Tuckman M, Bloom BR: Introduction of foreign DNA into mycobacteria using a shuttle phasmid. *Nature* 1987;327:532–535.

15. Lamb JR, Young DB: A novel approach to identification of T cell epitopes in Mycobacterium tuberculosis using human T lymphocyte clones. *Immunology* 1987;60:1–5.

16. Mehra V, Brennen PJ, Rada E, et al: Lymphocyte suppression in leprosy induced by unique M. leprae glycolipid. *Nature* 1984;308:194–196.

17. Mehra V, Convit J, Rubinstein A, Bloom BR: Activated suppressor T cells in leprosy. *J Immunol* 1982;129:1946–1951.

18. Mehra V, Mason LH, Fields JP, Bloom BR: Lepromin-induced supressor cells in patients with leprosy. *J Immunol* 1979;123:1813–1817.

19. Mehra V, Mason LH, Rothman W, et al: Delineation of a human T cell subset responsible for lepromin-induced suppression in leprosy patients. *J Immunol* 1980;125:1183–1188.

20. Mehra V, Sweetser D, Young RA: Efficient mapping of protein antigenic determinants. *Proc Natl Acad Sci USA* 1986;83:7013–7017.

21. Melancon-Kaplan J, Hunter SW, McNeil M, et al: Immunological significance of Mycobacterium leprae cell walls. *Proc Natl Acad Sci USA* in press, (1988)85:1917–1921.

22. Modlin RL, Brenner MB, Krangel MS, et al: T-cell receptors of human suppressor cells. *Nature* 1987;329:541–545.

23. Modlin RL, Kato H, Mehra V, et al: Genetically restricted suppressor T-cell clones derived from lepromatous leprosy lesions. *Nature* 1986;322:459–461.

24. Modlin RL, Mehra V, Wong L, et al: Suppressor T lymphocytes from lepromatous leprosy skin lesions. *J Immunol* 1986;137:2831–2834.

25. Mustafa AS, Gill HK, Nerland A, et al: Human T-cell clones recognize a major M. leprae protein antigen expressed in E. coli. *Nature* 1986;319:63–66.

26. Mustafa AS, Oftung F, Deggerdal A, et al: Gene isolation with human T lymphocyte probes. *J Immunol* 1988;141:2729–2733.

27. Nath I, van Rood JJ, Mehra NK, Vaidya MC: Natural suppressor cells in human leprosy: the role of HLA-D-identical peripheral lymphocytes and macrophages in the in vitro modulation of lymphoproliferative responses. *Clin Exp Immunol* 1980;42:203–210.

28. Ottenhoff THM, Elferink DG, Klaster PR, de Vries RRP: Cloned suppressor T cells from a lepromatous leprosy patient suppress Mycobacterium leprae reactive helper T cells. *Nature* 1986;322:462–464.

29. Ottenhoff THM, Klatser PR, Ivanyi J, et al: Mycobacterium leprae-specific protein antigens defined by cloned human helper T cells. *Nature* 1986;319:66–68.
30. Salgame PR, Colston MJ, Mitchison NA: Murine T-cell reactivity to cloned Mycobacterium leprae antigens. *Lepr Rev* 1986;57(suppl 2):301–304.
31. Stoner GL, Touw J, Belehu A, Naafs B: In vitro lymphoproliferative response to Mycobacterium leprae of HLA-D identical siblings of lepromatous leprosy patients. *Lancet* 1978;2:543–547.
32. Young RA, Mehra V, Sweetser D, et al: Genes for the major protein antigens of M. leprae. *Nature* 1985;316:450–452.

CHAPTER 32

Immunodiagnosis of Leprosy

K.D. Moudgil, R. Mukherjee, Becky M. Itty, and G.P. Talwar

Leprosy is a major global health problem in tropical countries. A conservative estimate indicates that there are about 10 million to 12 million persons suffering from leprosy in the world. India has an estimated 4 million leprosy patients, with an overall prevalence of about 5 per 1000, although in some areas it is as high as 40 per 1000 (21). The total number of leprosy patients in India has registered a continuous increase since 1941 (8). Moreover, the disease has spread to areas formerly free of the disease. The Indian population living in high prevalence areas at considerable risk is about 400 million.

Man harboring *Mycobacterium leprae* is the source of infection as well as the major reservoir of the infective organism. The incubation period of the disease is long: It takes up to 5 years or more for the clinical features to appear after infection with the bacillus. During the incubation period the bacilli multiply in various organs of the host. A small percentage of the host population cannot overcome the infection with *M. leprae* and develop the lepromatous type of leprosy. These patients, with bacilliferous lesions of skin and mucosa, and the asymptomatic carriers constitute a source and foyer of transmission of infection to the healthy population. Early diagnosis of symptomatic and asymptomatic carriers is of crucial importance for the control and eventual eradication of leprosy. Once the diagnosis is made, the infection can be cleared with the help of potent antileprosy drugs, thereby preventing transmission to others. Thus there is need to have sensitive and specific diagnostic tests for leprosy.

Immunodiagnostic Tests for Leprosy

A number of immunoassays for diagnosis of leprosy have been reported. The titration of antimycobacterial antibodies in patient sera has been the basis of most of these assays. Detection of *M. leprae* antigens in serum, urine, and tissues has also been attempted.

Immunoassays Based on Antibody Detection

Leprosy patients develop antibodies directed against *M. leprae* antigens. Antibodies have been quantitated using different antigenic preparations of *M. leprae* [intact cells (1), sonicate (13), or phenolic glycolipid-I (5)] or other mycobacteria [e.g., *M. smegmatis* (10,17), *M. fortuitum* (27)] that share several antigens with *M. leprae*. Alternatively, monoclonal antibodies directed against *M. leprae*-specific antigens have been employed in antibody-competitive assays (15,25). An enzyme immunoassay (EIA) that quantitates antibodies directed against peripheral nerve antigens has been developed (Mukherjee, unpublished data). Salient features of these immunoassays are discussed below.

Fluorescent Leprosy Antibody-Absorption Test

The earliest serological test, the fluorescent leprosy antibody-absorption (FLA-ABS) test, was proposed by Abe et al in 1976 (1). The test is based on detection of anti-*M. leprae* antibodies in leprosy sera. The FLA-ABS test has been employed extensively and has given reliable results. In two independent studies (2,3) the percentage positivity for lepromatous leprosy (LL) varied from 94.7 to 99.3% in comparison to 76.5 to 86.1% for tuberculoid leprosy (TT). The specificity of the test was high. However, the patient serum needs absorption with cardiolipin, lecithin, and polysaccharides of tubercle bacilli and a sonicated suspension of BCG and *M. vaccae* to remove antibodies directed against cross-reactive antigens of *M. leprae*. Moreover, the interpretation of the test on a semiquantitative scale is based on the degree of fluorescence observed by microscopy. In a way, it is a cumbersome test and is not easily amenable to field use, although a central well-equipped laboratory can conduct assays on sera collected from the field.

Radioimmunoassay

Harboe et al (13) reported a radioimmunoassay (RIA) using radiolabeled *M. leprae* filtrate as the antigen. The assay was specific for leprosy, as patients with pulmonary tuberculosis were not detected. The readability of LL patients was 100% compared to 25% for tuberculoid leprosy patients. However, absorption of patient serum with BCG sonicate is required to render the assay specific for leprosy, and the assay suffers from the general drawbacks associated with the use of radioisotopes, e.g., short shelf life of the reagents, need for costly equipment (counters), problems of waste disposal, and health hazards.

Enzyme Immunoassays

A number of EIAs for immunodiagnosis of leprosy have been described. Of them, the assays based on phenolic glycolipid I (PGL-I) are highly specific for leprosy.

EIAs Based on PGL-I

Phenolic glycolipid I is an antigen unique to *M. leprae*. The terminal sugars of the glycolipid carry the immunodeterminant (4,5). Leprosy patients develop predominantly immunoglobulin M (IgM) type of antibodies against PGL-I, and the last two sugars bind to the antibodies. This disaccharide residue has been chemically synthesized (11). Both PGL-I (native or deacylated) as well as the synthetic disaccharide (linked to bovine serum albumin) have been employed in EIAs.

The assays are sensitive for detection of active LL/BL (borderline lepromatous leprosy) patients with high bacillary load. In three independent studies (6,22,28) the percentage positivity for LL patients varied from 86 to 100% compared to 29 to 41.6% for TT patients. The diagnosis of paucibacillary patients and treated LL patients with low bacillary load is fairly low. A false positivity of 5.60 to 7.41% with PGL-I (6,12) and 14.8% for the synthetic disaccharide (6) has also been reported.

The EIAs described above are microtiter plate assays. Plate assays are sensitive and enable quantitation of the titers in sera. However, they are essentially laboratory-based assays and need microtiter plates of high quality, multichannel pipettes, a plate-washer, and a spectrophotometer or EIA reader. These facilities are not available in most of the leprosy control centers, particularly in rural field areas. To overcome these limitations, attempts have been made to develop alternative simple approaches intended for eventual field use. Our laboratory has developed a dipstick dot EIA using the synthetic disaccharide as the antigen (16). The dipstick consists of two nitrocellulose pads stuck onto a plastic strip. The antigen is dotted in the center of the lower nitrocellulose pad; the upper pad serves as an internal reagent control. A positive result is indicated by a blue dot against a clear white background. The antigen-coated dipsticks stored at room temperature are stable for up to 4 months, and the assay can be completed in 30 minutes. Further simplification of the procedure may enable the field technician or paramedical personnel to perform the assay under field conditions. Modifications of the assay to achieve this goal are in progress. Young et al (29) have reported a spot test using PGL-I as the antigen and a polysulfone membrane as the solid phase.

EIAs Based on *M. smegmatis* and *M. fortuitum*

Enzyme immunoassays using *M. smegmatis* [intact cells (10) or arabinomannan (17)] have been employed for serodiagnosis of leprosy as well as for monitoring chemotherapy. With arabinomannan, the ability of the assay to diagnose LL patients was 95% whereas for TT patients it was only 27%. The assay is not specific for leprosy. An EIA using *M. fortuitum* as the antigen has also been reported (27) in which the patient serum required absorption with *M. marinum* to render the assay specific for leprosy.

Sera from patients with bacilliferous leprosy gave high reactivity, but there was an overlap with values of normal sera.

EIA Using Sonicate Supernatant of *Mycobacterium w*

We have developed an EIA titrating antibodies in leprosy sera using sonicate supernatant of a cultivable, atypical bacterium, *Mycobacterium w* (*M. w*) (24). *M. w* is a candidate vaccine for leprosy (26). The choice of *M. w* as antigen for diagnosis was made after screening soluble antigens of seven mycobacteria (*M. w, M. kansasii, M. smegmatis, M. vaccae, $H_{37}Rv$, $H_{37}R_A$* and *M. avium*) for relative reactivity with sera of leprosy patients and normal healthy controls (18). *M. w* sonicate supernatant gave very low reactivity with sera of normal healthy subjects, whereas reactivity with leprosy sera was high. These studies indicated the merits of *M. w* sonicate for serodiagnosis of leprosy. The *M. w*-based EIA was found to be sensitive for detection of LL patients, and the readability of the assay extended toward the paucibacillary end of the spectrum. The assay could also diagnose pulmonary tuberculosis. Using the same antigen, we have developed a latex agglutination test that is sensitive, simple, and rapid (2 minutes) and can diagnose both pulmonary tuberculosis and leprosy.

EIA Using Peripheral Nerve Antigens

Sensory loss is one of the cardinal signs of leprosy. It is used clinically to discriminate leprosy lesions from other nonspecific dermatological disorders. Histopathological studies have revealed that the peripheral nerves are sites of predilection for *M. leprae* and are involved in all forms of leprosy. An enzyme immunoassay titrating antibodies against the peripheral nerve antigens has been developed (Mukherjee et al, unpublished data). Sera from leprosy patients classified as LL, BL, BB, BT, and TT, according to the Ridley-Jopling scale (23), were examined. The antineural antibody test was positive in all patients irrespective of their position in the spectrum of the disease. The sera of healthy control subjects and patients with tuberculosis and other dermatological conditions were negative in the assay. This test therefore could be used to (a) monitor peripheral nerve damage due to leprosy and (b) diagnose leprosy. This test has not missed a single case among 330 leprosy sera examined so far.

Assays Using *M. leprae*-Specific Monoclonal Antibodies

The availability of *M. leprae*-specific monoclonal antibodies (mAbs) has provided a new thrust to the development of these immunoassays (15,25). Sinha et al (25) reported a serum antibody competitive test (SACT) using a mAb (MLO4) directed against the My2a determinant of *M. leprae*. The assay is based on inhibition of binding of ^{125}I-labeled MLO4 antibodies to *M. leprae* sonicate (coated on a microtiter plate) by antibodies in leprosy sera. The assay is highly specific for leprosy. The readability of the assay for LL/BL patients was 100% compared to 46.7% for paucibacillary (TT/BT) patients. Sera from normal healthy individuals and patients with pulmonary tuberculosis or diseases such as cancer and autoimmune diseases were negative in the assay. Klaster et al (15) described an ELISA inhibition test using an enzyme-labeled monoclonal antibody recognizing 36-kilodalton (kDa) protein of *M. leprae*. Using this assay, seropositivity was found in 100% of multibacillary and 91% of paucibacillary patients. The false positivity rate was 5%.

Assays Based on Antigen Detection

By and large, most of the antibody-based assays described above are sensitive for detection of multibacillary (LL/BL) patients with a high bacillary load. However, with the exception of the FLA-ABS and ELISA-inhibition tests, the utility of the assays for diagnosis of paucibacillary (TT/BT) leprosy is fairly low. In addition, high antibody titers cannot discriminate a person with exposure to *M. leprae* from a person with clinical features of leprosy. Thus there is need to identify a more specific parameter for immunodiagnosis of leprosy. Detection of *M. leprae* antigens in the clinical specimen (serum, urine, etc.) would be better indicators of infection.

Young et al (30) have reported the detection of PGL-I in serum and urine of LL and BL patients. Lipid extracts of serum samples were submitted to silicic acid chromatography. PGL-I in the sample was determined using anti-PGL-I–monoclonal antibody and radiolabeled second antibody via autoradiography on a polysulfone chromatogram. The detection limit of the assay was about 50 ng PGL-I/ml serum. The antigen was demonstrable in 8 of 17 LL and 2 of 5 BL patients. Sera from normal individuals and from patients with TT leprosy or pulmonary tuberculosis were negative. It was observed that the level of circulating antigen decreased shortly (within a few weeks) after initiation of therapy. The levels of the antigen declined earlier than the decline of the antibody titers.

Cho et al (7) reported the detection of PGL-I in serum and urine of LL/BL patients. Serum samples were extracted with chloroform/methanol and fractionated on a column of silicic acid. PGL-I in the sample was revealed by chemical [thin-layer/high performance liquid chromatography (TLC/HPLC)] as well as immunological (ELISA) methods. The ELISA had high sensitivity (500 pg). PGL-I was detected in serum and urine of untreated LL/BL patients. In patients undergoing chemotherapy, the amount of PGL-I declined sooner than the decline in anti-PGL-I antibody titers.

Olcen et al (19) have also reported the detection of *M. leprae* antigens in urine of LL patients by an inhibitory RIA. With this test the presence of *M. leprae* antigens in the sample is reflected in inhibition of the binding of radiolabeled *M. leprae* antigen (sonicate supernatant) to a standard antibody. The 24-hour urine sample was concentrated 100 times and analyzed for the antigen. The detection limit of the antigen was 20 μg. The assay detected the antigen in 2 of 23 paucibacillary and 11 of 23 multibacillary patients (20). A significant correlation between bacterial index (BI) and antigen concentration was found. The amount of antigen excreted in urine decreased with effective chemotherapy.

Application of Recombinant DNA Technology in Immunodiagnosis of Leprosy

Genomic libraries of *M. leprae* have been constructed using various strategies (9,14,31). Several *M. leprae*-specific antigens have been identified by screening the DNA library with monoclonal antibodies specific for *M. leprae* (9,31) or with polyvalent anti-*M. leprae* rabbit or patient sera (14). Recombinant DNA technology hopefully will enable the production of *M. leprae*-specific antigens in large amounts. Some antigens thus prepared may be used in specific and sensitive immunoassays for diagnosis of leprosy. Moreover, DNA probes representing *M. leprae*-specific genes would be useful for detection of *M. leprae* DNA in tissue specimens (e.g., skin biopsy) and nasal droppings.

References

1. Abe M, Izumi S, Saito T, et al: Early serodiagnosis of leprosy by indirect immunofluorescence. *Lepr India* 1976;48:272–276.
2. Abe M, Minagawa F, Yoshino Y, et al: Fluorescent leprosy antibody absorption (FLA-ABS) test for detecting subclinical infection with Mycobacterium leprae. *Int J Lepr* 1980;48:109–119.
3. Bharadwaj VP, Ramu G, Desikan KV: Fluorescent leprosy antibody absorption (FLA-ABS) test for early serodiagnosis of leprosy. *Lepr India* 1981;53:518–524.
4. Brett SJ, Draper P, Payne SN, et al: Serological activity of a characteristic phenolic glycolipid from Mycobacterium leprae in sera from patients with leprosy and tuberculosis. *Clin Exp Immunol* 1983;52:271–279.
5. Cho SN, Yanagihara DL, Hunter SW, et al: Serological specificity of phenolic glycolipid-I from Mycobacterium leprae and use in serodiagnosis of leprosy. *Infect Immun* 1983;41:1077–1083.
6. Cho SN, Fujiwara T, Hunter SW, et al: Use of an artificial antigen containing the 3,6-di-0-methyl-B-D-glucopyranosyl epitope for the serodiagnosis of leprosy. *J Infect Dis* 1984;150:311–322.
7. Cho SN, Hunter SW, Gelber RH, et al: Quantitation of the phenolic glycolipid of Mycobacterium leprae and relevance to glycolipid antigenemia in leprosy. *J Infect Dis* 1986;153:560–569.
8. Christian M: The epidemiological situation of leprosy in India. *Lepr Rev* 1981;52(suppl 1):35–42.
9. Clark-Curtiss JE, Jacob WR, Doucherty MA, et al: Molecular analysis of DNA and construction of genomic libraries of Mycobacterium leprae. *J Bacteriol* 1985;161:1093–1102.

10. Douglas JT, Naka SO, Lee JW: Development of an ELISA for detection of antibody in leprosy. *Int J Lepr* 1984;52:19–25.

11. Fujiwara T, Hunter SW, Cho SN, et al: Chemical synthesis and serology of the disaccharides and trisaccharides of phenolic glycolipid antigen from the leprosy bacillus and preparation of a disaccharide protein conjugate for leprosy. *Infect Immun* 1984;43:245–252.

12. Gonzalez-Abreu E, Gonzalez A: Seroreactivity against the Mycobacterium leprae phenolic glycolipid I in mycobacteria infected or stimulated groups of individuals. *Lepr Rev* 1987;58:149–154.

13. Harboe M, Closs O, Bjune G, et al: Mycobacterium leprae specific antibodies detected by radio-immunoassay. *Scand J Immunol* 1978;7:111–120.

14. Khandekar P, Munshi A, Sinha S, et al: Construction of genomic libraries of mycobacterial origin: identification of recombinants encoding mycobacterial proteins. *Int J Lepr* 1986;54:416–422.

15. Klaster PR, De Wit Madeleine YL, Kolk AHJ: An ELISA-inhibition test using monoclonal antibody for the serology of leprosy. *Clin Exp Immunol* 1985;62:468–473.

16. Kumar S, Moudgil KD, Band AH, et al: A dot enzyme immunoassay for detection of IgM antibodies against phenolic glycolipid-I in sera from leprosy patients. *Indian J Lepr* 1986;58:185–190.

17. Miller RA, Dissanayake S, Buchanan TM: Development of an enzyme-linked immunosorbent assay using arabinomannan from Mycobacterium smegmatis: a potentially useful screening test for the diagnosis of incubating leprosy. *Am J Trop Med Hyg* 1983;32:555–564.

18. Moudgil KD, Gupta SK, Srivastava LM, Mishra R, Talwar GP: Evaluation of an enzyme immunoassay based on sonicate supernatant antigens of *Mycobacterium W* for immunodiagnosis of leprosy. *Indian J Lepr* (In Press).

19. Olcen P, Harboe M, Warndorff T, et al: Antigens of *Mycobacterium leprae* and anti-*M. leprae* antibodies in the urine of leprosy patients. *Lepr Rev* 1983;54:203–216.

20. Olcen P, Harboe M, Warndorff Van Diepen T: Antigens of *Mycobacterium leprae* in urine during treatment of patients with lepromatous leprosy. *Lepr Rev* 1986;57:329–340.

21. Park JE, Park K: Leprosy, in *Textbook of Preventive and Social Medicine,* 9th ed. India, B. Bhanot, 1983, pp 269–280.

22. Ralhan R, Band AH, Roy A, et al: An enzyme immunoassay titrating IgM antibody against phenolic glycolipid for diagnosis of lepromatous leprosy. *Indian J Med Res* 1985;82:110–115.

23. Ridley DS, Jopling WH: Classification of leprosy according to immunity: a five group system. *Int J Lepr* 1966;34:255–273.

24. Saxena VK, Singh US, Singh AK: Bacteriological study of a rapidly growing strain of Mycobacterium. *Lepr India* 1978;50:588–596.

25. Sinha S, Sengupta U, Ramu G, et al: Serological survey of leprosy and control subjects by a monoclonal antibody based immunoassay. *Int J Lepr* 1985;53:33–38.

26. Talwar GP: Towards development of a vaccine against leprosy: introduction. *Lepr India* 1978;50:488–491.

27. Vithayasai V, Songsiri S, Vithayasai P, et al: Serological study of leprosy by enzyme-linked immunosorbent assay, in *Proceedings of the Workshop on Se-*

rological Tests for Detecting Subclinical Infection in Leprosy. Tokyo, Sasakawa Memorial Health Foundation, 1983, pp 79–84.

28. Young DB, Buchanan TM: A serological test for leprosy with a glycolipid specific for Mycobacterium leprae. *Science* 1983;221:1057–1059.

29. Young DB, Fohn MJ, Khanolkar SR, et al: A spot test for detection of antibodies to phenolic glycolipid I. *Lepr Rev* 1985;56:193–198.

30. Young DB, Harnisch JP, Knight J, et al: Detection of phenolic glycolipid I in sera from patients with lepromatous leprosy. *J Infect Dis* 1985;152:1078–1081.

31. Young RA, Mehra V, Sweetser D, et al: Genes for the major protein antigens of leprosy parasite Mycobacterium leprae. *Nature* 1985;316:450–452.

CHAPTER 33

Immunodiagnostic Approaches to the Detection of *M. leprae* Infection in Leprosy

V.P. Bharadwaj and Kiran Katoch

Clinical symptoms aided by smear and histological examination are the conventional parameters used for the diagnosis of leprosy. These criteria, however, have self-imposed limitations, and it is sometimes difficult to diagnose the very early stage of the disease. Similarly, it is difficult to determine inactivity in some cases. Another important problem associated with epidemiologic studies of leprosy is the lack of reliable information on the occurrence of subclinical infection in leprosy, with the result that these studies are based mainly on the occurrence of manifest disease. There is sufficient evidence indicating that in leprosy, as in tuberculosis, there are far more persons with subclinical infection in any given area than those with overt disease. It is for this reason that the detection of subclinical leprosy infections are of paramount importance. Such detection may be useful for establishing the extent of infection and the course of the disease, as well as instituting proper immuno- or chemoprophylaxis.

Although a variety of approaches have been suggested (16,17,25,54,), the immunological approaches have proved to be useful for detection of disease and subclinical infections. In this chapter, we describe in detail the development of various immunonological tests and their application to the epidemiology of the disease.

Immunological Approaches

The immunological approaches may be classified into two broad groups: tests based on detection of delayed-type hypersensitivity (DTH) and tests based on detection of humoral responses.

Tests Based on Detection of DTH

Lepromin Test

The lepromin test has been found to measure the DTH to *Mycobacterium leprae*. Previously the most widely used test was the late Mitsuda lepromin test, but it has been found that Dharmendra's lepromin preparation gives equally good results. In clinical cases the lepromin test has been found to be useful for classifying the disease into various types (20). For sub-clinical cases the test is positive in some instances, but it has been found to be relatively insensitive. It has also been observed that there may be cross-reactions due to sensitization with other mycobacteria, as a positive response to lepromin has been observed in subjects from nonendemic areas, thereby giving rise to the problem of false positivity (46,50,56). The lepromin-negative contacts have been found to be at a much higher risk for developing disease than lepromin-positive contacts (20).

Bharadwaj et al (9,10) observed that in the young contacts of multi-bacillary forms of leprosy the infection rate, detected by the fluorescent leprosy-antibody absorption (FLA-ABS) test, is much higher than the positive lepromin response, which comes at a later stage, meaning that these cases would have been missed by lepromin testing alone.

Lymphocyte Transformation Test and Lymphocyte Migration Inhibition Test

Immune responsiveness to *M. leprae* has been studied throughout the histopathological spectrum of leprosy by both the lymphocyte transfor-mation test (LTT) and the lymphocyte migration inhibition test (LMIT), and it has been found that the responses continuously decreased from the polar tuberculoid leprosy group to the polar lepromatous group. This find-ing implies that the phenomenon underlying these tests is directly related to expression of T cell functions of leprosy patients. In a study conducted by Godal and associates (27;28) it was found that LTT and LMIT detected a large percentage of subclinical infection in medical attendants and household contacts of leprosy patients. However, the sensitivity of these tests has not been compared with that of the serological techniques, nor has a follow-up of these contacts been reported; therefore it is difficult to comment on the utility of these methods.

Tests for Detection of Humoral Responses to *M. leprae*

Detection of *M. leprae*-specific antibodies continues to be a fascinating approach to detecting humoral responses to *M. leprae* infection.

Immunodiffusion Test

Using a double diffusion test in gels, Stanford *et al* (53) reported four antigens specific to the leprosy bacillus. This technique was found to be of taxonomic importance; however, it has not been used for the detection of *M. leprae*-specific antibodies in cases or contacts. Caldwell et al (15) described an immunodiffusion test based on the principle of immuno-chemical structural integrity of surface antigens. The acetone-killed *M.*

leprae separated from infected armadillo liver without the use of proteases were treated with lithium acetate; and concentrated antigen extract has been used for detection of *M. leprae*-specific antibodies. These authors emphasized that if the specific surface antigen of *M. leprae* can be produced on a large scale, this test can be used for the diagnosis of subclinical infection. The test has not been tested in cases or contacts of leprosy patients, and thus it will be difficult to comment on its merits. Immunodiffusion techniques as such have not been found to be sensitive to the detection of antibodies or antigen for any of the infectious diseases. However, other sensitive assays based on the above antigen might be useful for detecting clinical and subclinical infection.

Crossed Immunoelectrophoresis

Crossed immunoelectrophoresis (CIE) has been tried by Rojas Espinosa et al (48) and Harboe et al (29), who described it as a potent technique for precise identification of immunogenic components of *M. leprae* antigens. More than 20 such antigens of *M. leprae* have been defined using this system. Antigen 21B by CIE has been thought to be *M. leprae*-specific (35). Though this technique is definitely more sensitive than immunodiffusion, its utility for diagnosing active disease and detecting subclinical infection is not known and has not yet been established.

Fluorescent Leprosy-Antibody Absorption Test

The antibody-combining specificity of *M. leprae* has been established by indirect immunofluorescence tests using antihuman gamma globulin fluorescent antibody. Cross-reactive components and antibodies are removed by absorbing the sera with cardiolipin, lecithin, BCG, and *M. vaccae* (2–4,7).

The FLA-ABS test is now being used as a routine method in our laboratory (JALMA Institute, Agra); and in our previous studies (8) on multi- and paucibacillary forms of leprosy, the test was found to be highly sensitive and specific. Using this test *M. leprae*-specific antibodies could be demonstrated in 95% of the multibacillary types and in 80% of the paucibacillary cases (6,8,13). These studies have revealed the presence of *M. leprae*-specific antibodies in all types of leprosy cases irrespective of the type and duration of disease. In a follow-up study it has been observed that the percent of positive FLA-ABS tests falls gradually after subsidense of the disease (13). The percent positivity of FLA-ABS in lepromatous cases was 36% even after 5 years of subsidense. On the other hand, the positivity percentage fell from 80% to 10% after 5 years in paucibacillary cases. Thus the test, like other available immunoassays, cannot be used (if positive) to confirm the inactivity of the disease. This test has been tried by other investigators as well. Kim and Lee (34) reported 100% positivity in lepromatous/(LL/BL) leprosy cases and 50% positivity in tuberculoid/(TT/BT) leprosy cases. Samuel and Adiga (49) found 47% positivity in indeterminate cases of leprosy. As the test is positive in a high number of healthy contacts, it appears to have a greater value in the detection of subclinical infection than in the serodiagnosis of the disease.

Abe et al (3) found high positivity in household contacts of leprosy patients in endemic areas. Bharadwaj et al (10,11) and Ramu et al (45) carried out the test in household contacts of multi- and paucibacillary forms of leprosy. The lepromin test (Dharmendra lepromin) was used to determine the delayed-type hypersensitivity (DTH) response in these individuals. The following interesting findings were observed:

1. Lepromin positivity is lowest in the 0 to 5 years and 6 to 10 years age groups of household contacts of multibacillary forms of leprosy, whereas FLA-ABS positivity in these age groups is high (Table 33.1). This finding means that a DTH response develops after the antibody response in many cases. Hence the test is sensitive in this group for detection of subclinical infection. The positivity in this group also could be used to monitor the effect of various control programs on the transmission of disease in the community. Follow-up studies are in progress with the thought that it may also help us understand the temporal relation between humoral and cell-mediated immune (CMI) responses.
2. FLA-ABS test positivity was greater among the contacts of lepromatous patients (69%) (6,9–12).
3. Using the results of the FLA-ABS test and the lepromin reaction, there were four possible combinations of groups (6,9–12). The significance and importance of these four groups is emphasized below (6,10–12,19,20).

Lepromin + ve and FLA-ABS + ve: Infected with *M. leprae* but showing good DTH.

Lepromin + ve and FLA-ABS − ve: A positive lepromin reaction may be due to nonspecific antigens or to the failure of the FLA-ABS test.

Lepromin − ve and FLA-ABS + ve: This combination indicates infection with *M. leprae* along with a poor CMI response in the host; thus there is a high risk for the individual to get the disease. A follow-up study

Table 33.1. Age-wise distribution of the contacts of multi-bacillary forms of leprosy.

Age group (years)	Total no. of cases	Lepromin		FLA-ABS	
		(+ve)	(−ve)	(+ve)	(−ve)
0–5	55	20	35	41	14
6–10	100	52	48	85	15
11–15	82	40	42	68	14
15	118	58	60	102	16
Total	355	170	185	296	59

has revealed that of the 38 contacts who developed the disease 34 (89.4%) belonged to this group. However, the lepromin status changes as the disease evolves. Thus the combination of the FLA-ABS and lepromin tests definitely helps in the identification of individuals at high risk. These individuals may need immunoprophylaxis or chemoprophylaxis, or both.

Lepromin − ve and FLA-ABS − ve: This finding indicates infection with neither *M. leprae* nor any other *Mycobacterium* sharing antigens with *M. leprae.* These individuals also could be genetically determined to be "poor responders" to the antigenic stimulation of *M. leprae.*

Using FLA-ABS and lepromin testing, detailed immunoepidemiological studies are in progress at our institute and at several other laboratories around the world. The results are fascinating and informative. Though the equipment and subjective assessments are the limiting factors, this test is of proved specificity and sensitivity, and it can be used on a large scale for various immunoepidemiological studies.

Radioimmunoassay

Harboe et al (29) developed a specific radioimmunoassay (RIA) for demonstration and quantitative estimation of antibody against BCG antigen 60. This antigen 60 is cross-reactive with *M. leprae* antigen 7. An RIA was described by Melsom et al (41) using this antigen 7 of *M. leprae.* A solid-phase RIA (sRIA) has been developed by Melsom et al (37–40) to detect class-specific antibodies to antigen 7 and have shown that immunoglobulin A (IgA). IgG, and IgM antibody response to this antigen occurs in contacts and all types of leprosy patients.

Using the RIA, Melsom et al found that antibody activity decreased in lepromatous leprosy during the first 3 years of treatment. Employing this assay, Yoder et al (57) reported similar findings. As it is a cross-reactive component, it is less likely to be used for specific detection of subclinical infection as infection due to other mycobacteria may also give false-positive results. Harboe et al (30) later described an RIA for detection of *M. leprae*-specific antibodies. With this test sera are recommended to be absorbed with BCG, *M. avium,* and *M. nonchromogenicum* to make the test specific. This test was found to be specific for *M. leprae* antibodies according to the authors' experience. Wherever facilities are available, this test could be useful for detection of subclinical infection, but the results can be confirmed only after using it in contacts on a large scale.

Enzyme-Linked Immunosorbent Assay

Reggiardo et al (47) described for the first time an ELISA test for detection of antimycobacterial antibodies in which mycobacterial glycolipids were recommended for use as antigens. Using the glycolipids from BCG, the authors reported 95% in lepromatous leprosy cases and 30% positivity in healthy contacts of leprosy patients.

Phenolic glycolipid I (PGL-I) has been purified from leprosy bacillus and has been shown to be chemically unique for *M. leprae* (31). An ELISA using this specific lipid has been developed in several laboratories (14,44,58,59; Talwar et al, unpublished data). From these studies it is clear that the active cases of multibacillary leprosy are highly positive for the test. Brett et al (14) and Cho et al (18) found a rapid decrease in antibody titers in patients within 2 to 3 years of chemotherapy. Young et al (60) developed a visual assay in a paper strip for diagnosis of leprosy. Kumar et al (36) have described a dot enzyme immunoassay for detection of IgM antibodies against PGL-I in the sera from leprosy patients. The terminal sugar of the phenolic glycolipid has been synthesized and is available from the World Health Organization (WHO). Qinixue et al (43) compared ELISA tests based on PGL-I as well as synthetic antigens and reported them to be equally sensitive and specific.

Another ELISA using whole *M. leprae* and other mycobacteria as antigens has been tried by Vithayasai et al (55) and Douglas et al (24). Douglas et al (22) noted that, of six mycobacterial species, whole *M. smegmatis* (autoclaved) was most effective as an antigen source in their ELISA system. Hence Douglas and Worth (23) used *M. smegmatis* in their field studies. Their preliminary data showed that the test can identify 70% of the preclinical cases at a 2-year screening interval and more than 90% (including all multibacillary cases) at the 1-year interval. In addition, these authors showed that, with treatment, high initial titers came down slowly to a lower level in multibacillary cases. On the other hand, antibody levels that were initially low in paucibacillary cases dropped rapidly after treatment. By employing the test with *M. smegmatis,* the sensitivity of the test is definitely reduced because the antibody (test serum) reacts four to ten times less than with *M. leprae.* The low specificity of the test is obvious as a different *Mycobacterium* has been used as an antigen.

Leproagglutination Test

The leproagglutination test has been developed in relation to serodiagnosis of syphilis and is based on the principle that biological false-positive reactions are due to antibodies reacting with the mixture of cardiolipin/lecithin in a 1:10 ratio, whereas in leprosy it is positive in a 1:1 ratio (1). Though this test has not been evaluated for detection of subclinical infection, it is likely to be less useful owing to the nonspecific nature of the antigens. Similarly, because of the same nonspecificity, the test would be of limited interest for active disease.

Indirect Hemagglutination Test

An indirect hemagglutination (IHA) test for leprosy was described by Jagannath and Sengupta (33). Petchelai et al (42), using the procedure, found that there is a high degree of cross-reaction with tuberculosis. This test needs to be made specific by proper absorption of sera. An IHA test based on specific antigens would be useful for the seroepidemiology of this disease.

Tests Using Monoclonal Antibodies

Monoclonal antibodies (mABs) produced by hybridoma cells are specific because the antibody produced in vitro is directed against a single epitope of a complex antigen. At present, several laboratories have raised mABs against *M. leprae* (5,26,32,60). Except for a few, these antibodies have been found to be cross-reactive. They have been used mostly for identification of antigens in *M. leprae*. However, for the first time Sinha et al (51) used a quasispecific antibody (MLO4) to detect antibodies in clinical and healthy contacts of leprosy; and based on their findings, a serum antibody competition test (SACT) has been developed at our institute (52). This test has detected 100% of active BL/LL cases. However, the positivity in BB and TT/BT patients was 87.5% and 47.0%, respectively. The test has also been applied to the detection of infection in contacts and is discussed elsewhere in this volume (see Chap. 35).

For the present, therefore, the practical approach is to use the FLA-ABS test until another more sensitive, more specific test is available for use on a large scale (19,21). It is clear from the above grouping that by employing the FLA-ABS and lepromin tests it may be possible to identify healthy persons who are at a high risk to develop the disease.

With the advances in hybridoma technology and the application of DNA recombination techniques, it may be possible to identify many more specific antigens and to produce them in vitro. In can be hoped, then, that other specific tests based on these techniques will be developed in the future. It is also possible that for the seroepidemiology of the disease more than one serological test may be necessary, as a particular test may have more diagnostic value and other tests may be more sensitive in detecting subclinical infection. It is an open field with a vast potential for the future.

References

1. Abe M: Specific serodiagnostic tests for leprosy, in Chatterjee BR (ed): *Window on Leprosy*. Wardha, India, Gandhi Memorial Leprosy Foundation, 1978, pp 235–241.
2. Abe M, Izumi S, Saito T, Mathur SK: Early serodiagnosis of leprosy by indirect immunofluorescence. *Lepr India* 1976;48:272–276
3. Abe M, Minagawa F, Yoshino Y, et al: Fluorescent antibody absorption (FLA-ABS) test for detecting subclinical infection with Mycobacterium leprae. *Int J Lepr* 1980;48:109–119.
4. Abe M, Yoshino Y: Antigenic specificity of *M. leprae* by indirect immunofluorescence. *Int J Lepr* 1978;46:119.

5. Atlaw T, Kozbor D, Roder JC: Human monoclonal antibodies against Mycobacterium leprae. *Infect Immun* 1985;49:104–110.
6. Bharadwaj VP: Immunodiagnostic approaches to the early detection of subclinical infection in leprosy. *Ann Natl Acad Med Sci (India)* 1985;21:128–139.
7. Bharadwaj VP, Desikan KV: Fluorescent leprosy antibody absorption (FLA-ABS) test for early serodiagnosis of leprosy. *Int J Lepr* 1979;47(suppl):391.
8. Bharadwaj VP, Ramu G, Desikan KV: Fluorescent leprosy antibody absorption (FLA-ABS) test for early serodiagnosis of leprosy. *Lepr India* 1981;53:518–524.
9. Bharadwaj VP, Ramu G, Desikan KV: A preliminary report on subclinical infection in leprosy. *Lepr India* 1982;54:220–227.
10. Bharadwaj VP, Ramu G, Desikan KV: Immunoepidemiological studies on subclinical infection in leprosy in the household contacts in India—a preliminary report. Working paper for the 7th Joint US–Japan Conference on Leprosy Research. Sendai, Japan, 25–27 August 1982.
11. Bharadwaj VP, Ramu G, Desikan KV, Katoch K: Studies on subclinical infection in leprosy, in *Proceedings of the Workshop on Serological Tests for Detecting Sub-clinical Infection in Leprosy*. Tokyo, Sasakawa Memorial Health Foundation, 1983, pp 29–35.
12. Bharadwaj VP, Ramu G, Desikan KV, Katoch K: Extended studies on subclinical infection in leprosy. *Indian J Lepr* 1984;56:807–812.
13. Bharadwaj VP, Katoch K, Ramu G, et al: Seroepidemiological studies in subsided cases of multi-and paucibacillary types of leprosy using FLA-ABS test. *Indian J Lepr* 1987;59:30–35.
14. Brett SJ, Draper P, Payne SN, Rees RJW: Serological activity of a characteristic phenolic glycolipid from Mycobacterium leprae in sera from patients with leprosy and tuberculosis. *Clin Exp Immunol* 1983;52:271–279.
15. Caldwell HD, Kirchheimer WF, Buchanan FM: Identification of Mycobacterium leprae specific protein antigen and its possible application for serodiagnosis of leprosy. *Int J Lepr* 1979;47:477–488.
16. Chatterjee BR: Is early diagnosis of leprosy in an Indian village not possible? *Lepr India* 1973;45:225–227.
17. Chatterjee BR, Thomas J, Taylor CE, Naidu GN: Epidemiological findings of a longitudinal study in a defined population in Jhalda, West Bengal, in *Proceedings of All India Leprosy Workers Conference: Silver Jubilee and Centenary of Hansen's Bacillus Discovery*, Sevagram, India, abstract, 1973, p 5.
18. Cho SN, Yanagihara DL, Hunter SW, et al: Serological specificity of phenolic glycolipid I from Mycobacterium leprae and use in serodiagnosis of leprosy. *Infect Immun* 1983;41:1077–1083.
19. Dharmendra: Detection of sub-clinical infection in leprosy (editorial). *Lepr India* 1982;54:193–203.
20. Dharmendra: Lepromin test, in Dharmendra (ed): *Leprosy*, Vol 2. Bombay, Samant, 1985, pp 999–1066.
21. Dharmendra: Immuno-epidemiological surveys for detection of subclinical infection, in Dharmendra (ed): *Leprosy*, Vol 2. Bombay, Samant, 1985, pp 1207–1222.
22. Douglas JT, Naka SO, Lee JW: Development of an ELISA for detection of antibody in leprosy. *Int J Lepr* 1984;52:19–25.

23. Douglas JT, Worth RM: Field evaluation of an ELISA test to detect antibody in leprosy patients and their contacts. *Int J Lepr* 1984;52:26–33.

24. Douglas JT, Worth RM, Murray CJ, et al: ELISA techniques with application to leprosy, in *Proceedings of the Workshop on Serological Tests for Detecting Sub-clinical Infection in Leprosy*. Tokyo, Sasakawa Memorial Health-Foundation, 1983, pp 85–90.

25. Figuerdo N, Desai SD: Positive bacillary findings in the skin of contacts of leprosy patients. *Indian J Med Sci* 1949;4:253–262.

26. Gillis TP, Buchanan TM: Production and partial characterisation of monoclonal antibodies to Mycobacterium leprae. *Infect Immun* 1982;37:172–178.

27. Godal T: Immunological detection of sub-clinical infection in leprosy. *Lepr India* 1975;47:30–41.

28. Godal T, Negassi K: Subclinical infection in leprosy. *Br Med J*. 1973;2:557–560.

29. Harboe M, Closs O, Bjorvatn B, et al: Antibody responses in rabbits to immunization with Mycobacterium leprae. *Infect Immun* 1977;18:792–805.

30. Harboe M, Closs O, Bjune G, et al: Mycobacterium leprae specific antibodies detected by radio-immunoassay. *Scand J Immunol* 1978;7:11–120.

31. Hunter SW, Brennan PJ: A novel phenolic glycolipid from Mycobacterium leprae possibly involved in immunogenicity and pathogenicity *J Bacteriol* 1981;147:728–735.

32. Ivanyi J, Sinha S, Aston R, et al: Definition of species specific and cross reacting antigenic determinants of M. leprae using monoclonal antibodies. *Clin Exp Immunol* 1983;52:528–536.

33. Jagannath C, Sengupta DN: Serology of leprosy. I. Indirect haemagglutination test with stabilised, sensitized red cells. *Lepr India* 1981;53:507–512.

34. Kim Do II, Lee H-Y: Some experimental studies on the fluorescent leprosy antibody absorption (FLA-ABS) tests for leprosy patients and household contacts, in *Proceedings of the Workshop on Serological Tests for Detecting Subclinical Infection in Leprosy*. Tokyo., Sasakawa Memorial Health Foundation, 1983, pp 49–54.

35. Kronvall G, Stanford JL, Walsch GP: Studies of mycobacterial antigens with special reference to Mycobacterium leprae. *Infect Immun* 1976;13:1132–1139.

36. Kumar S, Moudgill KD, Band AH, et al: A dot enzyme immunoassay for detection of IgM antibodies against phenolic glycolipid-I in sera from leprosy patients. *Indian J Lepr* 1986;58:185–190.

37. Melsom R, Harboe M, Duncan ME, Bergsiwick M: IgA and IgM antibodies against Mycobacterium leprae in the cord sera and in patients with leprosy: an indicator of intrauterine infection in leprosy. *Scand J Immunol* 1981;14:342–352.

38. Melsom R, Harboe M, Duncan ME: IgA, IgM and IgG anti M. leprae antibodies in babies of leprosy mothers during the first two years of life. *Clin Exp Immunol* 1982:49:532–542.

39. Melsom R, Harboe M, Myrvang B, et al: Immunoglobulin class specific antibodies to M. leprae in leprosy patients, including the indeterminate group and healthy contacts as a step in the development of methods for serodiagnosis of leprosy. *Clin Exp Immunol* 1982;47:225–233.

40. Melsom R, Harboe M, Naafs B: Class specific anti-M. leprae antibodies assay in lepromatous (BL-LL) leprosy patients, during the first two to four years of DDS treatment. *Int J Lepr* 1982;50:271–281.

41. Melsom R, Naafs B, Harboe M, Closs O: Antibody activity against Mycobacterium leprae antigen 7 during the first year of DDS treatment in lepromatous (BL-LL) leprosy. *Lepr Rev* 1978;49:17–29.

42. Petchelai B, Khupulsup KS, Sampattavanich S, et al: Serodiagnosis of leprosy by an indirect haemagglutination test, in *Proceedings of the Workshop on Serological Tests for Detecting Subclinical Infection in Leprosy*. Tokyo, Sasakawa Memorial Health Foundation, 1983, pp 77–78.

43. Qinixue WU, Ganyun YE, Xinyu L, et al: A preliminary study on serological activity of a phenolic glycolipid from Mycobacterium leprae in sera from patients with leprosy, tuberculosis and normal controls. *Lepr Rev* 1986;57:129–136.

44. Ralhan R, Band AH, Roy A, et al: An enzyme immunoassay titrating IgM antibody against phenolic glycolipid for diagnosis of lepromatous leprosy. *Indian J Med Res* 1985;82:110–115.

45. Ramu G, Bharadwaj VP, Katoch K, Desikan KV: Studies on the healthy contacts of leprosy patients, in *Proceedings of the Workshop on Serological Tests for Detecting Subclinical Infection in Leprosy*. Tokyo, Sasakawa Memorial Health Foundation, 1983, pp 37–43.

46. Rees RJW: The significance of lepromin reaction in man. *Prog Allergy* 1964;8:224–237.

47. Reggiardo Z, Vazquez E, Schnaper L: ELISA tests for antibodies against mycobacterial glycolipids. *J Immunol Methods* 1980;34:55–60.

48. Rojas Espinosa O, Estrada-Para S, Serranomirado SA, Latapi E: Antimycobacterial antibodies in diffused lepromatous leprosy detected by counterimmunoelectrophoresis. *Int J Lepr* 1976;44:448–452.

49. Samuel NM, Adiga RB: Detection of antibodies in the sera of leprosy patients and contacts by immunoabsorbent assay (ELISA), in *Proceedings of the Workshop on Serological Tests for Detecting Subclinical Infection in Leprosy*. Tokyo, Sasakawa Memorial Health Foundation, 1983, pp 55–68.

50. Shepard CC, Saitz EW: Lepromin and tuberculin reactivity in adults not exposed to leprosy. *J Immunol* 1967;99:637–645.

51. Sinha S, Sengupta U, Ramu G, Ivanyi J: A serological test for leprosy based on competitive inhibition of monoclonal antibody binding to MY2a determinant of M. leprae. *Trans R Soc Trop Med Hyg* 1983;77:869–871.

52. Sinha S, Sengupta U, Ramu G, Ivanyi J: Serological survey of leprosy and control subjects by a monoclonal antibody based immunoassay. *Int J Lepr* 1985;53:33–38.

53. Stanford JL, Rook GAW, Convit J, et al: Preliminary taxonomic studies on the leprosy bacillus. *Br J Exp Pathol* 1975;56:570–585.

54. Taylor CE, Elliston EP, Gideon H: Asymptomatic infection in leprosy. *Int J Lepr* 1965;33:716–731.

55. Vithayasai V, Songsiri S, Vithayasai P, Nelson K: Serological study of leprosy by enzyme linked immunoabsorbent assay, in *Proceedings of the Workshop on Serological Tests for Detecting Subclinical Infection in Leprosy*. Tokyo, Sasakawa Memorial Health Foundation, 1983, pp 79–86.

56. Waters MFR: Significance of the lepromin test in tuberculin negative volunteers permanently resident in a leprosy free area. Presented at the Tenth International Leproy Congress, Bergen, 1973, p 81 (abstract).
57. Yoder L, Naffs B, Harboe M, Bjune G: Antibody activity against Mycobacterium leprae antigens in leprosy: studies on the variation of antibody content throughout the spectrum and the effect of DDS treatment and relapse in BT leprosy. *Lepr Rev* 1979;50:113–121.
58. Young DB, Buchanan TM: The phenolic glycolipid from Mycobacterium leprae: use in serological tests, in *Proceedings of the Workshop on Serological Tests for Detecting Subclinical Infection in Leprosy*. Tokyo, Sasakawa Memorial Health Foundation, 1983; pp 19–24.
59. Young DB, Buchanan TM: A serological test for leprosy using a glycolipid specific for Mycobacterium leprae. *Science* 1983;221:1057–1059.
60. Young DB, Fohn MJ, Khanolkar SR, Buchanan TM: A spot test for detection of antibodies to phenolic glycolipid I. *Lepr Rev* 1985;56:193–198.
61. Young DB, Khanolkar SR, Barg LL, Buchanan TM: Generation and characterisation of monoclonal antibodies to the phenolic glycolipid of Mycobacterium leprae. *Infect Immun* 1984;43:183–188.

CHAPTER 34

Serological Survey of Leprosy Using a Monoclonal Antibody-Based Immunoassay and Phenolic Glycolipid ELISA

U. Sengupta, Sudhir Sinha, S.A. Patil, B.K. Girdhar, and G. Ramu

The present global estimate of leprosy patients is 11.5 million (21), one-fourth of whom live in India (6). The spread of infection can be arrested effectively by diagnosis and treatment of infectious subjects at an early ("preclinical") stage (21). However, the available diagnostic tools are often inadequate for this purpose. Three of the *Mycobacterium leprae*-specific serological tests—the fluorescent leprosy-antibody absorption test (FLA-ABS), the phenolic glycolipid I-based enzyme-linked immunosorbent assay (PG-ELISA) and a monoclonal antibody–serum antibody competition test (SACT) have shown promising results when applied to the study of healthy contacts of leprosy patients (1,3,8,17,19). The present review is mainly attributed to the work done using SACT.

Technique

The serum antibody competition test is an inhibition immunoassay wherein antibodies present in the test sample (serum) inhibit the binding of tracer-labeled specific monoclonal antibodies to a proteinaceous antigen in a crude extract of *M. leprae*. The quasispecific monoclonal antibody MLO4 (12), which binds to an epitope on the 35-kilodalton (kDa) soluble protein of *M. leprae* (11), was found suitable for the development of this test (18). The standardized methodology of SACT has also been described (19).

Briefly, 96-well PVC microtiter plates are coated with the whole soluble extract of *M. leprae* (20) purified from the infected armadillo tissues (9) at a concentration of 50 μg protein/ml (50 μl/well) and blocked with 2% bovine serum albumin (BSA-PBS). A 25-μl aliquot of each of the serial dilutions of test samples (in BSA-PBS) is preincubated in an antigen well for 1 hour at 37°C before adding ^{125}I-MLO4 ($\approx$60,000 cpm/25 μl/well) and incubating for another 2 hours at 37°C. The washed wells are cut and counted in a gamma counter, and percent bound radioactivies are obtained for each serum dilution (100% being the binding in the absence of competing test serum). Under the defined test conditions, a (serum) sample causing ≥50% inhibition of ^{125}I-MLO4 binding to the antigen well at a dilution of ≥1:5 (i.e., ID_{50}≥1:5) is regarded as positive for specific antibody. This radioimmunoassay has also been adopted as an enzyme immunoassay using peroxidase-labeled monoclonal antibody.

Study of Untreated Patients and Controls

Initially, specificity and sensitivity of the assay system (SACT) was evaluated using sera from different classes (15) of untreated leprosy patients, patients with other diseases, and healthy controls from geographical areas either endemic (India) or nonendemic (Europe) for leprosy. The results are compiled in Figure 34.1. The observed negative results in an array of control sera indicate specificity of SACT for leprosy. The test was found to be highly sensitive for the detection of multibacillary leprosy showing 100% positivity with high titers (ID_{50} ≥1:625 mostly) in LL/BL (lepromatous leprosy/Borderline leprosy) cases. However, the positivity in TT/

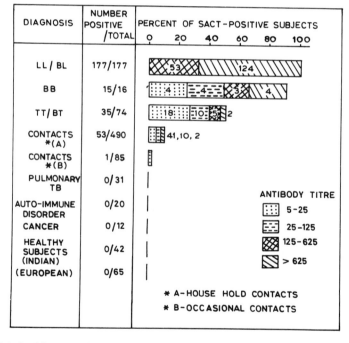

Fig. 34.1. Incidence and titer of *M. leprae*-specific antibody in leprosy patients, healthy contacts, and control subjects assayed by SACT. Sera giving ≥ 50% inhibition of ^{125}I-MLO4 binding at ≥1:5 dilution were scored as positive. Different antibody titers and the corresponding number of test subjects are indicated within the bars.

BT (Tuberculoid leprosy/Borderline tuberculoid leprosy) cases was only about 50%, that too mostly with low titers. Thus it could be speculated that the MLO4 binding determinant on 35-kDa antigen is predominantly immunogenic in individuals who are comparatively more susceptible to *M. leprae* (e.g., LL/BL patitents).

Figure 34.2 shows the titration curves obtained with representative SACT-positive ($ID_{50} \geq 1:5$) sera from leprosy patients and their contacts and SACT-negative ($ID_{50} < 1:5$) control sera that showed higher limits of inhibition. Although four serum dilutions (1:5, 1:25, 1:125, and 1:625) have been tested routinely, it is evident from this figure that an ID_{50} titer of more than 1:625 (obtained mostly with LL/BL sera) could be as high as 1:10,000.

Effect of Chemotherapy

One of the uses of a serological test may be to monitor the efficacy of treatment in patients and to evolve more potent therapeutic regimens. Thus SACT was performed in 62 lepromatous leprosy patients who had

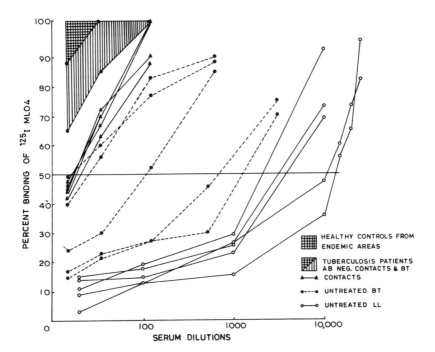

Fig. 34.2. Representative SACT titration curves of sera from antibody-positive or antibody-negative leprosy patients and their contacts, as well as maximally inhibitory (antibody-negative) control sera.

been receiving chemotherapy for 0 to 5 years (14 patients), 5 to 10 years (27 patients), or 10 to 15 years (21 patients). Skin smears for acid-fast bacilli (AFB) were done for all the patients routinely. Results of this study, compared to the results with 50 untreated BL/LL patients, are shown in Figure 34.3. The antibody titers decreased progressively but slowly over the years of treatment, with six patients eventually becoming SACT-negative after 10 years of treatment. There was no difference in the scatter of antibody titers of patients receiving dapsone (DDS) monotherapy or multidrug therapy. Persistence of anti-*M. leprae* antibodies in the blood for such long periods is not unexpected, as it takes prolonged chemotherapy for the lepromatous leprosy patients to show negativity for AFB (7). As seen in Figure 34.3, most of the patients remained bacteriologically positive for up to 5 years of treatment. Moreover, *M. leprae* is also known to persist in some protected sites in the human body (14). Thus the antigenic components of disintegrated bacilli are expected to remain in the body for long periods after clinical subsidence of the disease. However, the patients showing consistently high titers even after 10 years of chemotherapy (Fig. 34.3) could also be suspected of being refractory to treat-

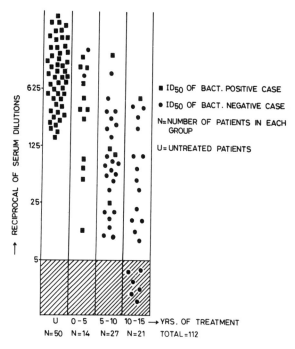

Fig. 34.3. Effect of chemotherapy on SACT titers in lepromatous leprosy patients. Hatched area represents antibody-negative sera. The bacteriologically positive case in the 10–15 years group was a case of relapse.

ment. Two such patients have already undergone a relapse (becoming bacteriologically positive again after remaining negative).

Study of Contacts

As mentioned at the outset, a test is needed more for the diagnosis of "preclinical" leprosy than for clinically established disease. SACT positivity with high antibody titers was observed in all patients who are totally anergic to *M. leprae* infection (lepromatous type). Such individuals, if not diagnosed early (or preclinically) and rendered noninfectious with the help of chemotherapy, are potential transmitters of infection. Therefore along with some other serological tests, the efficiency of SACT in diagnosing such individuals at an early stage is being evaluated. In the initial pilot study, about 30% of the household contacts of leprosy patients were found to be SACT-positive (19), a finding that encouraged us to undertake large-scale hospital-based (2) and field-based prospective studies in apparently healthy children in contact with leprosy patients. The antibody positivities in different categories of contacts studied so far are compiled in Figure 34.1. In this large sample the rate of positivity in the household contacts has decreased to about 10% (from the initially observed 30%), which is a more realistic figure considering the leprosy prevalence rates in these parts of the country (6). The preliminary data on the follow-up of 93 contacts are given in Table 34.1. Within a period of 2 years, 35% of the SACT-positive contacts developed clinical signs of leprosy, compared to 5% of the SACT-negative contacts who developed the disease. Notably, the initial presentation of the disease was in the form of multiple lesions in all contact-turned patients from the SACT-positive group.

Adaptation for Field Studies

The summer temperature in certain parts of tropical countries such as India may exceed even 45°C, and most often it is not possible to work in a field area under controlled temperatures. Moreover, storage space for a large number of serum samples at low temperatures may be difficult to

Table 34.1. Results of 2 years' follow-up of contacts.

Type	No.	Developed disease so far		Initial presentation of disease
		No.	%	
SACT-positive	17	6	35.3	Multiple lesions in all
SACT-negative	76	4	5.3	Single lesion in all
Total	93	10	10.8	

provide. Thus an attempt was made to apply SACT on the blood samples collected and dried on filter paper.

The standardized technique is as follows. Spots of blood (intravenous or finger prick) are collected on Whatman No. 3 chromatography paper (diameter of each spot $\simeq$ 2 cm) and dried at ambient temperature. Discs of 16 mm diameter (50 µl whole blood) are punched out from these spots. Each disc is placed in a small tube and eluted in an appropriate volume of PBS pH 7.4 by keeping it at 4°C overnight. Serial dilutions of filter paper blood (assuming a hematocrit value of 50%, the serum volume of each spot was considered to be 25 µl) and corresponding serum dilutions were subjected to SACT.

Initially, the effect of storage of dried spots at 4°, 37°, or 40°C for 2 weeks was assessed. SACT titers remained unaltered under these storage conditions.

Figure 34.4 shows the results of SACT in different serum samples and the corresponding elutes from blood spots dried at ambient temperature and stored at 40°C for 2 weeks. A significant correlation was observed between the two antibody titers.

Appraisal of Serological Tests for Leprosy

The various serological tests for leprosy that have been developed so far have shown varying degrees of success in diagnosing clinically established or "preclinical" infection. Thus it is imperative to evaluate the presently available as well as forthcoming tests in a comparative manner so as to know which test (or a combination of tests) would ultimately be suitable for large-scale field trials. We have initiated such an attempt by performing SACT and PG-ELISA on a set of serum samples. For this purpose the ELISA for antibodies to the phenolic glycolipid I of *M. leprae* (PG), as described by Cho et al (5), was adopted. In this assay, as observed by earlier workers, a serum dilution of 1:300 produced a minimum of false-positive and false-negative results. However, the optical densities of negative control sera were higher than those reported earlier, which may be

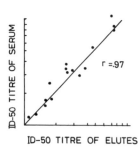

Fig. 34.4. Titers of sera and eluates of corresponding blood blots stored at 40°C, by antibody competition test.

due to the fact that our nonleprosy controls belonged to a leprosy endemic area (16).

The comparative results obtained so far have been compiled in Table 34.2. Although all of the bacteriologically positive (skin smears positive for AFB) lepromatous leprosy patients (untreated or treated) were SACT-positive, their (n = 38) PG-ELISA positivity was only 68%. Both these tests detected positivity in tuberculoid leprosy patients broadly to the same extent (30 to 40%), which has been observed in earlier studies as well. An important observation in the present study was the segregation of SACT-positive and PG-ELISA-positive subjects among household contacts of leprosy patients. Although both tests were positive in about 27% of the contacts, the cumulative positivity was 47%. Moreover, about 50% of the PG-ELISA-positive contacts showed positivity for only IgG-type antibodies, whereas it is conventionally believed that the IgM-type antibody response is the one seen predominantly in humans.

Presently some preliminary data are available regarding efficiency of both tests in detecting "preclinical" leprosy (mentioned above) (8). However, more work is required for proper appraisal of the tests.

Prospects

If we consider the problem of "preclinical" diagnosis of leprosy specifically, it is likely that a serological test would tend to be positive in many healthy subjects living in a leprosy endemic area, most of whom will not contract the disease (21). The observed positivity rates of the contacts of

Table 34.2. Positivity of different groups of test sera for SACT and PG-ELISA: comparative analysis.

Subjects[a]	No.	Positive for[b]			Cumulative positivity
		SACT alone	PG-ELISA alone	Both	
L/BL (u+)	14	2 (14%)	0	12 (86%) (M5,M,G7)	14 (100%)
L/BL (t+)	24	10 (42%)	0	14 (58%) (M8,G2,M,G4)	24 (100%)
L/BL (t−)	20	11 (55%)	0	7 (35%) (G4,M,G3)	18 (90%)
T/BT (u−)	18	1 (6%)	2 (11%) (M,G2)	4 (22%) (G1,M,G3)	7 (39%)
Contacts	55	10 (18%)	12 (22%) (M5,G5,M,G2)	4 (7%)	26 (47%)

[a]L = lepromatous. T = tuberculoid. B = borderline. t = treated. u = untreated. + = smear positive for AFB. − = smear negative for AFB.
[b]M = IgM and G = IgG classes of anti-PG antibodies. Numbers indicate test sera positive for IgM, IgG, or both types of antibody.

leprosy patients with SACT and PG-ELISA (10 to 25%) indicate such a situation. Nevertheless, the application of these tests may considerably reduce the workload for the purpose of close surveillance of the population at risk, provided that all or most of the prospective patients emerge from the test-positive groups. In this context, it would be important to monitor the study population periodically, which may help in identification of consistently test-positive subjects who are probably incubating a progressive infection. Furthermore, by simultaneous use of *M. leprae*-specific cell-mediated immunity (CMI) tests, a group of antibody-positive and CMI-negative individuals, at higher risk of developing serious forms of the disease, may also be identified.

The antibody detection tests mentioned above have fallen short of diagnosing more than 50% of the tuberculoid (or paucibacillary) type of leprosy. One of the future antibody detection tests or, more likely, a test based on the detection of *M. leprae*-specific CMI may help in diagnosing such cases efficiently.

The detection of antigen, rather than antibody, is considered a better indication of the active infection. However, the sensitivity of antigen detection tests is likely to be compromised owing to antibody excess situations as observed in leprosy (10,13). A method to demonstrate *M. leprae*-specific phenolic glycolipid I antigen in the serum and urine of active lepromatous leprosy patients has not demonstrated desirable sensitivity (4,22). Nevertheless, the detection of specific antigens in the skin, which is the site of predilection and multiplication of *M. leprae* in humans, may be more promising. The *M. leprae*-specific genetic probes are presently being developed for this purpose (23).

Acknowledgments. We are grateful to Dr. J. Ivanyi, Director, MRC unit for tuberculosis and related infections, Hammersmith Hospital, London; Dr. R.J.W. Rees, Clinical Research Centre, Harrow, U.K.; and Dr. K.V. Desikan, Director, Central JALMA Institute for Leprosy, Agra for their help and suggestions throughout this study. The work was supported by ad hoc grants from the Indian Council of Medical Research.

References

1. Abe M, Minagawa F, Yoshino Y, et al: Fluorescent leprosy antibody absorption (FLA-ABS) test for detecting subclinical infection of M. leprae. *Int J Lepr* 1980;48:109–119.
2. Ashworth M, Sinha S, Patil SA, et al: The detection of subclinical leprosy using a monoclonal antibody based radioimmunoassay. *Lepr Rev* 1986;57:237–242.
3. Bharadwaj VP, Ramu G, Desikan KV: A preliminary report on subclinical infection in leprosy. *Lepr India* 1982;54:220–227.

4. Cho SN, Hunter SW, Gelber RH, Brennan PJ: Quantitation of phenolic glycolipid of M. leprae. *J Infect Dis* 1986;153:560–569.
5. Cho SN, Yanagihara PL, Hunter SW, et al: Serological specificity of phenolic glycolipid I from M. leprae and use in serodiagnosis of leprosy. *Infect Immun* 1983;41:1077–1083.
6. Christian M: The epidemiological situation of leprosy in India. *Lepr Rev* 1981;52(suppl 1):35–42.
7. Dharmendra: Treatment of leprosy, in Dharmendra (ed): *Leprosy,* Vol 1. Bombay, Kothari Medical Publishers, 1978, pp 359–473.
8. Dissanayake S, Young DB, Khanolkar SR, et al: Serodiagnosis of subclinical leprosy: evaluation of an ELISA using phenolic glycolipid antigen to detect infection with M. leprae, in *Proceedings of the XII International Leprosy Congress,* New Delhi, 1984, pp 79–83.
9. Draper P: *Protocol 1/79: Purification of M. leprae.* Annex 1 to the report of the enlarged Steering Committee for IMMLEP Meeting of 7-8 February 1979. Geneva, World Health Organization, 1979.
10. Harboe M: Significance of antibody studies in leprosy and experimental models of the disease (editorial). *Int J Lepr* 1982;50:342–350.
11. Ivanyi J, Morris JA, Keen M: Studies with monoclonal antibodies to mycobacteria, in Macario AJL, Macario EC (eds): *Monoclonal Antibodies Against Bacteria.* New York, Academic Press, 1985.
12. Ivanyi J, Sinha S, Aston R, et al: Definition of species specific and cross-reactive antigenic determinants of M. leprae using monoclonal antibodies. *Clin Exp Immunol* 1983;52:528–536.
13. Patil SA, Sinha S, Sengupta U: Detection of mycobacterial antigens in immune complex of leprosy sera. *J Clin Microbiol* 1986;24:169–171.
14. Ramu G, Desikan KV: A study of scrotal biopsy in subsided cases of lepromatous leprosy. *Lepr India* 1979;51:341–346.
15. Ridley DS, Jopling WH: Classification of leprosy according to immunity: a five group system. *Int J Lepr* 1966;34:255–273.
16. Sinha S, Patil SA, Girdhar BK, et al: A comparative study of phenolic glycolipid ELISA (PG-ELISA) and serum antibody competition test (SACT) on a set of serum samples, in *Proceedings Indo-U.K. Symposium on Leprosy.* ed. Katoch VM. The coronation Press, Agra (India), 1987, pp 45–51.
17. Sinha S, Ramu G, Girdhar BK, Sengupta U: A monoclonal antibody based competition radioimmunoassay for monitoring anti-M. leprae antibodies in leprosy patients and contacts. *Ann Natl Acad Med Sci (India)* 1985;21:109–117.
18. Sinha S, Sengupta U, Ramu G, Ivanyi J: A serological test for leprosy based on competitive inhibition of monoclonal antibody binding to the MY2a determinant of M. leprae. *Trans R Soc Trop Med Hyg* 1983;77:869–871.
19. Sinha S, Sengupta U, Ramu G, Ivanyi J: Serological survey of leprosy and control subjects by a monoclonal antibody based immunoassay. *Int J Lepr* 1985;53:33–38.
20. Smelt AHM, Rees RJW, Liew FY: Induction of delayed type hypersensitivity to M. leprae in healthy individuals. *Clin Exp Immunol* 1981;44:501–506.
21. World Health Organization: Epidemiology of leprosy in relation to control. *WHO Techn Rep Ser* No. 716, 1985.

22. Young DB, Harnisch JP, Knight J, Buchanan TM: Detection of phenolic gly-
    colipid I in sera from patients with lepromatous leprosy. *J Infect Dis*
    1985;152:1078–1081.
23. Young RA, Mehra V, Sweetser D, et al: Genes for the major protein antigens
    of the leprosy parasite Mycobacterium leprae. *Nature* 1985;316:450–452.

Part IX
Malaria and Leishmaniasis

CHAPTER 35

Antisporozoite Malaria Vaccine Development Based on Circumsporozoite Protein

Altaf A. Lal, Vidal F. de la Cruz, Judith A. Welsh, and Thomas F. McCutchan

Malaria remains prevalent in the tropical and subtropical areas of the world, infecting hundreds of millions of people each year. The problem of malaria is compounded by the spread of insecticide resistance in mosquitos and by the appearance of *Plasmodium falciparum* resistance to drugs such as chloroquine and pyrimethamine. More than 100 *Plasmodium* species have been described that cause malaria in a wide range of vertebrate hosts. Four species are recognized that commonly infect humans: *P. falciparum, P. vivax, P. malariae,* and *P. ovale. P. falciparum* is responsible for almost all malaria-related deaths, and *P. vivax* causes considerable morbidity.

The *Plasmodium* spp. life cycle is complex, alternating between vertebrate host and insect vector. Infection is initiated by the inoculation of sporozoites during the bite of an infected mosquito. Within a few minutes, sporozoites become localized, primarily in hepatocytes. These liver (exoerythrocytic) stages then develop into the blood-form stages of the life cycle, some of which give rise to gametocytes, which may be taken up with a blood meal by a biting mosquito. In the mosquito these gametocytes further differentiate, undergo fertilization, and eventually give rise to sporozoites, whereby the cycle in the mammalian host can be reinitiated.

Each stage in the parasite life cycle is morphologically and for the most part antigenically distinct. Several of these antigenic targets are on different stage-specific molecules and could be employed for immunization. Ideally an antimalarial vaccine will be a multicomponent vaccine incorporating targets from several of the life cycle stages.

The circumsporozoite protein covers the surface of the sporozoite (11,26). Immunization with irradiated sporozoites of *Plasmodium* results in protection against further sporozoite challenge (3,5,14,20,25). In these rodent, simian, and human trials the humoral response was primarily directed against the circumsporozoite protein. Sporozoites incubated with immune sera in vitro develop a precipitate on the surface, which usually leads to neutralization of infectivity. It is known as the circumsporozoite precipitin reaction (6). Antibodies to the circumsporozoite protein have been implicated in protection against malaria (19). Thus current efforts for an antisporozoite vaccine are focused on the circumsporozoite protein and its gene. In the design of any subunit vaccine, the relative roles of antibody dependent and independent mechanisms, the role of malarial epitopes recognized by B and T cells, and the role of lymphokines in protective immunity must be considered.

Circumsporozoite Protein Gene and Immunological Considerations for Vaccine Design

Cirumsporozoite (CS) protein genes have been cloned and sequenced from several species of *Plasmodium* (1,7,9,10,15,17). All of the CS genes sequenced to date exhibit similar general features. The central one-third of the protein is composed of immunodominant repetitive sequences rich in glycine, proline, glutamine, asparagine, and alanine (Fig. 35.1). Flanking the central repeat domain are two sequences, generally conserved among the genes and referred to as region I and region II (Fig. 35.2), and these regions are in turn flanked by charged regions. A hydrophobic signal sequence and a putative hydrophobic anchor domain are found at the amino- and carboxyl-termini, respectively. The CS protein of the *Plasmodium yoelii* and *P. berghei* rodent malarias are different from other CS proteins in that the central repeat region is made up of two sets of repeats. In *P. yoelii,* for example, a 5′, 18- nucleotide motif repeated 15 times codes for QGPGAP, followed by eight tetrapeptide repeats. The first and last of these tetrapeptide repeats are variant, whereas the remaining six are QQPP. Another feature of the CS protein gene is a set of repeats upstream of region I, called the preregion I repeats, which have thus far been described only in the *P. yoelii* (15) and some *P. falciparum* genes (8,16) (Fig. 35.3). Even though sporozoites reside in the circulation for a few minutes before entering the hepatocytes, vaccination of rodents, monkeys, birds, and humans with attenuated sporozoites provides complete protection against malaria. Acquisition of resistance to malaria with age is correlated with higher titers of serum antibodies to sporozoites. However, a negative correlation between antibody level to the CS protein of *P. falciparum* and parasitemia and splenic enlargement has been observed (18). Newborn rodents suckled by sporozoite-immunized foster mothers were resistant to sporozoite-induced malaria (22).

Interferons may also play a role in addition to antibodies in the protective immunity mediated by x-irradiated sporozoites. The relevant observations are that spleen cells of mice vaccinated with x-irradiated sporozoites release high levels of γ-interferon when challenged in vitro with parasite extracts (21). Passive transfer of 10 μg of a monoclonal antibody to a mouse induces complete protection against a low-level sporozoite challenge (24). In vitro neutralization of sporozoite infectivity has been ob-

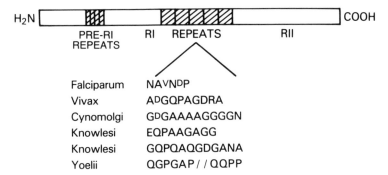

Fig. 35.1. Map of CS protein from various species of *Plasmodium:* CS protein repeats.

tained with monovalent Fab fragments, implying that simple binding of antibodies to the parasite surface interferes with infectivity (24).

Other observations in the murine system indicate that T cell immunity, independent of antibody, is protective (2). Also, in human volunteers immunized with irradiated sporozoites, specific protection was evident before the appearance of humoral antibody (4). T cell immunity is of course required for humoral as well as antibody-independent immunity. In view of difficulties in obtaining CS protein from sporozoites in quantities sufficient for vaccination, synthetic and genetically engineered vaccines have been developed for *P. falciparum*. The former is a conjugate of a protein carrier with a synthetic repeat of the *P. falciparum* gene, NANP. Rabbits and

Region I

P.K. (N) EEPKKPNENKLKQPEQP

P.K. (H) EEPKKPNENKLKQPNEG

P.V.   AEPKNPRENKLKQPGDR

P.F.   EKLRKPKHKKLKQPGDG

P.Y.   DDPPKEAQNKLNQPVVA

Region II

P.K. (N) SVTTEWTPCSVTCGNGVR

P.K. (H) SVTTEWTPCSVTCGNGVR

P.V.   TVGTEWTPCSVTCGVGVR

P.F.   SISTEWSPCSVTCGNGIQ

P.Y.   QLTEEWSQCSVTCGSGVR

Fig. 35.2. Comparison of the amino acid sequences of the conserved regions (RI and RII) in CS protein.

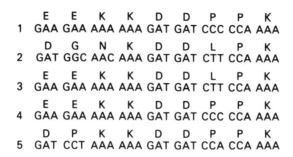

```
       E   E   K   K   D   D   P   P   K
   1 GAA GAA AAA AAA GAT GAT CCC CCA AAA

       D   G   N   K   D   D   L   P   K
   2 GAT GGC AAC AAA GAT GAT CTT CCA AAA

       E   E   K   K   D   D   L   P   K
   3 GAA GAA AAA AAA GAT GAT CTT CCA AAA

       E   E   K   K   D   D   P   P   K
   4 GAA GAA AAA AAA GAT GAT CCC CCA AAA

       D   P   K   K   D   D   P   P   K
   5 GAT CCT AAA AAA GAT GAT CCA CCA AAA
```

```
         Q   G   P   G   A   P
    1 CCA GGG CCA GGA GCA CCA
    2 --- --- --- --- --- ---
    3 --- --A --- --- --- ---
    4 --G --T --- --- --- ---
    5 --- --A --- --- --- ---
    6 --- --A --- --- --- ---
    7 --- --T --- --- --- ---
    8 --G --T --- --- --- ---
    9 --G --T --- --- --- ---
   10 --- --A --- --- --- ---
   11 --G --- --- --- --- ---
   12 --- --A --- --- --- ---
   13 --- --A --- --- --- ---
   14 --G --- --- --- --- ---
   15 --- --- --- --- --- ---
```

```
         Q   Q   P   P
    1 CAA CAA CCA CCC
    2 --- --- --- --A
    3 --- --G --- --A
    4 --- --G --- --A
    5 --- --G --- --A
    6 --- --G --- --A
```

Fig. 35.3. Nucleotide and amino acid sequence of the repeating units in the CS protein of *P. yoelii*.

mice immunized with this formulation produce high titers of antipeptide antibodies that recognize the CS protein and neutralize parasite infectivity (28). The vaccine made by recombinant DNA technology (R32Tet32) is a fusion protein produced in *Escherichia coli*. In mice this product induces antibodies that react with CS protein and block sporozoite invasion of human hepatoma cells in vitro (27).

If high-titer antibodies to the CS repeat region are required for protective immunity, a carrier conjugated peptide may be sufficient to induce the immune response. However, if the antibodies are required to persist over a long period of time, as would be desirable, T cell sites derived from the CS protein would be useful, as there would be natural boosting of antibody production from an infected mosquito bite. From murine studies, Good et al have identified two T-helper epitopes on the *P. falciparum* CS protein molecule, one contained within the central repeat region of the molecule and recognized by mice bearing the I-A^b gene (12) and the other located on the N-terminal to region II and recognized by mice bearing the I-A^k gene (13). From examining a large panel of H-2 congenic mice, it appears that these two T sites may be the major T helper sites on the molecule. Variation of T epitopes may influence both the initial response to infection and the ability to boost the prior memory. Antigenic variation may be a mechanism for the parasite to evade immune pressure, and so it has to be considered in vaccine design.

Rodent Model to Examine the Efficacy of Subunit Anti-CS Vaccine

There are several mouse malaria species on which an animal model might be based. *P. yoelii* was used as a model for various reasons. (a) It is more genetically diverse than species such as *P. berghei*. Moreover, isolates of *P. berghei* are often mixtures of *P. berghei* and *P. yoelii* (23), and it is difficult to differentiate them because of morphological similarities. (b) Cloned lines of *P. yoelii* are available that are infective to mosquitoes. (c) The laboratory procedures for infecting mosquitoes require fewer facilities. The sequence of the gene encoding the CS protein of *P. yoelii* has been reported (15). There are polymorphic forms of the CS gene in the isolate of *P. yoelii*, as evidenced by multiple bands appearing on Southern blots of parasite DNA.

For vaccination of mice three chemically synthesized peptides were cross-linked by glutaraldehyde to keyhole limpet hemocyanin (KLH). (QGPGAP) × 3 represents three repeats from the putative immunodominant portion of the molecule. (QQPP) × 4 is derived from the second set of repeats to the carboxy-terminal of the central repeats. (EEKKDDPPKDPKKDDPPKEA) represents the repetitive sequence amino-terminal of the central repeats (i.e., the *P. yoelii* preregion I repeats) (Fig. 35.3). Three strains of mice, BALB/c, B10.BR, and B6.D2, were grouped as follows: unimmunized, immunized with KLH emulsified in Freund's adjuvant, and immunized with the peptide–KLH conjugate emulsified in Freund's adjuvant. The details of the immunization schedule and the antibody titers are given in Table 35.1. Initially all three strains of mice responded poorly, and it took five boosts and about 90 days for a high titer of antibodies to develop (Fig. 35.4). Table 35.1 shows the immune response to the dominant repeat (QGPGAP). Against the other two peptide–KLH conjugates, development of the immune response followed the same pattern. Immunized and unimmunized mice were challenged with 500 sporozoites of *P. yoelii* (obtained by dissection of infected

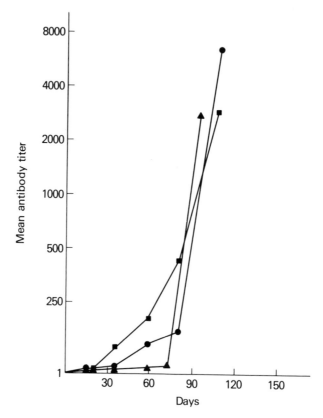

Fig. 35.4. Immune response to synthetic peptide (QGPGAP) ×3 conjugated to KLH in mice. ●—● BALB/c. ■—■ B10.BR. ▲—▲ B6.D2. Each point indicates the day of boost.

mosquitoes) by an intravenous route. Taken as a group, 22 of 24 unimmunized mice became infected, 18 of 28 immunized with KLH alone became infected, and 19 of 24 immunized with the KLH-(QGPGAP) × 3 conjugate became infected. There was no significant relation between antibody titer and infection. At the extremes, an animal with a titer of 512 was protected, whereas another with a titer of 8000 was infected. Similar results were obtained when saponin was used as an adjuvant. Also, unimmunized and immunized animals showed 0.5% parasitemia on approximately the same day.

It is clear from these studies that high-titer antibodies against the immunodominant repeat of CS protein do not neutralize sporozoites necessarily in vivo as they do in vitro assays. This finding raises two questions. (a) What is the basis of successful vaccination against irradiated sporozoites? (b) Why does passive transfer of monoclonal antibodies against the repeats of the CS protein gene protect rodents against sporozoite chal-

Table 35.1. Immunization of mice with KLH-(QGPGAP) $\times 3$ in Freund's adjuvant and challenge with sporozoites.

Immunization schedule	No. of animals challenged	No. of animals infected	Average day of 0.5% parasitemia	Range of antibody titer of infected animals	Range of antibody titer of protected animals
		PART A[a]			
Unimmunized	24	22	7	—	—
KLH + CFA	28	18	6	—	—
KLH-(QGPGAP) $\times 3$ + CFA	24	19	7	128–8000	512–32,000
		PART B[b]			
BALB/c					
Unimmunized	12	11	7	—	—
KLH + CFA	9	7	7	—	—
KLH-(QGPGAP) $\times 3$ + CFA	6	3	7	128–8000	1,024–32,000
B10.BR					
Unimmunized	6	5	7	—	—
KLH + CFA	10	6	7	—	—
KLH-(QGPGAP) $\times 3$ + CFA	8	8	6	128–4096	—
B6.D2					
Unimmunized	6	6	6	—	—
KLH + CFA	9	5	7	—	—
KLH-(QGPGAP) $\times 3$ + CFA	10	8	7	1024–4096	512–8000

Mice (BALB/c, B10.BR, and B6.D2) were immunized with 100 µg of the conjugate intraperitoneally. The conjugate in PBS was emulsified in complete Freund's adjuvant. The mice were boosted about every 2 weeks with 50 µg of the conjugate in incomplete Freund's adjuvant. Two days before boosting the animals were bled by the tail artery and serum-tested for anti-QGPGAP antibodies by ELISA.
[a]Part A shows the immune response and protection in unimmunized and immunized group of animals.
[b]Part B shows the same data when the animals are divided according to their haplotype.

lenge? Some possible answers are that irradiated sporozoites remain invasive and elicit a cytotoxic T cell response. In this case the cytotoxic T cell response would be involved in protection. Alternatively, there may be a role for T cells in an antibody-independent induction of lymphokine secretion that ultimately leads to protection. These hypotheses could be more easily tested in a rodent model system.

References

1. Arnot DE, Barnwell JW, Tam JP, et al: Circumsporozoite protein of immunodominant epitope. *Science* 1985;230:815–818.

2. Chen DH, Tigelaar RE, Weinbaum FI: Immunity to sporozoite induced malaria in mice. I. The effect of immunization of T- and B-cell-deficient mice. *J Immunol* 1977;118:1322–1327.
3. Clyde DF, Most H, McCarthy V, Vanderberg JC: Immunization against sporozoite-induced falciparum malaria. *Am J Med Sci* 1973;266:169–177.
4. Clyde DF, McCarthy VC, Miller RM, Hornick RB: Specificity of protection of man immunized against sporozoite-induced falciparum malaria. *Am J Med Sci* 1973;266:398–403.
5. Clyde DF, McCarthy VC, Miller RM, Woodward WE: Immunization of man against falciparum and vivax malaria by use of attenuated sporozoites. *Am J Trop Med Hyg* 1975;24:397–401.
6. Cochrane AH, Nussenzweig RS, Nardin EH: Immunization against sporozoites, in Kreier JP (ed): *Malaria,* Vol 3. New York, Academic Press, 1980, pp 163–202.
7. Dame JB, Williams JL, McCutchan TF, et al: Structure of the gene encoding the immunodominant surface antigen on the sporozoite of the human malaria parasite Plasmodium falciparum. *Science* 1984;225:593–599.
8. de la Cruz VF, McCutchan TF: Heterogenity at the 5' end of the circumsporozoite protein gene of Plasmodium falciparum is due to a previously undescribed repeat sequence. *Nucl Acid Res* 1986;14:4695.
9. Eichinger DJ, Arnot DE, Tam JP, et al: Circumsporozoite protein of Plasmodium berghei: gene cloning and identification of the immunodominant epitopes. *Mol Cell Biol* 1986;6:3965–3972.
10. Ellis J, Ozaki LS, Gwadz RW, et al: Cloning and expression in E. coli of the malarial sporozoite surface antigen gene from Plasmodium knowlesi. *Nature* 19xx;302:536–538.
11. Fine E, Aikawa M, Cochrane AH, Nussenzweig RS: Immunoelectron microscopic observation on Plasmodium knowlesi sporozoites: localization of protective antigens and its precursors. *Am J Trop Med Hyg* 1984;33:220–226.
12. Good MF, Berzofsky JA, Maloy WL, et al: Genetic control of the immune response to a Plasmodium falciparum sporozoite vaccine: widespread nonresponsiveness to a single malaria T-epitope in highly repetitive vaccines. *J Exp Med* 1986;164:655–660.
13. Good MF, Maloy WL, Lunde MN, et al: Construction of synthetic immunogen: use of new T-helper epitope on malaria circumsporozoite protein. *Science* 1987;235:1059–1062.
14. Gwadz RW, Cochrane AH, Nussenzweig V, Nussenzweig RS: Preliminary studies on vaccination of rhesus monkeys with irradiated sporozoites of Plasmodium knowlesi and characterization of surface antigens of these parasites. *Bull WHO* 1979;57(suppl 1):165–173.
15. Lal AA, de la Cruz VF, Welsh JA, et al: Structure of the gene encoding the circumsporozoite protein of Plasmodium yoelii: a rodent model for examining anti-malarial sporozoite vaccines. *J Biol Chem* 1987;262:2937–2940.
16. Lockyer MJ, Schwarz RT: Strain variation in the circumsporozoite gene of Plasmodium falciparum. *Mol Biochem Parasitol* 1987;22:101–108.
17. McCutchan TF, Lal AA, de la Cruz VF, et al: Sequence of the immunodominant epitope for the surface protein of sporozoites of Plasmodium vivax. *Science* 1985;230:1381–1383.

18. Miller LH, Howard RJ, Carter R, et al: Research towards malaria vaccine. *Science* 1986;234:1349–1356.
19. Nardin EH, Nussenzweig RS, Bryan J, McGregor IA: Antisporozoite antibodies: their frequent occurrence in individuals living in an area of hyperendemic malaria. *Science* 1979;206:597–599.
20. Nussenzweig RS, Vanderberg J, Most H, Orton C: Protected immunity produced by injection of x-irradiated sporozoites of Plasmodium berghei. *Nature* 1967;216:160–162.
21. Ojo-Amaize EA, Vilcek J, Cochrane AH, Nussenzweig RS: Plasmodium berghei sporozoites are mitogenic for murine T-cells, induce interferon, and activate natural killer cells. *J Immunol* 1984;133:1005–1009.
22. Orjih AU, Cochrane AH, Nussenzweig RS: Active immunization and passive transfer of resistance against sporozoite-induced malaria in infant mice. *Nature* 1981;291:331–332.
23. Peters W, Chance ML, Lissner R, et al: The chemotherapy of rodent malaria. XXX. The enigmas of the "NS lines" of P. berghei. *Ann Trop Med Parasitol* 1978;72:23–36.
24. Potocnjak P, Yoshida N, Nussenzweig RS, Nussenzweig V: Monovalent fragments (Fab) of monoclonal antibodies to sporozoite surface antigen (Pb44) protect mice against malarial infection. *J Exp Med* 1980;151:1504–1513.
25. Rieckman KH, Carson PE, Beaudoin RL, et al: Sporozoites induce immunity in man against an Ethiopian strain of P. falciparum. *Trans R Soc Trop Med Hyg* 1974;60:258–259.
26. Yoshida N, Nussenzweig RS, Potocnjak P, et al: Hybridoma produces protective antibodies directed against the sporozoite stage of malaria parasite. *Science* 1980;207:71–73.
27. Young JF, Hockmeyer WT, Gross M, et al: Expression of Plasmodium falciparum circumsporozoite proteins in Escherichia coli for potential use in human malaria vaccine. *Science* 1985;228:958–962.
28. Zavala F, Tam JP, Hollingdale MR, et al: Rationale for development of a synthetic vaccine against Plasmodium falciparum malaria. *Science* 1985;228:1436–1440.

CHAPTER 36

Malaria Sporozoite Vaccine Development: Recent Progress

W. Ripley Ballou

Control of falciparum malaria remains one of the world's greatest health challenges. Global malaria eradication and control programs conducted over the past 50 years have been seriously hampered by the widespread development of drug resistance by the parasite and insecticide resistance by the mosquito vector. As a result, there is great interest in developing vaccines to prevent malaria, and there has been dramatic progress in the development of vaccines directed against the sporozoite stage of *Plasmodium falciparum*.

Sporozoites are considered to be suitable vaccine targets because they are injected into the blood by mosquitoes and are thus exposed to antibodies for a brief period of time before they invade hepatocytes. To be efficacious, a vaccine against sporozoites must neutralize all infectious parasites prior to the time they invade liver cells, for if even one sporozoite survives to invade, mature, and release merozoites, clinical malaria results. The feasibility of immunization against the sporozoite stage was first demonstrated by Nussenzweig et al (13) using radiation-attenuated sporozoites in rodents, and later analogous studies were conducted in humans (5,15). However, immunization required the exposure of volunteers to many hundreds of infected mosquitoes over a period of several weeks to months, and the limited availability of purified sporozoites made such an approach impractical for vaccine development.

Analysis of sera from sporozoite-immunized mice demonstrated high levels of antibodies that were almost exclusively directed against a protein that covered the parasite surface, the circumsporozoite (CS) protein (16). With the development of monoclonal antibody and gene cloning techniques, it was shown that the general structures of CS proteins are remarkably similar among those malaria species studied to date (10). The most striking feature of CS proteins is that more than 40% of their primary structure is made up of multiple tandem repeats of relatively short amino acid sequences. More importantly, antibodies directed against CS proteins recognize epitopes contained within these repeat sequences and are biologically active against sporozoites in vivo and in vitro (1,14). We demonstrated the feasibility of immunization against sporozoites with CS protein subunit vaccines using the *P. berghei* rodent malaria model (7). Rapid progress toward developing a human malaria vaccine resulted from cloning the CS gene of *P. falciparum* (6) and identifying its CS repeat epitopes as targets for functional antibodies (1). A series of recombinant DNA *P. falciparum* CS proteins expressed in *Escherichia coli* (19) were identified as potential sporozoite vaccine candidates (18), and one of them, alum-adjuvanted R32tet$_{32}$ (FSV-1), was selected to undergo phase I trials of safety and immunogenicity, as well as a preliminary study of efficacy in a small number of human volunteers (2). These sporozoite vaccine studies were conducted in collaboration with D.M. Gordon, I. Schneider, R.A. Wirtz, W.T. Hockmeyer, and J.D. Chulay at the Departments of Immunology and Entomology, WRAIR; S.L. Hoffman at the Malaria Branch, NMRI, Bethesda, MD; J.F. Young and G.F. Wasserman at Smith Kline and French Laboratories, Swedeland, PA; J.A. Sherwood and F.A. Neva at the National Institutes of Health; and M.R. Hollingdale at the Bethesda Research Institute, Rockville, MD.

Safety and Immunogenicity of FSV-1

Fifteen healthy men aged 22 to 50 with no prior exposure to malaria and without detectable antibodies to sporozoite or blood-stage parasites of *P. falciparum* gave informed consent to be immunized with the recombinant vaccine. The vaccine consisted of single-dose ampules of sterile R32tet$_{32}$ [MDP(NANP)$_{15}$NVDP(NANP)$_{15}$NVDPtet$_{32}$, where tet$_{32}$ refers to the first 32 amino acids encoded by a tetracycline resistance gene, read out of frame] in aqueous saline containing 0.5 mg Al^{3+} (as aluminum hydroxide gel) per 0.5 ml dose with thimerosal added as a preservative. FSV-1 was prepared at five concentrations (10, 30, 100, 300, and 800 μg R32tet$_{32}$ per 0.5 ml dose). The vaccine was administered intramuscularly to each of three volunteers at each of the five doses at week 0, with booster doses at weeks 4 and 8.

FSV-1 was safe and well tolerated at all five doses. Mild local pain associated with the actual injection of vaccine occurred in seven of nine volunteers receiving doses of 100 μg or greater. One volunteer who received 800-μg doses developed repeated sneezing and urticaria within 5 minutes after a third dose of FSV-1. These symptoms resolved spontaneously within 30 minutes and did not recur after a fourth dose was administered several months later.

To measure antibodies directed only against the CS repeat portion of R32tet$_{32}$, we used as antigen in the ELISA a related recombinant protein containing the identical CS repeat portion but only the first two amino acids of the tet$_{32}$ tail (R32LR). Antigen-specific immunoglobulin G (IgG) was detected as early as 2 weeks after the primary immunization and was dose-dependent between 100 and 800 μg of R32tet$_{32}$. Twelve of fifteen (80%) volunteers had antibody titers of 1:50 or greater, with a maximal titer of 1:800 in one individual. Maximal antibody responses were sustained for 3 to 4 weeks, and titers returned to baseline by 4 months after the third dose. Immunoglobulin class determinations revealed IgM, IgA, and IgG antibodies to the antigen in all positive sera, with IgG antibodies predominating. IgG$_1$ was the major isotype found and comprised 69 to 94% of total antigen-specific IgG. IgG$_2$ and IgG$_3$ antibodies were found in nearly equal amounts. IgG$_4$ antibodies were detectable in only one volunteer. Total antigen-specific IgG, estimated by summing isotype concentrations, range from less than 1 to more than 50 μg/ml.

When week 0 and week 10 antibodies were tested by immunofluorescence using air-dried *P. falciparum* sporozoites, only week 10 sera from the three volunteers receiving 800-μg doses were positive. Two sera gave 2+ reactions at a 1:100 dilution, and a third gave a 3+ reaction at a 1:400 dilution. Lymphocyte proliferation assays performed at weeks 4 and 8 demonstrated blastogenic responses to R32tet$_{32}$ in 11 of 13 volunteers, but there was no apparent correlation between the magnitude of proliferative responses and vaccine dose or antibody levels.

Preliminary Study of Efficacy

Although it was clear that the vaccine was less immunogenic for humans than preclinical studies in rodents had suggested, we believed it was important to determine if a subsequent booster might increase antibody titers and if these antibodies were protective. Thus 12 months after the start of the study, six of the original volunteers received a fourth dose of FSV-1. Antibody to CS epitopes increased above baseline in four subjects, but all titers were less than the maximal titers achieved during the primary immunization. Three weeks later these subjects and two nonimmunized controls were challenged using a method we had previously shown to be safe and reliable (4). Five mosquitoes infected with the chloroquine-sensitive NF54 strain of *P. falciparum* were allowed to feed on the volunteers. Seven of the eight volunteers developed clinical malaria 9 to 13 days later. Parasitemia never developed in the individual who had the highest antibody response, and the incubation and prepatent periods were prolonged in the two subjects with the higher antibody titers among the subjects who became parasitemic (Table 36.1). Lymphocyte blastogenic responses to R32LR were positive in those volunteers who developed antibodies, but the magnitude of the responses did not correlate with either the antibody titer or protection. The clinical manifestations of malaria were not modified by a delay in parasitemia.

Table 36.1. Immunogenicity and efficacy of a fourth dose of FSV-1.

Subject	O.D.[a] (1:50)	Dose (μg)	S.I.[b]	IFA (1:100)	Prepatent period (days)
Ctl 1	0.195	—	ND	—	9
Ctl 2	0.137	—	ND	—	10
11	0.122	300	1.08	—	10
7	0.147	100	1.40	—	10
10	0.264	300	2.95	—	10
8	0.294	100	9.88	—	12
13	0.613	800	7.07	2 +	13
14	1.058	800	4.04	2 +	>30

[a]ELISA absorbance (optical density).
[b]Lymphycote blastogenesis to R32LR; mean stimulation index (S.I.) of four nonimmunized controls was 1.71. ND = not done. Adapted from Ballou et al (2).

Implications for Future Development

These studies confirmed the hypothesis that humans can be successfully protected from *P. falciparum* sporozoite challenge by immunization with a highly purified recombinant CS protein subunit vaccine. The current vaccine, however, cannot be considered a candidate for larger field trials. Protection was seen in only one volunteer, and very high doses of vaccine were required to elicit a protective level of antibody. This level was similar to those found in naturally immunized individuals from highly endemic malarious areas, but such levels do not protect these populations from malaria under conditions of intense sporozoite transmission (12). It is clear that higher and more sustained levels of antibodies in a large proportion of volunteers will be needed before a CS subunit vaccine can be considered ready for large-scale clinical trials.

Among the approaches under consideration to increase the immunogenicity of this vaccine are the use of higher doses, alternative routes or schedules of immunization, novel adjuvants, and incorporation of the vaccine into liposomes. Alternatively, the recombinant protein itself may need to be modified as a new fusion protein or combined with other carrier proteins. Studies of antibody responses and lymphocyte blastogenesis using R32tet$_{32}$ in mice have identified both T and B cell epitopes on the repeat portion of the molecule, and mice primed with these CS epitopes can be boosted by intact sporozoites (11). Eighty percent of the recombinant CS protein in FSV-1 consists of multiply-repeated units of only four amino acids. This region of the molecule contains but a single murine T cell epitope (8) and may thus be inefficient in eliciting boosting responses in humans under natural exposure to sporozoites. The identification of additional T epitopes on the CS gene and their incorporation into new vaccines (9) or the addition of nonsporozoite T epitopes using alternate carrier proteins or fusion products may significantly enhance CS antibody production.

Finally, the fundamental role of cellular immune responses afforded by immunization with irradiated sporozoites must be considered in the development of second-generation sporozoite vaccines (3,7,17). However, target antigens for cell-mediated immune responses to preerythrocytic parasite stages are unknown and may be independent of the CS protein. In summary, these studies have demonstrated that humans can be protected with a purified recombinant sporozoite vaccine, but much work is required to develop a vaccine candidate suitable for field trials.

References

1. Ballou WR, Rothbard J, Wirtz RA, et al: Immunogenicity of synthetic peptides from circumsporozoite protein of Plasmodium falciparum. *Science* 1985;228:996–999.
2. Ballou WR, Hoffman SL, Sherwood JA, et al: Safety and efficacy of a recombinant DNA Plasmodium falciparum sporozoite vaccine. *Lancet* 1987;1:1277–1281.
3. Chen DH, Tigelaar RE, Weinbaum FI: Immunity to sporozoite induced malaria infection in mice. I. The effect of immunization of T and B cell deficient mice. *J Immunol* 1977;188:1322–1327.
4. Chulay JD, Schneider I, Cosgriff TM, et al: Malaria transmitted to humans by mosquitoes infected from cultured Plasmodium falciparum. *Am J Trop Med Hyg* 1986;35:66–68.
5. Clyde DF, McCarthy VC, Miller RM, Woodward WE: Immunization of man against falciparum and vivax malaria by use of attenuated sporozoites. *Am J Trop Med Hyg* 1975;24:397–401.
6. Dame JB, Williams JL, McCutchan TF, et al: Structure of the gene encoding the immunodominant surface antigen on the sporozoite of the human malaria parasite Plasmodium falciparum. *Science* 1984;225:593–599.
7. Egan JE, Weber JL, Ballou WR, et al: Efficacy of murine malaria sporozoite vaccines: implications for human vaccine development. *Science* 1987;236:453–456.
8. Good MF, Berzofsky JA, Maloy WL, et al: Genetic control of the immune response in mice to Plasmodium falciparum sporozoite vaccine: widespread nonresponsiveness to single malaria T epitope in highly repetitive vaccine. *J Exp Med* 1986;164:655–660.
9. Good MF, Maloy WL, Lunde MN, et al: Construction of synthetic immunogen: use of new T-helper epitope on malaria circumsporozoite protein. *Science* 1987;235:1059–1062.
10. Hockmeyer WT, Ballou WR, Young JF: Sporozoite malaria vaccines, in Chedid L, Hadden JW, Spreafico F, et al (eds): *Advances in Pharmacology*. Oxford, Pergamon Press, 1985, pp 357–361.
11. Hoffman SL, Cannon LT, Berzofsky JA: Plasmodium falciparum: a T cell epitope on sporozoite vaccine allows sporozoite boosting of immunity. *Exp Parasitol* 64:64–70(1987)

12. Hoffman SL, Oster CN, Plowe CV, et al: Naturally acquired antibodies to sporozoites do not prevent malaria: Vaccine Development Implications *Science* 237:639–642, 1987.

13. Nussenzweig R, Vanderberg J, Most H: Protective immunity produced by the injection of x-irradiated sporozoites of Plasmodium berghei. *Milit Med* 1969;134:1176–1182.

14. Potocnjak P, Yoshida N, Nussenzweig R, Nussensweig V: Monovalent fragments (Fab) of monoclonal antibodies to a sporozoite surface antigen (Pb44) protect mice against malaria infection. *J Exp Med* 1980;151:1504–1513.

15. Rieckmann KH, Beaudoin RL, Cassells JS, Sell KW: Use of attenuated sporozoites in the immunization of human volunteers against falciparum malaria. *Bull WHO* 1979;57(suppl 1):261–265.

16. Vanderberg J, Nussenzweig R, Most H: Protective immunity produced by the injection of x-irradiated sporozoites of Plasmodium berghei. V. In vito effects of immune serum on sporozoites. *Milit Med* 1969;134:1183–1190.

17. Verhave JP, Strickland GT, Jaffe HA, Ahmed A: Studies on the transfer of protective immunity with lymphoid cells from mice immune to malaria sporozoites. *J Immunol* 1978;121:1031–1033.

18. Wirtz RA, Ballou WR, Schneider I, et al: Plasmodium falciparum: immunogenicity of circumsporozoite protein constructs produced in Escherichia coli. *Exp Parasitol* 1987;63:166–172.

19. Young JF, Hockmeyer WT, Gross MT, et al: Expression of Plasmodium falciparum circumsporozoite proteins in Escherichia coli for potential use in a human malaria vaccine. *Science* 1985;228:958–962.

CHAPTER 37

Community-Based Integrated Vector Control of Malaria in India

V.P. Sharma and R.C. Sharma

The National Malaria Eradication Programme (NMEP) of the Government of India in collaboration with the state health departments is responsible for malaria control in the country. The strategy of malaria control in rural areas is to interrupt transmission by spraying residual insecticides such as DDT, HCH, or malathion and in urban areas mainly by application of antilarval methods, i.e., larvicidal application and destruction of breeding sites. With particular reference to rural malaria, an innovative approach to the control of malaria is being demonstrated in Gujarat by involving communities that suffer from the disease. This approach is in contrast to a spraying strategy in which participation of the villagers is passive, if it exists at all. Most of the time the populace is completely indifferent to spraying, as this measure is the responsibility of the government.

In an alternate strategy the main emphasis is on eliciting a positive response from all sections of the society, involving them in the simple methods of mosquito control. This strategy has become important in view of the innumerable problems faced by the spray squads all over the country: As a result of continuous spraying since the mid-1950s, malaria mosquitoes have become resistant, no longer dying in the same numbers as they once did. Added to this problem is the fact that several collateral benefits of spraying, such as the control of bedbugs and flies, are no longer visible; and, in fact, because of the excito-repellent action of DDT, the bedbug nuisance increases considerably soon after spraying. The spraying also has the inherent disadvantage of long persistence in the environment, with the possibility of an adverse ecological impact. In some areas refusal rates of spraying in the households are high, and the spraying is restricted to the cattle sheds. It has had a counterproductive impact, as the mosquitoes are driven toward the houses, thereby enhancing transmission. Cultural and agricultural practices also seriously impede the spraying operations. In the former case, in many areas walls are mud-plastered soon after spraying, and in the latter case spraying is not accepted because of the possible destruction of cottage industries, e.g., silk culture and beekeeping. There are also instances of large-scale misuse of insecticides that find their way into agriculture. In many areas spraying has little value, as the malaria mosquitoes rest and feed outdoors, avoiding contact with the sprayed walls.

The economics of spraying is alarming. The government of India is spending about 45% of the health budget annually on malaria control, and an equal amount is to be shared by state governments. Many states find it difficult to provide matching grants, and so in some areas malaria control is put aside pending the availability of funds. Rural malaria control alone costs the government about Rs. 160 crores annually. The epidemiology of the disease is such that if neglected for a year or two it may assume epidemic proportions, resulting in high morbidity.

Although large sums of money are spent annually on malaria control, there is no monitoring of the program. The data collected on the incidence are the indirect result of fortnightly surveillance to prevent morbidity due to malaria. As a result of this feedback, at least 2 million parasite-positive cases are detected every year, 30% of which are due to *P. falciparum*. There is no further true decline in malaria cases, although the government is doing almost everything possible (3). Focal studies have shown a considerable increase in falciparum malaria in many parts of the country, with fulminating epidemics resulting in reports of deaths from Jalpaiguri (WB), areas in Assam & Punjab, and U.P. (Shahjahanpur, and Bareilly) to name only a few (1,2). The problem is compounded by the *P. falciparum* parasite becoming resistant to the commonly used antimalarial chloroquine (5). Malaria control therefore is in a state of crisis and at the crossroads, waiting for some alternatives to emerge.

Because most of the problems of spraying relate to the behavior and response of the villagers, it is absolutely essential that any long-term malaria control must heavily depend on their active participation. For the method to be successful, the techniques should be simple and appropriate for local situations, and it should also result in long-term benefits to the villagers. With this background a demonstration cum feasibility study was launched in 1983 in a 26,000 population (seven villages) in Nadiad taluka, Kheda district, Gujarat. The methods employed in tackling breeding problems were simple and inexpensive. Surveys, both inside and outside the villages, were carried out to map all water collection sites and areas that can support mosquito breeding. The second step was to terminate breeding in such areas. The most common method was to turn the pots with water upside down, cover the water receptacles, and fill the borrow pits with soil. While performing these activities, villagers were shown the mosquito breeding and how simple it was to destroy these potential sources of mosquito production. Major potential breeding sites were ponds and wells. Breeding in ponds was controlled by the introduction of fish that eat mosquito larvae. These fish, found indigenously, were collected and bred in abandoned cement tanks and ponds. The margins of the ponds were cleaned periodically so that the fish could reach the grassy margins, which are the common breeding spots for mosquitoes (8).

Breeding in the wells generally takes place when they are not used regularly. In most of the wells, expanded polystyrene (EPS) beads which are composed of an inexpensive, nonbiodegradable, lasting material, were introduced, and they formed a mat on the water surface. Wells treated with the beads do not support mosquito breeding for at least 3 years, which is the maximum period under observation (4). Some wells that are only occasionally used by the villagers were not treated with beads; instead, a larvivorous fish ("guppy") was introduced. In any habitat with larvivorous fish the breeding of mosquitoes is prevented so long as the fish continue to survive in that habitat.

The villagers thus participated actively in eliminating the mosquito breeding sites, introducing the larvivorous fishes, and destroying intradomestic breeding in containers and any other water collection vessels that may support mosquito breeding. Constant health education and interaction with all sections of society was the key to communication.

The villages generally have innumerable borrow pits and other depressions as well as uneven land. Most of these sites become potential breeding sites after the rains. These places were mapped, and the villagers were mobilized to fill in and level the areas with soil. This maneuver is being done on a routine basis by organizing Shram Dans; at present four tractors fitted with trolleys are constantly working under the guidance of village panchayats in leveling the land inside and outside the villages.

An important and unhygienic problem was the flow of waste water on the streets that used to collect in pits. It was an ideal place for *Culex* mosquitoes to breed. Although they are not the vectors of malaria but of filaria, transmission may become established as had happened in many towns of eastern U.P. including Terai region (9). At first the drains were cleaned with the help of villagers, but now soak pits have been introduced. The soak pits, which cost about US$4 have become popular, and there is no stagnant water on the streets. The demand for soak pits has increased tremendously, making it difficult to cope with the installation program.

Another area of importance that has been successfully exploited is the composite culture of edible fishes with the larvivorous fishes. During the first year, although there was large-scale poaching and major thefts in eight ponds that were taken up for fish production, the auction earned 100 times more money than was usually earned by the ponds. If no poaching was done, the income would have easily reached 500 times. This situation was so when three ponds dried up because of drought. It has been estimated that larvivorous fishes grown in 20 villages can produce enough guppies for the entire taluka of 100 villages. Moreover, food fish culture in all the ponds in Nadiad taluka can make the Panchayats economically viable, so they can initiate mosquito and malaria control from their own resources (10). The movement has increased tremendously, and now there is a demand for similar experiments in all the ponds of Nadiad taluka.

The second economical scheme to boost the village economy was the social forestry scheme. There were large areas in villages that were either low-lying or marshy. In all such areas eucalyptus trees have been planted. This measure will dry the marshy areas and prevent mosquito breeding. Surveys also revealed that each village had large areas of waste land. In all these areas large scale plantations are being encouraged with the help of the social forestry department. During the first 2 years about 0.2 million plantations have been created, but the demand of Nadiad taluka has exceeded 5 million, a target that was impossible to achieve through intersectoral collaboration. Therefore in-house nurseries and village nurseries are being encouraged so that villagers can plant trees of their own liking, and there shall be no expenditure by the government. The additional advantage of care by the villagers would protect plantations and increase survival. This tremendous resource, which was otherwise going to waste, can produce so much income that one year's earnings can control malaria in the entire taluka for a few years.

At this point it is important to mention that the key to success in the implementation of integrated control of a malaria program was community participation from all sections of the society. Gujarat state is endemic for malaria, and Kheda district recorded the highest incidence for several years. Nadiad taluka was the worst affected area, with a high incidence of falciparum malaria and deaths due to malaria. Mosquito breeding sites were innumerable, and spraying of insecticides was not producing any tangible impact in the reduction of disease transmission. Intensive surveillance and prompt radical treatment was the first line of action that reduced morbidity and completely eliminated mortality. It also infused tremendous confidence in the villagers.

Because of the spectacular success in malaria control in the seven villages, the study was extended to another group of 14 villages, giving a total population of 60,000 in the 21 villages. The vector densities of A. culicifacies and A. stephensi almost completely disappeared from most of the villages. The incidence of malaria started to go down in the experimental villages during successive years. The study was extended to the entire taluka comprising 100 villages with a population of 350,000 (Table 37.1). The results in the entire taluka were equally spectacular, with general improvement in the environment and enhanced awareness about the water-borne and water-based diseases (7). The experiment is still in progress, as it is a continuing process; it was to be extended to towns and other talukas to cover about 1 million population by the end of 1988. In this innovative approach to malaria control only noninsecticidal methods were used that were environmentally safe and that enriched the ecology of the area.

It was also realized that as the incidence of malaria and mosquito nuis-

Table 37.1. Incidence of malaria in the experimental villages.

Year	History of spray	Jan–Mar	Apr–Jun	Jul–Sep	Oct–Dec	Total
			Quarterly record of malaria cases			
	Complex A: 7 Villages; Population 26,000					
1981	HCH/mal. sprayed	374	955	914	1,797	4,040
1982	HCH/mal. sprayed	982	849	417	107	2,355
1983	DDT sprayed	64	75	187	85	411
1984[a]	Integrated control	26	36	61	18	141
1985[a]	Integrated control	6	8	44	10	68
1986[a]	Integrated control	2	8	30	22	62
	Complex B: 14 Villages; population 34,000					
1981	HCH/mal. sprayed	214	586	754	477	2,031
1982	HCH/mal. sprayed	396	908	549	115	1,968
1983	DDT sprayed	86	149	185	52	472
1984	DDT sprayed	17	33	72	21	143
1985[a]	Integrated control	4	29	48	5	86
1986[a]	Integrated control	6	12	33	17	68
	Complex C: 79 villages; population 290,000					
1981	HCH/mal. sprayed	1,174	3,998	4,036	2,417	11,625
1982	HCH/mal. sprayed	1,667	5,170	3,474	931	11,242
1983	DDT sprayed	556	761	1,392	543	3,252
1984	DDT sprayed	210	381	645	138	1,374
1985	DDT sprayed	87	253	394	172	906
1986[a]	Integrated control	30	109	229	128	496

[a]No insecticides used.

ance falls below the tolerance threshold, the villagers may start losing interest and consider it as but one of the ongoing routine activities. In order to sustain interest, other schemes of improved chulhas, biogas plants, solar cookers, and other improved agricultural practices are being promoted and are resulting in the holistic development of the village.

A study of this nature would be incomplete if the cost of the mosquito control operations were not affordable and equal to or less than the current expenditure incurred on malaria control using DDT. DDT spraying cost is Rs. 34 lakhs for one million population. The integrated control expenditure would come to about Rs. 37 lakhs for the same population. The expenditure to control malaria using malathion would be Rs. 2 crores. As the size of the population increases, the expenditure for integrated control would not increase proportionately. Calculations show that, to control malaria in the entire rural population of Kheda district (2.7 million population), the expenditure for DDT, malathion, and integrated control would be Rs. 92 lakh, Rs. 537 lakh, and Rs. 85 lakh, respectively. Therefore the strategy for malaria control without the use of insecticides and with the use of other safe methods and the help of the community is the most inexpensive and lasting method of malaria control (6). If we take into account income generated as a result of fish culture, social forestry, etc., not only is malaria control free but it also provides considerable income to the village panchayats, thus making them self-sufficient and strengthening the panchayat movement. It would also bring to reality the dream of the father of the nation, Mahatma Gandhi.

Integrated control of malaria has many advantages over insecticidal methods. Some of these advantages are that semipermanent to permanent changes are incorporated in the environment, thus making the areas unsuitable for mosquito breeding on a long-term basis. These changes are brought about by maintaining the ecological integrity of the area, keeping the developments in harmony with nature. Malaria control is carried out with the active participation of the village community, thereby reducing the cost and bringing about social transformation of members of the society, who become conscious of their surroundings and keep them clean. The method inculcates scientific temper in these rural communities, which were often left behind and were unable cope with the changing environment. Because the methods are labor-intensive, the money is spent to provide employment to unemployed youths. This money would otherwise have gone to the purchase of insecticides.

Malaria control along with other developments opens a window for the introduction of new developmental schemes. Because of the rapport established with the villagers and the confidence gained, it becomes easy to convince them of new developments that will benefit the communities. For example, it was easy to introduce several alternate energy sources, such as the improved chulhas, solar cookers, and biogas plants. These units save energy and provide a healthy environment.

One of the major advantages is that the method automatically tackles the intractable problem of insecticide resistance. The single factor of insecticide resistance has made the spraying strategy a failure, and the huge costs incurred are not producing desirable results, thus questioning the performance of the malaria control departments. The introduction of replacement insecticides increases the cost manyfold, with no real hope in the near future. It is nearly impossible to allot such huge funds for malaria control alone, particularly when this allotment must be made year after year with the usual escalation of costs of insecticides. It is envisaged that by the end of the seventh plan the entire Kheda district will be brought under integrated control of malaria; thus we would stop the use of insecticides in at least one endemic area with a problem of multiple resistance.

Based on the spectacular success of this experiment, the study has been extended to other malaria endemic areas such as Hardwar, Shahjahanpur, Haldwani, and Allahabad (U.P.); Berhampur (Orissa); Madras (T.N.); Mandla (M.P.); and Sonapur (Assam). It is envisaged that at least one or two Primary Health Centres (PHCs) would be brought under integrated control at each site. It would stimulate interest in other parts of the country and would provide an opportunity to develop situation-specific alternate strategies for each area for its adoption in the malaria control program of the country.

References

1. Ansari MA, Batra CP, Sharma VP: Outbreak of malaria in villages of Bareilly, District U.P. *Indian J Malariol* 1984;21:121–123.
2. Chandrahas RK, Sharma VP: Malaria epidemic in Shahjahanpur. *Indian J Malariol* 1983;20:163–166.
3. Sharma GK: Review of malaria and its control in India, in Sharma VP (ed): *Proceedings Indo-UK. Workshop on Malaria*. Delhi, Malaria Research Centre, 1984, pp 13–40.
4. Sharma RC, Yadav RS, Sharma VP: Field trails on the application of expanded polystyrene (EPS) beads in mosquito control. *Indian J Malariol* 1985;22:107–109.
5. Sharma VP: Drug resistant P. falciparum malaria in İndia, in Sharma VP (ed): *Proceedings Indo-U.K. Workshop on Malaria*. Delhi, Malaria Research Centre, 1984, pp 169–184.
6. Sharma VP: Cost-effectiveness of the bioenvironmental control of malaria in Kheda district, Gujarat. *Indian J Malariol* 1986;23:141–145.
7. Sharma VP, Sharma RC: Bio-environmental control of malaria in Nadia, district Kheda, Gujarat. *Indian J Malariol* 1986;23:95–117.
8. Sharma VP, Sharma RC: Review of the integrated control of malaria in Kheda district, Gujarat, in *Proceedings of the ICMR/WHO Workshop to Review Research Results on Community Participation for Disease Vector Control*. 3–9 February 1986.
9. Sharma VP, Malhotra MS Mani TR In: facets of Environmental problems (Five case studies) Krishna Murti CR (ed): Malaria: Entomological and Epidemiological studies in Terai, District Nainital, U.P. INSA - ICSU 1983, pp 35–46.
10. Sharma RC, Gupta DK, Sharma VP: Studies on the role of indigenous fishes in the control of mosquito breeding. *Indian J Malariol* 1987;24: 73–77.

CHAPTER 38

Leishmaniasis and Malaria: New Tools for Epidemiological Analysis

Dyann F. Wirth, William O. Rogers, Robert Barker,
Heitor Dourado, Laksami Suesebang,
and Bernadino Albuquerque

Parasitic diseases are still prevalent in many parts of the world, causing both human suffering and economic loss. Major efforts to control and even eradicate parasitic diseases have met with some success (e.g., elimination of malaria from the southern United States and Cuba), but in many developing tropical countries parasitic diseases remain major health problems. Such diseases can pose a significant barrier to economic development, and their control is an important goal for improved world health. Intensive research is being devoted to the development of new control measures for many parasitic diseases. These control measures include development of vaccines and new chemotherapeutic agents as well as improved vector control strategies. Previous experience has demonstrated the need for extensive baseline information before the introduction of any control program and the need for continued monitoring of the control program in order to assess its effectiveness. These diseases have complex life cycles involving the vectors of transmission and often the intermediate hosts, both of which have an impact on the transmission of disease and can affect the outcome of any control measure.

Developments in biotechnology, including the use of monoclonal antibodies and recombinant DNA, have provided new tools for the collection of information on these diseases and have the potential for providing more extensive and detailed information on the parasite in the infected human and in insect vectors. This chapter focuses on the potential impact of these new methodologies on epidemiological studies of two parasitic diseases, leishmaniasis and malaria, for which new methods of detection, in man and insect vectors, have been developed. These diseases are caused by parasitic protozoa of different genera, *Leishmania* and *Plasmodium,* respectively, each of which has a unique life cycle. Each poses a separate set of problems for epidemiological studies and eventual control.

Leishmaniasis

Human leishmaniasis is caused by at least 14 species and subspecies of the genus *Leishmania*. The clinical manifestations of the disease depend in part on the infecting *Leishmania* organism and fall into three general categories: simple cutaneous disease, which is often self-limiting; mucocutaneous disease, which involves the destruction of nasal tissue; and visceral disease, a systemic infection that is often fatal if untreated. In fact, the diseases share relatively few properties except that they are caused by organisms of the same genus and that in humans the parasite grows in the phagolysosomal vesicles of macrophages. Epidemiological studies over the last 25 years have shown that, in general, leishmaniasis is a zoonotic disease; the parasite is transmitted to man from a reservoir mammalian host by a *Phlebotomus* (sandfly) vector during a blood meal. Of the numerous *Leishmania* species infective for mammals, only a subset can infect and cause disease in humans. Presumably, this selectivity is due to a combination of factors, including the intrinsic susceptibility of humans and the feeding habits of the sandfly vector. Enormous effort has been devoted to the isolation and characterization of *Leishmania* organisms that infect man and to identification of the principal mammalian reservoirs and species of *Phlebotomus* vector (26,28). The result of many such studies has been the correlation of particular clinical manifestations with certain species or subspecies of the parasite (see Table 38.1 for a summary).

The identification of *Leishmania* species is based on a variety of ecological, biological, biochemical, and immunological criteria (11). Each cultured isolate of the parasite has been analyzed by the use of one or more of these criteria and categorized as to species and subspecies. There remain certain controversies as to whether organisms isolated in distant geographic locations but sharing certain common properties belong to the same or distinct subspecies of *Leishmania*. For example, *L. mexicana garnhami* (52), isolated in Venezuela, is similar to *L. mexicana amazonensis,* isolated in Brazil, as determined by isoenzyme profiles and monoclonal antibodies (25,40,44), but there are conflicting results as to its growth characteristics in sandfly vectors (27). Whether these organisms represent strains of the same subspecies or distinct subspecies cannot be resolved because there is no single generally accepted method for species identification in the genus *Leishmania*. This uncertainty complicates comparison of the disease epidemiology in distinct geographic locations and represents a potential limitation on the transfer of control measures from one geographic location to another. The World Health Organization (WHO) has addressed this problem by establishing a set of reference strains for the various species and subspecies to be used for comparison and classification of new isolates.

The diverse nature of leishmaniasis clearly requires a diverse control program with specific targets for each focus of the disease. This goal, in

Table 38.1. Major *Leishmania* species causing human diseases.

Primary geographic disease[a] and species	Location
Cutaneous leishmaniasis	
L. mexicana mexicana	Mexico, Central America
L. mexicana amazonensis	Brazil Amazon region
L. mexicana pifoni	Venezuela
L. major	Southern USSR, Middle East
L. tropica	Asia, southern Europe, northern and western Africa
L. braziliensis guyanensis	Northern, southern America
L. braziliensis panamensis	Central America
L. braziliensis peruviana	Peru
Diffuse cutaneous disease	
L. mexicana amazonensis	Brazil, Amazon region
L. aethiopica	Ethiopia and Kenya
Mucocutaneous disease	
L. braziliensis braziliensis	Western and northern South America
Visceral disease	
L. donovani	India
L. donovani infantum	Mediterranean area
L. chagasi	Northern South America

These results are from 57 patients at the *Leishmania* clinic, Instituto de Medicina, Tropical de Manaus during January–February 1984. Each patient was given a Montenegro test, and those with positive results in the test underwent further diagnostic tests that included a tissue biopsy (4 mm) taken directly from the lesion. This biopsy was divided in half; one-half was placed directly in Schneider's and the other half touched to nitrocellulose, the remainder of this half of the specimen being processed for histopathology. The hybridization was performed by standard procedures, as described elsewhere (59). The *L. mexicana* probe was a combination of DNA extracted from strain WR303 (*L. mexicana amazonensis*) and L11 (*L. mexicana mexicana*). The *L. braziliensis* probe was isolated from strain WR2903. All hybridization results were read independently by two people.
[a]Based on Marinkelle (10).

turn, requires more extensive collection of baseline data with regard to the infecting *Leishmania* species in humans, the relevant sandfly vector, and the principal mammalian reservoir of the particular species or subspecies. These data must be collected at each focus for proper implementation of any control program directed at either the sandfly vector or the mammalian reservoir. For control measures that involve identification and treatment of patients, accurate, rapid diagnosis of leishmaniasis must be achieved before treatment with relatively toxic chemotherapeutic agents is begun (48).

One of the major problems of further analysis of the ecology and epidemiology of the disease is the laborious task of identifying the parasite. All of the work documenting the species and subspecies of *Leishmania* is dependent on the isolation of the organism either directly in culture or after passage through a susceptible laboratory animal, most commonly the hamster. This method has several limitations, i.e., the number of samples that can be handled at any one time, the time it takes to grow the parasites and subsequently type them, and, perhaps more important, the selection from an otherwise mixed population of parasites those that grow either in vitro or in experimental animals. A method of direct identification of *Leishmania* parasites from lesions, sandfly vectors, and intermediate mammalian reservoirs is necessary if broader epidemiological studies involving large numbers of samples are to be initiated.

Current diagnosis of leishmaniasis is achieved either by direct examination of a tissue biopsy or by means of a delayed-type hypersensitivity test referred to as the Montenegro test (45). Neither of these methods is able to distinguish *Leishmania* species or subspecies, and the Montenegro test cannot distinguish current from previous infections. For certain forms of leishmaniasis, the most effective control measures may be the direct treatment of infected patients (34) and will require specific diagnosis of *Leishmania* species or subspecies from lesion material in order to design treatment regimens that minimize morbidity and mortality (14). For example, cutaneous infection with *L. braziliensis* is often associated with subsequent mucocutaneous disease, and early diagnosis could facilitate treatment and perhaps reduce the frequency of mucocutaneous disease.

DNA Probes in the Diagnosis of Leishmaniasis

A new methodology based on DNA probes specific for the various *Leishmania* species was developed to provide a direct diagnosis of patients with leishmaniasis and to eliminate the need for culturing parasites before species identification (59). This methodology allows direct diagnosis from lesion material without requiring isolation of the parasite. Such direct diagnosis of *Leishmania* species, which had not been possible with any of the existing methodologies, provides the basis for the clinical management of this disease. In the initial studies the DNA probes, which were based on total kinetoplast DNA (kDNA), could differentiate the major species complexes in the New World: *L. mexicana* and *L. braziliensis*. Subsequent experiments in which recombinant DNA methodologies were used resulted in the development of DNA probes that can differentiate species, subspecies, and even distinct isolates of the parasite (7,8,23,30,59,60).

The basis for the DNA probes is the minicircle, which is a highly repeated small circular DNA molecule found within the mitochondria of the parasite. It has no apparent function and has an apparent high rate of DNA sequence divergence as measured by restriction site polymorphism and DNA hybridization studies (15,21a,24,31,54). In the New World *Leishmania* spp. the kDNA minicircles isolated from *L. mexicana* do not share any sequence homology with those isolated from *L. braziliensis* (7,59). These differences in DNA sequence have provided the basis for a

DNA probe that can distinguish the two *Leishmania* species directly when material obtained from a lesion is applied to nitrocellulose.

These DNA probes have been used to diagnose leishmaniasis in patients from the Institut du Tropical de Medicine de Manaus (51) (Table 38.2). In each case, the results of the DNA probe were compared to standard diagnostic tests, including the Montenegro test (see above), histopathology, and culturing and subsequent characterization of the parasite. As can be seen in Table 38.2, the DNA probe detected the disease in 32 of the 43 patients who had positive results in the Montenegro test. This observation may mean that in some patients the parasite density was below the detection limit of the assay, which, based on laboratory experiments, is as few as 50 organisms in a single spot on nitrocellulose. There are cases in which organisms were cultured from lesions that were originally negative. Another possibility is that the delayed-type hypersensitivity test detected either a previous infection or a cross-reacting antigen from another type of infection (3). A third possibility is that the parasite in the lesion is of a type not recognized by the kDNA probe. However, in each case where parasites have been isolated from a lesion, these parasites have reacted with either the *L. braziliensis* or *L. mexicana* kDNA probes.

The DNA probe clearly detects infections in more patients than either histopathology or culturing. Although this high detection rate probably results from the greater sensitivity of the DNA probe compared to the other methods, it may also represent false positives in the DNA probe method. Because there is no single "gold standard" for the detection of parasites, the exact determination of false positives is not possible. However, in ten cases in which histopathology failed to detect parasites and the result of the DNA probe was positive, culturing of parasites clearly demonstrated their presence. The question of the false-positive reaction with DNA probes has been tested experimentally. These kDNA probes did not hybridize with touch preparations of uninfected tissue from several animals and from a limited number of human lesions that subsequently proved not to be due to leishmaniasis. In addition, these DNA probes did not react with *Trypanosoma cruzi,* malaria, or *Escherichia coli.* Further

Table 38.2. Comparison of diagnostic tests for leishmaniasis.

Diagnostic test	No. of patients
Montenegro-positive	43
Hybridization-positive	32
L. mexicana DNA probe	2
L. braziliensis DNA probe	30
Culture-positive	27
Histopathology	17

evidence that the number of false positives with the kDNA probes is relatively low was obtained by testing duplicate touch preparations of each lesion with kDNA from both *L. mexicana* and *L. braziliensis*. In every case in which there was a reaction with the kDNA probe, it was specific for either *L. mexicana* or *L. braziliensis,* and the duplicate lesion showed no reaction above background. When parasites were isolated from these lesions and subsequently tested by kDNA hybridization, the original identification was confirmed. An alternative method of limiting the number of false-positive reactions due to nonspecific binding of labeled DNA to tissue or blood is to perform in situ hybridization and examine each preparation under the microscope as suggested by Barker and co-workers (8). It is a time-consuming and expert process not easily adapted to large numbers of samples.

One of the limitations of the kDNA minicircle as a hybridization probe is that, although it can distinguish the major species complexes of New World cutaneous leishmaniasis, the kDNA minicircle from each complex is homologous to all the subspecies, and it is therefore impossible to distinguish subspecies (7,59). This identification of subspecies is important clinically because certain manifestations of the disease are specific to the parasite subspecies as well as being important in the description of any intermediate host or insect vector carrying a particular subspecies. In addition, our work and the work of others has indicated that in the Old World *Leishmania* species causing cutaneous and visceral disease kDNA sequence homology occurs among different species (1,3,6,30,53,59,60). For example, kDNA isolated from *L. major* hybridizes with both of the other cutaneous species (*L. tropica* and *L. aethiopica*) and with the visceral strain *(L. donovani)* (60).

Therefore a new approach is required for the development of DNA probes that can distinguish subspecies in New World cutaneous leishmaniasis and differentiate the species complexes of the Old World isolates. Several groups have used recombinant DNA methods to develop DNA probes with these narrower specificities (Table 38.3) (6,8,20,23,29,30,50,60). The method has been to clone restriction fragments of a kDNA minicircle and to use these cloned subfragments as more specific probes for species, subspecies, and even isolates. The general observation from this body of work is that within the minicircle population there are DNA sequences that have undergone rapid sequence divergence and can thus serve to differentiate even closely related organisms (29). In addition, we have shown that this sequence divergence can occur within a single minicircle. A nested set of deletions of a single cloned minicircle fragment from *L. mexicana amazonensis* was generated and then tested for hybridization specificity. The full-length minicircle had a specificity similar to that of total kDNA, and two deletions demonstrated species- and isolate-specific hybridization patterns (50).

The next step for the utilization of DNA probes will be in the detection

Table 38.3. Hybridization specificities of cloned kinetoplast DNA fragments.

Source	Specificities	Ref.
L. donovani	Species	29
L. infantum	Species	29
L. donovani	Visceral complex	30
L. chagasi	Non-Indian visceral	30
L. major	Isolate	60
L. mexicana amazonensis	Species	50
	Subspecies	50
	Isolate	50
L. major	Isolate	23
L. tropica	Isolate	23
L. aethiopica	Isolate	23

of infected insect vectors and intermediate hosts. Preliminary laboratory experiments show that parasites can be detected in infected sandflies that have been squashed directly on nitrocellulose (47). This approach must now be tested in the field. Similarly, laboratory-based experiments have demonstrated that parasite infections can be detected in tissue touch preparations from animals experimentally infected (59); however, both the intensity of infection and the target tissue in natural hosts will be different and so the DNA probes must be tested directly in field-extracted material.

The DNA probe methodology should be readily adaptable to field situations. Once the tissue biopsy or sandfly vector is obtained, it is applied directly to nitrocellulose or other solid supports, and it is stable in this form indefinitely. Thus samples could be collected from distant sites and returned for processing. The major disadvantage of this methodology is the requirement for a radioisotope. Alternative methods of labeling DNA, which are being developed and tested for such biological specimens, should be useful in field situations.

Use of Monoclonal Antibodies in Detecting Leishmaniasis

A second method for the identification of *Leishmania* species and subspecies that should also facilitate the collection of epidemiological data is the development of monoclonal antibodies specific for *Leishmania* species and subspecies (18,22,37,39). Cultured promastigotes were used in most of the reported work with *Leishmania*-specific monoclonal antibodies, but in principle an immunofluorescent antibody (IFA) test, analogous to those described for several viral and bacterial systems, could be developed that could be used directly on lesion material from either human patients or other mammalian hosts. Preliminary work on the use of IFA in the identification of promastigotes directly from laboratory-infected sandflies has been reported (38).

Malaria

Human malaria is caused by the four major *Plasmodium* species: *P. falciparum, P. vivax, P. malaria,* and *P. ovale.* In most parts of the world the prevailing parasite species is *P. falciparum,* which causes the most severe form of the acute disease and is often fatal in children (12). *Plasmodium vivax,* the next most prevalent malarial infection, is characterized by relapses caused by parasites that remain in the liver in a latent form. The parasite is transmitted by various species of the anopheline mosquito to the human host. The sporozoite, the infectious form of the parasite released from mosquito salivary glands, initiates the exoerythrocytic cycle in the liver. Subsequently developed merozoites invade erythrocytes, and the asexual cycle continues through the course of the infection. A subset of the infected erythrocytes develop into gametocytes; this form can develop in mosquitoes and results in disease transmission. There is no significant animal reservoir for this disease.

Malaria is among one of the major infectious diseases in the world, with acute clinical malaria affecting some 90 million to 100 million people per year according to the WHO estimates (61). In addition, there is a large reservoir of chronic infection. The WHO estimates that more than 40% of the world's population is at risk of malaria infection and that some 365 million people live in areas where malaria endemicity has remained unchanged despite intensive world efforts at malaria eradication (61).

Elimination of malaria worldwide has proved to be a difficult goal, and thus current efforts are devoted to its control (56). During the eradication program, several problems arose that will have an impact on any control program. Among these problems is the widespread resistance of anopheline vectors to insecticides (57), the emergence of *P. falciparum* strains resistant to chloroquine (the primary chemotherapeutic agent), and the subsequent development of multidrug-resistant parasite strains (58). New approaches to control measures include improved conventional methods of vector control and chemotherapy and the development of innovative measures including vaccines for the malaria parasite and biological control of anopheline vectors.

The parameters of malaria transmission and disease prevalence have been studied for several decades in many parts of the world. Mathematical models (10,35,35a,46) of disease transmission have been developed based on both entomological factors and human factors such as immunity (43). In addition, epidemiological studies have demonstrated an association of malaria prevalence with certain variants in erythrocytes, sickle cell trait (16,21,55), glucose-6-phosphate dehydrogenase deficiency (32,32a), Duffy blood group antigens (41,42), and α-thalassemia (17). These studies have shown that malaria is a complex and dynamic disease that is varied throughout the world. Thus any control program must take into consideration the multiple factors that can affect malaria transmission and must measure these factors in each situation. For this goal to be achieved, it is imperative that new and efficient means of measuring both entomological and human factors on a large scale be implemented.

Detection of Malaria Infection

Both vaccine and future drug trials will require a sensitive and rapid method for detecting parasites. Currently, malaria infection is determined by use of a thick smear stained with Giemsa. This method is both specific and sensitive for the diagnosis of malaria but has severe limitations when large numbers of samples must be handled in a timely fashion, as will be the case for the collection of baseline data for many of the vaccine trials and chemotherapy studies. A trained microscopist is required for each determination, a time-consuming, tiring task that is potentially subject to reader bias, especially when large numbers of slides must be read within a short time period. Thus alternative methods for handling large numbers of samples are necessary.

DNA probes specific for human malaria have been developed by several groups (19,36,49), and work by Barker et al (9) has demonstrated that the DNA probe specific for *P. falciparum* can be used to detect malaria infection directly in fingerstick blood of infected patients. The DNA probes specific for *P. falciparum* are dispersed, highly repeated DNA sequences isolated from the *P. falciparum* genome using recombinant DNA technology. For both laboratory and field testing the pPF14 probe (9) is specific for *P. falciparum* and does not react with *P. vivax,* the other major cause of human malaria. The methodology compares favorably in sensitivity with routine microscopy, detecting parasite densities as low as 40 parasites per microliter of blood. As can be seen in Table 38.4, the DNA probe method detects *P. falciparum* infection in 129 of 632 patients compared with the 121 detected by routine examination of Giemsa-stained thick smears. Subsequent examination of duplicate slides by expert microscopists confirmed the DNA probe diagnosis in those patients missed by routine microscopy. The DNA probe method offers the advantage of a standardized procedure that can be used in a batchwise fashion on large numbers of samples. An important feature of this methodology is that it is reproducible over a large number of samples and should be less subject to reader bias. In addition, we have been able to correlate the intensity of DNA hybridization with parasite density, and thus the DNA probes may also provide information on the intensity of infection. One limitation of the correlation of hybridization intensity with parasite density is the potential for variation in the number of repeated target sequences in dif-

Table 38.4. Comparison of DNA probes specific for *P. falciparum.*

Diagnostic test	No. of patients
Positive results	
Microscopy: Giemsa stain of thick blood smears	121
DNA probe method	129
Negative results	
Microscopy: Giemsa stain of thick blood smears	511
DNA probe method	503

The patient population consisted of 632 patients examined at malaria clinics of the Malaria Division of the Thailand Ministry of Public Health in either Bangnamron or Chantiburi, Thailand in July 1985. Blood was collected by digital puncture into a heparinized capillary and treated as previously described (9). Malaria thick smears were prepared in the routine manner for diagnosis at the clinic. A separate set of slides, both thin and thick smears, was prepared and subsequently analyzed by malaria experts (9).

ferent *P. falciparum* strains. The DNA probe methodology now must be tested in an epidemiological study to assess its general usefulness.

Methods for the detection of malaria-specific antigens or antibodies have also been developed and tested (4,33). A major problem with these assays has been the presence of both antigens and antibodies after the malaria parasites have disappeared from the bloodstream. Malaria-specific antibodies can persist for long periods, so it is useful to assay initially infected naive individuals for antibodies, especially young children (this test is complicated by the presence of maternal antibodies), but the assay cannot be used to determine present infection with the parasite in individuals previously infected (5). Extensive work with monoclonal antibodies has led to the identification of many malaria antigens, and a major area for research in the future is the testing of specific antigens or antibodies and their correlation with active infection or protective immunity.

Detection of Infective Mosquito Vectors

An important parameter in malaria transmission is the inoculation rate, which is the number of infective mosquito bites per unit time. This rate is based on both the man-biting rate of the vector species and the fraction of infective mosquitoes. Zavala et al. (13,62) have developed an immunological method for the determination of infective mosquitoes that uses a monoclonal antibody specific for the major protein of the malaria sporozoite, the circumsporozoite protein. This method has been field-tested and compared with the existing method, which is the capture and dissection of mosquitoes to determine infection. The advantage of this new methodology is that it can determine the species of sporozoite in the infected mosquito. Another advantage is that large numbers of mosquitoes can be tested easily. Thus previously overlooked vectors that have a low rate of infection are now being discovered, and their contribution to malaria transmission is being determined (2).

Determination of Human Genetic Parameters

Epidemiological studies have demonstrated an association between malaria prevalence and certain variants in erythrocytes. Advances in the detection of human genetic variants with the use of restriction site polymorphisms of specific DNA fragments should allow more extensive investigation of these genetically inherited diseases as well as new diseases that have not yet been associated with malaria prevalence. An elegant study by Flint et al (17) demonstrated the potential of this technology. A single variant of α-thalassemia has been correlated with malaria prevalence in Melanesia.

Conclusion

The application of new methods of biotechnology to the epidemiology of leishmaniasis and malaria is in its initial phases. The new tools offer distinct advantages with regard to specificity, sensitivity, and ease of use for large numbers of samples compared to existing methodologies, and they have enormous potential for their contribution to new knowledge on the transmission and prevalence of these diseases. Before these methods are generally accepted for use, they must be extensively tested under field situations and modified to provide the relevant information important for epidemiological analysis.

Acknowledgments. We thank Ramona Gonski for her careful preparation of the manuscript and Carolyn McDowell for her proofreading. Work from the authors' laboratory received support from the UNDP/World Bank/ WHO Special Programme for Research and Training in Tropical Diseases, NIH (AI 21365, AI 19392), and The John D. and Catherine T. MacArthur Foundation. D.F.W. is a Burroughs-Wellcome Scholar in Molecular Parasitology.

References

1. Arnot DF, Barker DC: *Mol Biochem Parasitol* 1981;3:47.
2. Arrada M, et al: *Am J Trop Med Hyg* (in press).
3. Aston DC, Thornley AP: *Trans R Soc Trop Med Hyg* 1970;75:537.
4. Avraham H, et al: *J Immunol Methods* 1982;53:61.
5. Avraham H, et al: *Am J Trop Med Hyg* 1983;32:11.
6. Barker DC, Arnot DF: *Eur J Cell Biol* 1980;22:124.
7. Barker DC, Butcher J: *Trans R Soc Trop Med Hyg* 1983;77:285.
8. Barker DC, et al: *Parasitology* 1985;91:S139.

9. Barker RB, et al: *Science* 1986;231:1434.
10. Bruce-Chwatt JA: *Trop Geogr Med* 1976;28:1–8.
11. Chance ML, Walton BC (eds): *Biochemical Characterization of Leishmania.* Geneva, UNDP/World Bank/WHO, 1982.
12. Cohen S, Lambert PH: In Cohen S, Warren KS (eds): *Immunology of Parasitic Infections.* Oxford, Blackwell, 1982, p 422.
13. Collins FH, et al: *Am J Trop Med Hyg* 1982;33:538.
14. Deane LM, Grimaldi G: In Chang, Bray (eds): *Leishmaniasis.* Amsterdam, Elsevier, 1985.
15. Englund PT: *J Biol Chem* 1979;254:4895.
16. Fleming AF: *Ann Trop Med Parasitol* 1979;73:161.
17. Flint J, et al: *Nature* 1986;321:744.
18. Frankenberg, et al: *Am J Trop Med Hyg* 1985;34:266.
19. Franzen, et al: *Lancet* 1984;1:525.
20. Frasch ACC, et al: *Mol Biochem Parasitol* 1985;4:163.
21. Haldane JBS: *Hereditas* 1949;35(suppl):267–273.
21a. Jackson PR, et al: *Am J Trop Med Hyg* 1984;33:808.
22. Jaffe CC, et al: *J Immunol* 1984;133:440
23. Kennedy WPK: *Mol Biochem Parasitol* 1984;12:313.
24. Kidane GZ, Hughes D, Simpson L: *Gene* 1984;27:265.
25. Lainson R: In *Proceedings of the 3rd Venezuelan Congress of Microbiology and Symposium on Leishmaniasis.* Bouquisimeto, Venezuela, 1983.
26. Lainson R: *Trans R Soc Trop Med Hyg* 1983;77:569.
27. Lainson R, Shaw J: In Lunden WHR, Evans (eds): *Biology of Kinetoplastidae 2.* London, Academic Press, 1979, pp 1–116.
28. Lainson K, Shaw JJ: *Nature* 1978;273:595.
29. Lawrie JM, et al: *Am J Trop Med Hyg* 1985;34:257.
30. Lopes UG, Wirth DF: *Mol Biochem Parasitol* 1986;20:77–84.
31. Lopes UG, et al: *Parasitology* 1984;70:89.
32. Luzzato L, et al: *Bull WHO* 1974;50:195.
32a. Luzzato L: *Blood* 1979;54:961.
33. Mackey L, et al: *Bull WHO* 1980;60:69.
34. Marinkelle CJ: *Bull WHO* 1980;58:807.
35. McDonald G: *Trop Dis Bull* 1952;49:813.
35a. McDonald G: *Proc R Soc Trop Med* 1955;48:295.
36. McLaughlin E, et al: *Am J Trop Med Hyg* 1985;34:837.
37. McMahon-Pratt D, David J: *Nature* 1981;291:581
38. McMahon-Pratt D, et al: *Am J Trop Med Hyg* 1983;32:1268.
39. McMahon-Pratt D, et al: *J Immunol* 1985;134:1935.
40. Miles MA, et al: *Trans R Soc Trop Med Hyg* 1980;74:243.
41. Miller LH: et al: *N Engl J Med* 1976;295:302.
42. Miller LH, et al: *Am J Trop Med Hyg* 1978;27:1069.
43. Molineaux L, Gramiccia G: *The Garki Project.* Geneva, World Health Organization, 1980, pp 109–115.
44. Momen H, Grimaldi G: *Trans R Soc Trop Med Hyg* 1984;78:701.
45. Montenegro: *An Fac Med Univ Sao Paulo* 1926;1:323.
46. Najera JA: *Bull WHO* 1974;50:449–457.
47. Perkins P, Wirth DF: Unpublished observations.
48. Peters W, et al: *J R Soc Med* 1983;76:540.

49. Pollack Y, et al: *Am J Trop Med Hyg* 34:663.
50. Rogers WO, Wirth DF: *Proc Natl Acad Sci USA* 1987;84:565–569.
51. Rogers WO, et al: In preparation.
52. Scorza JU, et al: *Trans R Soc Trop Med Hyg* 1979;73:293.
53. Spithill TW, et al: *J Cell Biochem* 1984;24:103.
54. Steinert M, Van Assel S: *Plasmid* 1980;3:7.
55. Walker JH, Bruce-Chwatt LJ: *Trans R Soc Trop Med Hyg* 1956;50:511.
56. *WHO Tech Rep Ser* No. 640, 1979.
57. *WHO Tech Rep Ser* No. 655, 1980.
58. *WHO Tech Rep Ser* No. 711, 1984.
59. Wirth DF, McMahon-Pratt D: *Proc Natl Acad Sci USA* 1982;79:6999.
60. Wirth DF, Rogers WO: In Kingsbury D, Falkow S (eds): *Rapid Detection and Identification of Infectious Agents*. New York, Academic Press, 1985.
61. *World Health Stat Q* 1984;37:130–161.
62. Zavala F, et al: *Nature* 1982;299:737.

Part X
New Approaches to Vaccinology

CHAPTER 39

Vaccinia Virus Expression Vectors

Bernard Moss

Infectious diseases remain a major problem throughout the world. It is clear, at least for viruses, that prevention is our best defense. Most successful vaccines (e.g., smallpox, rubella, Sabin poliomyelitis, measles, mumps, yellow fever) have consisted of live attenuated viruses. In general, they provide long-lasting immunity, high potency, and economy of manufacture and delivery. A few vaccines (e.g., Salk poliomyelitis, influenza, rabies) are killed viruses. One vaccine (hepatitis B) consists of a purified subunit protein. Developments in recombinant DNA technology have opened new possibilities for vaccine production. Genetic engineering can be used to produce subunit proteins, attenuate viruses by gene deletion or modification, or construct live recombinant vectors. As a potential vector, vaccinia virus has impressive credentials, as it was responsible for the eradication of smallpox. Studies suggest that vaccinia virus can be used for expression of genes from other microorganisms and that such recombinants retain infectivity, induce humoral and cell-mediated immunity, and provide significant disease protection to experimental animals.

Formation of Recombinant Viruses

Poxviruses, of which vaccinia virus is the best characterized, have several distinctive features that directly affect their use as vectors. They include a large DNA genome, cytoplasmic site of replication, and a unique virus encoded and packaged transcription system. Foreign DNA has been inserted into vaccinia virus by homologous recombination (17,23). Plasmid vectors that facilitate the cloning and expression of foreign genes as well as the selection of recombinant viruses have been constructed (18). These plasmids contain several important elements: a cassette (composed of a vaccinia virus promoter and restriction endonuclease sites for insertion of foreign genes) that is flanked by vaccinia DNA. Recombination of the foreign gene is directed by the flanking vaccinia DNA. If infectivity is to be retained, the foreign gene must be inserted into a nonessential locus.

The thymidine kinase (TK) gene is a particularly suitable locus because insertional inactivation provides a powerful selection technique; TK^- recombinant viruses can form plaques in TK^- cells in the presence of 5-bromodeoxyuridine, whereas TK^+ parental vaccinia virus cannot. Addition of the *Escherichiacoli* β-galactosidase gene to the plasmid vector, in such a way that it will be integrated into the vaccinia genome along with the desired foreign gene, provides an additional visual way of screening recombinant virus plaques, which turn blue when an appropriate indicator dye is added to an agar overlay (6). The level of foreign gene expression depends on the intrinsic strength of the vaccinia promoter chosen, as well as suitable engineering to maximize translation. There may also be some influence of the surrounding vaccinia DNA at the site of insertion. Because vaccinia virus RNAs are not spliced and transcription of the vaccinia virus genome occurs in the cytoplasm, only continuous open reading frames can be properly expressed by vaccinia virus vectors. The amount of vaccinia virus DNA that can be inserted, however, is large, exceeding 25,000 basepairs (28), and several genes can be expressed from one vector (26).

Expression of Foreign Proteins

Apparently any continuous open reading frame, whether derived from prokaryotic or eukaryotic sources, can be expressed by vaccinia virus. Moreover, certain difficulties that can arise with a nuclear transcription system (e.g., the presence of cryptic splice sites) are not a problem with vaccinia virus. When the expressed gene is derived from a mammalian or mammalian virus source, correct posttranslational modifications (e.g., proteolytic processing, glycosylation, transport) occur normally. About 1 to 3 mg of desired protein per liter of infected cells has been obtained. The development of a new bacteriophage T7/vaccinia virus hybrid system may result in greatly increased yields in the near future (12). In this system the highly specific and efficient RNA polymerase gene of bacteriophage T7 has been inserted into the genome of vaccinia virus; this polymerase then transcribes genes that have T7 regulatory signals added.

Stimulation of Humoral and Cell-Mediated Immunity

It is important to determine the significant microbial targets of humoral and cell-mediated immunity in order to use recombinant DNA technology for production of vaccines. Vaccinia virus vectors provide a powerful tool for such research. When an animal is inoculated with vaccinia virus, it mounts a strong antibody and cytotoxic T cell response to vaccinia antigens. When a recombinant vaccinia virus is used, the immune response is directed toward the foreign antigen as well. By expressing individual genes from another virus or microorganism in vaccinia virus, the relative importance of these antigens can be measured in several ways. The serum can be checked for neutralizing antibody, the lymphocytes can be assayed for their ability to proliferate in response to the foreign antigen and/or to lyse cells expressing the foreign antigen, and the animals can be challenged with the infectious agent. Results obtained with several systems are briefly reviewed.

Immunity to Specific Viruses

Hepatitis B Virus

The first report (29A) of an immune response to a recombinant vaccinia virus was obtained with a vector that contained the hepatitis B virus surface antigen (HBsAg). The protein was assembled into characteristic 22-nm spherical particles and secreted from recombinant virus infected cells. Rabbits inoculated with the live recombinant virus produced high antibody titers. At present there is no in vitro system to determine neutralizing antibody. Chimpanzees inoculated with the same recombinant vaccinia virus had no or barely detectable antibody but nevertheless were protected against an intravenous challenge with hepatitis B virus (21). The data indicated that the vaccination provided a priming immune response, allowing the animals to rapidly mount a protective anti-HBsAg anamnestic response. Further work is needed to enhance the immune response prior to testing in humans. One approach, thus far tested only in small animals, is to incorporate both pre-S and S epitopes into vaccinia virus (8).

Herpes Viruses

Herpes simplex virus encodes a large number of envelope glycoproteins. An effective vaccine will probably require an immune response to several of these glycoproteins. Recombinant vaccinia viruses that express only the glycoprotein D, however, can protect mice against both lethal (25) and latent (9) infections. Neutralizing antibody to Epstein-Barr virus has been produced by immunization with a recombinant vaccinia virus that expresses the gp340 protein (16).

Influenza Virus

Recombinant vaccinia viruses that express influenza antigens have been particularly useful for determining the targets of cytotoxic T cells in both mice (1,2,3,33) and man (20). These experiments demonstrated that the internal virus proteins, the nucleoprotein in particular, are the major cross-reactive targets of cytotoxic T lymphocytes (CTL). Nevertheless, animals inoculated with the recombinant vaccinia virus expressing the hemagglutinin provides high neutralizing antibody and much better protection against respiratory influenza infection (1,24,29B). These results demonstrate the prime importance of type-specific neutralizing antibody against this disease. In addition, they demonstrate that a peripheral (e.g., skin) vaccination can protect against lower respiratory infection. To protect against upper respiratory influenza infection, however, it was necessary to administer the recombinant vaccinia virus intranasally (27).

Respiratory Syncytial Virus

At present there is no effective vaccine against respiratory syncytial virus (RSV), a major cause of lower respiratory infection in infants. Two glycoproteins, G and F, are located on the surface of RSV. In order to identify which glycoprotein was the major target of neutralizing antibody, separate recombinant vaccinia viruses containing the gene for F or G were constructed. Both recombinants induced the production of neutralizing antibody in cotton rats, but better protection was achieved with glycoprotein F (11,22). Recombinant vaccinia virus expressing glycoprotein G was also shown to protect mice against RSV (30). Probably both proteins should be produced for optimal protection. In view of the need for an RSV vaccine to be given to infants, there is great concern regarding safety.

Rhabdoviruses

Vesicular stomatitis virus (VSV) causes a disease of some veterinary importance. Recombinant vaccinia viruses that express the envelope glycoprotein G and the internal protein N have been constructed (19). The G recombinant has been shown to induce neutralizing antibody and to protectively immunize cattle against an intralingual challenge with VSV. Interestingly, the N protein is a better cytotoxic T cell target (32) but does not provide protective immunity.

Rabies is caused by another member of the rhabdovirus family. Recombinant vaccinia viruses that express the rabies G protein induce high levels of neutralizing antibody in animals (14,31). The ability of oral administration of such a vaccine to protect foxes has raised the possibility of its use as a wild-life vaccine (4). Studies to determine the environmental impact of such use is needed.

Retroviruses

Vaccinia virus vectors are being used to learn more about the immune response to retroviruses. A recombinant vaccinia virus that expresses the envelope gene of Friend murine leukemia virus (FMuLV) was able to protect mice against virus-induced leukemia (10). After vaccination, however, neither neutralizing antibody nor cytotoxic T lymphocytes were detected; the only indication of an immune response was an in vitro T lymphoproliferative response to FMuLV antigen. Following FMuLV challenge, neutralizing antibody and cytotoxic T lymphocytes rapidly appeared, indicating that the animals were primed.

Recombinant vaccinia viruses that express the *env* gene of human immunodeficiency virus have been constructed (7,13) and used to study gp120-induced fusion of T4 lymphocytes (15). Monkeys vaccinated with the recombinant virus produced antibody that neutralized human immunodeficiency virus (HIV) in vitro (Moss, unpublished). It is necessary to immunize chimpanzees, as they are the susceptible animal other than man.

Further Work

Efforts are currently being made to improve expression of foreign genes by vaccinia virus vectors and to simplify and accelerate the construction of recombinants. The safety of live recombinant viruses is of particular concern. Studies indicate that it is possible to further attenuate vaccinia virus by inactivating or deleting certain genes, including thymidine kinase and a group with unknown function located near the left end of the genome (5). The prospects for constructing effective and safe vector-based vaccines are excellent.

References

1. Andrew ME, Coupar, BEH, et al: Cell mediated immune response to influenza virus antigens expressed by vaccinia virus recombinants. *Microbiol Pathol* 1986;1:443–452.
2. Bennink JR, Yewdell JW, Smith GL, et al: Recombinant vaccinia virus primes and stimulates influenza virus HA-specific CTL. *Nature* 1984;311:578–579.
3. Bennink JR, Yewdell JW, Smith GL, Moss B: Recognition of cloned influenza virus hemagglutinin gene products by cytotoxic T lymphocytes. *J Virol* 1985;57:786–791.
4. Blancou J, Kieny MP, Lathe R, et al: Oral vaccination of the fox against rabies using a live recombinant vaccinia virus. *Nature* 1986;322:373–375.
5. Buller RML, Smith GL, Cremer K, et al: Decreased virulence of recombinant vaccinia virus expression vectors is associated with a thymidine kinase negative phenotype. *Nature* 1985;317:813–815.
6. Chakrabarti S, Brechling K, Moss B: Vaccinia virus expression vector: coexpression of β-galactosidase provides visual screening of recombinant virus plaques. *Mol Cell Biol* 1985;5:3403–3409.
7. Chakrabarti S, Robert-Guroff M, Wong-Stall F, et al: Expression of the HTLV-III envelope gene by a recombinant vaccinia virus. *Nature* 1986;320:535–537.

8. Cheng K-C, Smith GL, Moss B: Hepatitis B large surface protein is not secreted but is immunogenic when selectively expressed by recombinant vacccinia virus. *J Virol* 1986;60:337–344.

9. Cremer KJ, Mackett M, Wohlenberg C, et al: Vaccinia virus recombinant expressing herpes simplex virus type 1 glycoprotein D prevents latent herpes in mice. *Science* 1985;228:737–740.

10. Earl PL, Moss B, Wehrly K, et al: T cell priming and protection against Friend murine leukemia by a recombinant vaccinia virus expressing env gene. *Science* 1986;234:728–731.

11. Elango N, Prince GA, Murphy BR, et al: Resistance to human respiratory syncytial virus (RSV) infection induced by immunization of cotton rats with a recombinant vaccinia virus expressing the RSV G glycoprotein. *Proc Natl Acad Sci USA* 1986;83:1906–1910.

12. Fuerst TR, Niles EG, Studier FW, Moss B: Eukaryotic transient expression system based on recombinant vaccinia virus that synthesizes bacteriophage T7 RNA polymerase. *Proc Natl Acad Sci USA* 1986;83:8122–8126.

13. Hu S, Kosowski SG, Dalyrimple JM: Expression of AIDS virus envelope gene in recombinant vaccinia viruses. *Nature* 1986;320:537–540.

14. Kieny MP, Lathe R, Drillien R, et al: Expression of rabies virus glycoprotein from a recombinant vaccinia virus. *Nature* 1984;312:163–166.

15. Lifson JD, Feinberg MB, Reyes GR, et al: Induction of CD4-dependent cell fusion by the HTLV-III/LAV envelope glycoprotein. *Nature* 1986;323:725–728.

16. Mackett M, Arrand JR: Recombinant vaccinia virus induces neutralizing antibodies in rabbits against Epstein-Barr virus membrane antigen gp340. *EMBO J* 1985;4:3229–3234.

17. Mackett M, Smith GL, Moss B: Vaccinia virus: a selectable eukaryotic cloning and expression vector. *Proc Natl Acad Sci USA* 1982;79:7415–7419.

18. Mackett M, Smith GL, Moss B: A general method for the production and selection of infectious vaccinia virus recombinants expressing foreign genes. *J Virol* 1984;49:857–864.

19. Mackett M, Yilma T, Rose JK, Moss B: Vaccinia virus recombinants: expression of VSV genes and protective immunization of mice and cattle. *Science* 1985;227:433–435.

20. McMichael AJ, Michie CA, Gotch FM, et al: Recognition of influenza A virus nucleoprotein by cytotoxic T lymphocytes. *J Gen Virol* 1986;67:719–726.

21. Moss B, Smith GL, Gerin JL, Purcell R: Live recombinant vaccinia virus protects chimpanzees against hepatitis B. *Nature* 1984;311:578–579.

22. Olmsted RA, Elango, N, Prince GA, et al: Expression of the F glycoprotein of respiratory syncytial virus by a recombininat vaccinia virus: comparison of the individual contributions of the F and G glycoproteins to host immunity. *Proc Natl Acad Sci USA* 1986;83:7462–7466.

23. Panicali D, Paoletti E: Construction of poxviruses as cloning vectors: insertion of the thymidine kinase gene from herpes simplex virus into the DNA of infectious vaccinia virus. *Proc Natl Acad Sci USA* 1982;79:4927–4931.

24. Panicali D, Davis SW, Weinberg RL, Paoletti E: Construction of live vaccines by using genetically engineered poxviruses: biological activity of recombinant vaccinia virus expressing influenza virus hemagglutinin. *Proc Natl Acad Sci USA* 1983;80:5364–5368.

25. Paoletti E, Lipinskas BR, Samsonoff C, et al: Construction of live vaccines using genetically engineered poxviruses: biological activity of vaccinia virus recombinants expressing the hepatitis B virus surface antigen and the herpes simplex virus glycoprotein D. *Proc Natl Acad Sci USA* 81:193–197.

26. Perkus ME, Piccini A, Lipinskas BR, Paoletti E: Recombinant vaccinia virus: immunization against multiple pathogens. *Science* 1985;229:981–984.

27. Small PA, Smith GL, Moss B: Intranasal vaccination with a recombinant vaccinia virus containing influenza hemagglutinin prevents both influenza virus pneumonia and nasal infection: intradermal vaccination prevents only viral pneumonia, in Lerner RA, Chanock RM, Brown F (eds): *Vaccines 85. Molecular and Chemical Basis of Resistance to Parasitic, Bacterial and Viral Diseases.* Cold Spring Harbor, New York, Cold Spring Harbor Laboratory, 1985, pp 175–176.

28. Smith GL, Moss B: Infectious poxvirus vectors have capacity for at least 25,000 base pairs of foreign DNA. *Gene* 1983;25:21–28.

29a. Smith, G.L., Mackett, M., and Moss, B. Infectious vaccinia virus recombinants that express hepatitis B virus surface antigen. *Nature* 302:490–495, 1983.

29b. Smith GL, Murphy BR, Moss B: Construction and characterization of an infectious vaccinia virus recombinant that expresses the influenza hemagglutinin gene and induces resistance to influenza virus infection in hamsters. *Proc Natl Acad Sci USA* 1983;80:7155–7159.

30. Stott EJ, Ball LA, Young KK, et al: Human respiratory syncytial virus glycoprotein G expressed from a recombinant vaccina virus vector protects mice against live-virus challenge. *J Virol* 1986;60:607–613.

31. Wiktor TJ, MacFarlan RI, Reagan KJ, et al: Protection from rabies by a vaccina virus recombinant containing the rabies virus glycoprotein. *Proc Natl Acad Sci USA* 1984;81:7194–7198.

32. Yewdell JW, Bennink JR, Mackett M, et al: Recognition of cloned vesicular stomatitis virus internal and external gene products by cytotoxic T lymphocytes. *J Exp Med* 1986;163:1529–1539.

33. Yewdell JW, Bennink JR, Smith GL, Moss B: Influenza A virus nucleoprotein is a major target antigen for cross-reactive antiinfluenza A virus cytotoxic T lymphocytes. *Proc Natl Acad Sci USA* 1985;82:1785–1789.

rather than antiviral immunity. In the absence of an appropriate animal model for the induction of tumors by HPV, we turned instead to the distantly related papova virus polyoma (PY). PY induces a variety of tumors in rodents, and PY-transformed cells express a novel TSA from the early region of the viral genome. However, establishing the precise relation between TSA and protein species synthesized from the early region of the PY genome has been complicated by the existence of three distinct early polypeptides, referred to as large-T (LT), middle-T (MT), and small-T (ST) in reference to their molecular size. We therefore elected to separately express the three T coding sequences in independent live recombinant viruses.

DNA sequences separately containing the three intact coding sequences have previously been constructed by intron excision in vitro (16). We introduced these coding sequences downstream of a vaccinia promoter and separately transferred the three coding sequences to the VV genome, generating VVpyLT, VVpyMT, and VVpyST (10).

Subcellular distribution of the three proteins was examined by immunofluorescence staining of infected tissue culture cells. The LT protein was observed exclusively within the nucleus, whereas the ST protein was cytoplasmic. MT protein appeared to form distinct aggregates in the cytoplasm, and additional perinuclear fluorescence suggested an association with the Golgi apparatus or other intracellular membranes, as other studies have indicated (3). Because no cell-surface fluorescence was observed we proceeded in parallel with the three recombinant viruses to examine their potential for eliciting tumor immunity.

Rats inoculated with PY-transformed syngeneic rat cells rapidly develop tumors. We sought to determine if prior administration of the vaccinia/polyoma recombinants might block the development of transplanted tumor cells. Groups of rats were inoculated subcutaneously or intradermally with recombinant virus, boosted, and seeded (subcutaneously) with syngeneic PY-transformed cells. All unvaccinated animals developed localized tumors, and no spontaneous regression was observed for the duration of the experiments.

In contrast, 50 to 60% of animals vaccinated subcutaneously with either VVpyLT or VVpyMT developed small tumors that rapidly regressed and were eliminated. The route of vaccination proved important, and intradermal rather than subcutaneous inoculation significantly improved the rejection frequency (90% for the VVpyMT vaccine). Animals vaccinated with VVpyST developed tumors that were maintained throughout the course of the experiment.

We next examined the possibility that animals already bearing tumors might be induced to reject their tumor cells by vaccination with the appropriate recombinant. Groups of rats were inoculated with PY-transformed cells as before and allowed to develop tumors. When all animals presented tumors 2 to 4 mm in diameter, groups were separately inoculated

CHAPTER 40

Vaccinia Recombinants Expressing Foreign Antigens: Antiviral and Antitumor Immunity

M.P. Kieny and R. Lathe

Virus infection elicits a variety of symptoms. In man infection can provoke immunosuppression [e.g., human immunodeficiency virus (HIV)], behavioral changes (e.g., rabies), or neoplastic transformation [e.g., human lymphotrophic virus type I (HTLVI), papilloma viruses]; and these changes are often linked with enhanced transmission and high morbidity. In many cases no appropriate vaccines yet exist against these pathogens, and vaccine development remains of the highest priority. We have adopted the live recombinant virus approach to the elaboration of vaccines against viral diseases, and we discuss here the efficacy of such vaccines for eliciting antivirus and antitumor immunity.

Vaccinia Virus

Vaccinia virus (VV), extensively used for the control and eradication of smallpox, is a large (180 kilobases) DNA virus that replicates in the cytoplasm of infected cells. In consequence of its cytoplasmic replication, VV is unable to take advantage of host nuclear components and, instead, carries in the virion those enzymes necessary for the onset of viral gene expression. For this reason VV DNA is noninfective. Hence cloning of foreign gene segments into VV has relied on the transfer of material to the VV genome by recombinational exchange in vivo (12,13). A number of groups have exploited this transfer procedure to develop VV as a vector for the expression of virus or parasite antigens, and inoculation with live recombinant viruses is more often than not effective in conferring protective immunity (for review, see ref.17). We hence adopted VV as an appropriate vector for the elaboration of new recombinant vaccines.

Antivirus Immunity: Rabies

Although a vaccine against rabies has been available for more than a century, the disease remains a problem. In developed countries rabies is predominantly a disease of wild animals, whereas in South America it is responsible for considerable loss of cattle. In addition, thousands of human rabies deaths occur every year in Asia, where infection of man occurs primarily through the bites of stray dogs.

Vaccines based on rabies virus grown on nervous tissue have proved effective but can be responsible for severe side effects. New vaccines, based on rabies virus grown in tissue culture, elicit protective immunity in man and animals, but their use is restricted because of the high cost and technical problems associated with the establishment of high-technology production plants in the developing world.

Rabies control could be achieved by mass vaccination of wild or stray animals, but for this purpose only oral administration of vaccine (in bait) is practicable. It is of note that all traditional vaccines based on inactivated rabies virus are unable to induce protection against the disease when administered orally. Attenuated rabies virus has been used for the oral vaccination of wild foxes in Switzerland, Germany, and Canada; and field trials have met with success in eradicating rabies from localized areas. Nevertheless, only a few animal species can be immunized with this vaccine, as dogs, skunks, and raccoons are to some extent refractory to it. In addition, the use of live attenuated rabies virus vaccine bears the attendant risk of reversion to virulence. There is hence a need for a rabies vaccine able to satisfy the following requirements: It should be inexpensive, effective for all target animals even when administered orally, stable, and safe. For this reason we set out to develop a recombinant vaccinia virus expressing the major rabies antigen.

Rabies virus, a typical rhabdovirus related to vesicular stomatitis virus, possesses a negative strand RNA genome encoding five polypeptides. One of these proteins, the glycoprotein (G), is the major target antigen of rabies virus and is the major protein capable of eliciting production of neutralizing antibodies or of conferring protection against the disease (reviewed in ref. 8). The G protein was thus chosen for expression in a recombinant vaccinia virus.

The cDNA of rabies glycoprotein (ERA strain) encodes a mature polypeptide of 505 amino acids. This cDNA was modified in vitro to remove structural alterations likely to prejudice expression of the antigen, and in vivo recombination was used to transfer the G cDNA, under the control of the Vaccinia Virus 7.5k promoter, to the VV genome, yielding recombinant VVTGgRAB (6,9). The recombinant G protein synthesized in VVTGgRAB-infected cells was recognized by both polyclonal and monoclonal antibodies raised against the natural G protein of rabies virus.

Upon inoculation into mice and rabbits, VVTGgRAB elicited production of a high level of rabies neutralizing antibodies and protected these animals against severe challenge with street rabies virus. In addition to the parenteral routes of administration it was shown that oral administration of the recombinant virus to laboratory animals also induced resistance to rabies challenge (20). This result prompted us to investigate if wild target animals could be vaccinated orally with the recombinant.

Young adult foxes were vaccinated either parenterally or orally with VVTGgRAB. Inoculation or ingestion of the live recombinant did not provoke any adverse reaction in the vaccinated animals. All routes of administration induced the production of neutralizing antibodies, and the recombinant virus was able to protect foxes against rabies challenge (2). We have also demonstrated that the minimum oral dose of recombinant virus required to protect 50% of the animals was $10^{5.6}$ pfu of virus.

To test the feasibility of a vaccine that could be delivered to animals in the wild we introduced the recombinant virus into baits. The latter consisted of a plastic capsule containing VVTGgRAB inserted into the beaks of chicken heads. Animals presented with this vaccine also raised a strong immune response and were protected against challenge with rabies virus (2).

Similar experiments on raccoons have been conducted with VVTGgRAB (21), and effective immunization was demonstrated. The recombinant vaccinia-rabies virus has thus proved safe and effective for all animals tested to date and may be considered as the prototype for a new vaccine against rabies for wild and domestic animals.

Antitumor Immunity: Papova Viruses

Rabies appears to be unique in that infection can be successfully treated by intensive vaccination. The protective effect in these cases has been attributed to the rapid elaboration of a cellular immune response (see ref 8 for review). The same mechanism appears to be involved in the rejection of tumors (4), and we examined the possibility that live recombinant viruses expressing tumor antigen might elicit antitumor immunity rather than antivirus immunity.

In man several rather innocuous viruses may elicit the formation of tumors. Neoplastic transformation can be responsible for high morbidity even when virus proliferation is successfully controlled by the immune system. Particularly striking examples are the link between hepatitis B virus and primary liver cancer or between human papilloma virus (HPV) and cervical cancer (15). The increased frequency of such tumors [e.g., 0.4 to 3.9% of women throughout the world develop cervical cancer (7)] makes the development of an antitumor vaccine of the highest priority.

Tumor cells often present novel tumor-specific antigens (TSA) that potentially present targets for the immune system. However, in many cases the TSA species are only poorly characterized, and immunization with antiidiotypic antibodies presenting the internal image of TSA has been attempted (5,11). Such methods are limited by the quantity of antigen material available and because an effective cellular immune response demands co-presentation of antigen with host histocompatibility determinants. To avoid these limitations we explored the possibility that TSA expressed from a live recombinant vaccinia virus might elicit antitumor with the vaccinia recombinants and boosted 4 days later. In all animals tumor growth continued to a diameter exceeding 15 mm. Significant tumor regression was observed only with VVpyMT (intradermal), and in 50% of vaccinated animals complete tumor elimination was observed.

It is of note that none of the early PY proteins appears to possess signals for cell surface presentation, although the primary structure of MT reveals a C-terminal stretch of hydrophobic amino acids that could reflect a membrane-spanning domain (18). However, it is now becoming clear that de facto "internal" proteins can be processed for presentation at the cell surface (1,19,22).

Discussion

Vaccinia virus is a potent vehicle for the expression of viral antigens. We have demonstrated that vaccinia recombinants expressing the rabies glycoprotein can elicit protective immunity in laboratory and wild animals, even when presented in bait as an oral vaccine. This vaccine has great potential for the wide-scale control of rabies.

In addition, we have shown that administration of live recombinant vaccinia virus bearing tumor-specific antigen can efficiently prevent the development of transplanted tumor cells, and tumor-bearing animals could be induced to reject their tumors by vaccination. Appropriate live viral recombinants may eventually be used in the management of human tumors presenting novel antigens.

It is of note that vaccinia recombinants may be of utility in the control of human AIDS (7), although the use of recombinant vaccinia viruses, in either animals or man, has not yet been approved.

Acknowledgments. We thank P. Chambon and J.P. Lecocq for their continued interest in this work.

References

1. Bennink JR, Yewdell JW, Smith GL, et al: Recombinant vaccinia primes and stimulates influenza hemagglutinin-specific cytotoxic T-cells. *Nature* 1984;296:75–76.
2. Blancou J, Kieny MP, Lathe R, et al: Oral vaccination of the fox against rabies using a live recombinant vaccinia virus. *Nature* 1986;322:373–375.
3. Dilworth SM, Hanson HA, Darnfors C, et al: Subcellular localization of the middle and large T-antigens of polyoma virus. *EMBO J* 1986;5:491–499.
4. Heberman RB: Cell-mediated immunity to tumor cells. *Adv Cancer Res* 1974;19:207–263.
5. Herlyn D, Ross AH, Koprowski H: Anti-idiotypic antibodies bear the internal image of a human tumor antigen. *Science* 1986;232:100–102.

6. Kieny MP, Lathe R, Drillien R, et al: Expression of rabies glycoprotein from a recombinant vaccinia virus. *Nature* 1984;213:163–166.

7. Kieny MP, Rautmann G, Schmitt D, et al: AIDS virus env protein expressed from a recombinant vaccinia virus. *Biotechnology* 1986;4:790–795.

8. Kieny MP, Desmettre P, Soulebot JP, Lathe R: Rabies vaccine: traditional and novel approaches. *Prog Vet Microbiol Immunol* 1987;3:73–111.

9. Lathe R, Kieny MP, Lecocq JP, et al: Immunization against rabies using a vaccinia-rabies recombinant virus expressing the surface glycoprotein, in Lerner RA, Chanock RM, Brown F (eds): *Vaccines 85*. Cold Spring Harbor, New York, Cold Spring Harbor Laboratory, 1985, pp 157–162.

10. Lathe R, Kieny MP, Gerlinger P, et al: Polyoma tumor antigens expressed from vaccinia virus: a model for tumor immunity, in Lerner RA, Chanock RM, Brown F (eds): *Vaccines 87*. Cold Spring Harbor, New York, Cold Spring Harbor Laboratory, 1987; pp. 242–249.

11. Lee VK, Harriott TG, Kuchroo VK, et al: Monoclonal anti-idiotypic antibodies related to murine oncofetal bladder tumor antigen induce specific cell-mediated tumor immunity. *Proc Natl Acad Sci USA* 1985;82:6286–6290.

12. Mackett M, Smith GL, Moss B: Vaccinia virus: a selectable eukaryotic cloning and expression vector. *Proc Natl Acad Sci USA* 1982;79:7415–7419.

13. Panicali D, Paoletti E: Construction of poxviruses as cloning vectors: insertion of the thymidine kinase gene from herpes simplex virus into the DNA of infectious vaccinia virus. *Proc Natl Acad Sci USA* 1982;79:4927–4931.

14. Peto R: Introduction: geographic patterns and trends. *Banbury Rep* 1986;21:3–15.

15. Peto R, Zur Hausen H (eds): *Viral Etiology of Cervical Cancer*. Banbury Report 21. Cold Spring Harbor, New York, Cold Spring Harbor Laboratory, 1986.

16. Rassoulzadegan M, Cowie A, Carr A, et al: The role of individual polyoma virus early proteins in oncogenic transformation. *Nature* 1982;300:713–718.

17. Smith GL, Mackett M, Moss B: Recombinant vaccinia viruses as new live vaccines. *Biotechnol Genet Eng Rev* 1984;2:383–407.

18. Tooze J: *DNA Tumor Viruses*. New York, Cold Spring Harbor Press, 1981.

19. Townsend ARM, McMichael AJ, Carter NP, et al: Cytotoxic T cell recognition of the influenza nucleoprotein and hemagglutinin expressed in transfected mouse L cells. *Cell* 1984;39:13–25.

20. Wiktor TJ, Macfarlan RI, Reagan KJ, et al: Protection from rabies by a vaccinia virus recombinant containing the rabies virus glycoprotein gene. *Proc Natl Acad Sci USA* 1984;81:7194–7198.

21. Wiktor TJ, Kieny MP, Lathe R: New generation of rabies vaccine: vaccinia-rabies glycoprotein recombinant virus. *Appl Virol Res* Vol. 1. Plenum Publishing Corporation, New York, pp. 69–90.

22. Yewdell JW, Bennink JR, Smith GL, Moss B: Influenza A virus nucleoprotein is a major target antigen for cross-reactive anti-influenza A virus cytotoxic T lymphocytes. *Proc Natl Acad Sci USA* 1985;82:1785–1789.

CHAPTER 41

Liposomes as Carriers of Vaccines

Carl R. Alving

The modern concept of liposomes as models of lipid bilayers of cell membranes was established in a landmark paper by Bangham et al (8). Shortly thereafter, starting in 1968, a related field was established that dealt with a specialized area of liposome research that has been loosely referred to as"immunologic aspects of liposomes." Based on an analysis of the literature through 1980 it was projected that the publication rate in the field of immunology of liposomes would reach 200 papers per year in 1983 (4). Extrapolation of the growth rate published in 1983 now leads to the further prediction that if growth were maintained at the same rate 670 papers would appear in 1987 and 910 papers in 1988.

The degree of interest in all of the fields involving immunologic aspects of liposomes has been maintained at a high level, and one of the areas that has undergone rapid growth has been research on liposomes for inducing immunity and as carriers of vaccines (3). Numerous candidate vaccines have been designed for human and veterinary use, and it is likely that liposome-based vaccines will find clinical use in the near future. Table 41.1 lists many of the protein antigens that have been employed as immunogens in liposomes.

Role of Macrophages in Processing Liposomes

The rationale for using liposomes for carrying antigens for immunization was originally based on the known propensity of parenterally injected liposomes to be ingested by macrophages. The natural targeting of liposomes to macrophages has served as the basis for utilizing liposomes as drug carriers for a variety of immunologic and nonimmunologic practical applications (2). Macrophages have a truly prodigious appetite for liposomes. Uptake by macrophages can be stimulated even further by opsonizing (coating) the liposomes with either immunoglobulin G (IgG) antibody for binding to the Fc receptor (12–14,17,18,25), IgM antibody and complement (C) for binding to the C3b receptor (22,29,30), fibronectin (14), or "mod-

Table 41.1. Protein antigens used in liposomes to induce immune responses.[a]

Viruses
>   Hepatitis B virus antigen
>   Adenovirus subunit antigen
>   Gross virus (tumor-associated antigen)
>   Influenza virus
>   Epstein-Barr virus glycoprotein
>   Rabies virus glycoprotein
>   Encephalomyocarditis virus
>   Semliki Forest virus
>   Herpes simplex virus antigen
>   Rubella virus protein

Parasites
>   Malaria merozoite antigen
>   Malaria sporozoite antigen
>   *Trypanosoma brucei* glycoprotein

Bacteria
>   Streptococcal cell wall antigen
>   *Brucella abortus* antigen
>   Gonococcal protein

Toxins
>   Diphtheria toxin
>   Cholera toxin
>   Snake venoms

Tumor Antigens

Enzymes
>   Horseradish peroxidase
>   Glucuronidase
>   Lysozyme

Serum proteins
>   Albumin
>   Gamma globulin

Miscellaneous
>   Various synthetic peptides
>   Thyroxine
>   Birth control antigens
>   Transplantation antigens

[a] Alving (3) should be consulted for references for this table.

ified" low density lipoproteins (LDL) for targeting to the receptor for modified LDL (15). As shown in Figure 41.1, virtually all of the cytoplasmic space of macrophages can be occupied by C-opsonized liposomes. Calculations from the experiment shown in Figure 41.1 revealed that an average of more than 1500 C-opsonized liposomes in the diameter range of 1 to 6 μm were ingested per mouse peritoneal macrophage.

Other factors that influence the uptake of liposomes are the charge, size, and degree of lipid unsaturation of the particles. Increasing the negative charge increases the uptake of unopsonized liposomes (14), but negative charge suppresses the uptake of C-opsonized liposomes (22). Upon

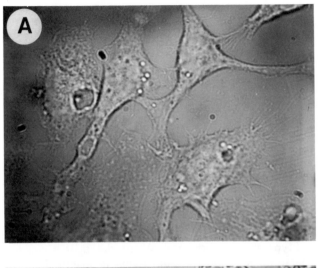

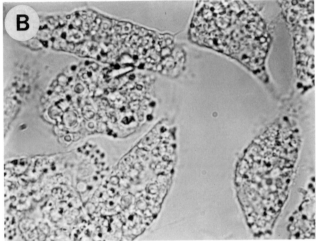

Fig. 41.1. Complement-dependent phagocytosis of liposomes. Liposomes containing Forssman antigen were incubated with mouse peritoneal macrophages in the presence of IgM anti-Forssman antibodies and complement. **A:** Light microscopy prior to adding liposomes. **B:** At 24 hours after adding liposomes. [From Alving (1), with permission.]

parenteral injection of liposomes, the blood circulation time of highly saturated small unilamellar liposomes before uptake by macrophages or other cells is much longer than that of large liposomes or unsaturated liposomes (23).

When liposomes are injected parenterally they undergo dynamic interactions with plasma proteins, and it is thought that numerous types of opsonizing agent might become attached to circulating liposomes (16).

The attachment of plasma proteins as opsonizing agents could explain why liposomes are removed so avidly from the circulation by macrophages. An excellent and comprehensive review on the fate of liposomes in vivo, the roles of plasma proteins, and the biochemical and biophysical factors that influence liposome distribution has been published by Senior (23).

From an immunologic standpoint, the macrophage, and particularly the Kupffer cell, is a major antigen-processing cell (APC) and plays a fundamental role in the production of T-cell-dependent humoral immunity (26,27). The importance of macrophages in the processing of liposome-encapsulated protein antigens has been shown in two ways. In the first type of study macrophage function was depleted in vivo by intraperitoneal administration of carrageenan (24). Carrageenan is a substance that selectively inhibits macrophage function without adversely affecting T or B lymphocytes. This type of pretreatment essentially abolished the humoral immune response to liposome-encapsulated bovine serum albumin (BSA). In the second type of study liposome-encapsulated BSA was ingested by peritoneal macrophages in vitro, and the macrophages themselves were then injected into mice. The anti-BSA response to the latter injection procedure was similar to that obtained by direct injection of liposome-encapsulated BSA into mice (9).

The latter study also clearly demonstrated that liposome-encapsulated BSA is a T-cell-dependent antigen. An anti-BSA response did not occur in T-cell-deficient athymic nude (nu^+/nu^+) mice. Humoral immunity was restored by reconstitution of the mice with T cells.

Although the way in which liposomes interact with macrophages and T cells to cause increased humoral immunity is not clear, it is possible that the process might include a role for class II major histocompatibility (MHC) antigen molecules. Liposomes that carried both class II MHC molecules and a foreign antigen induced antigen-specific MHC-restricted proliferation and interleukin 2 production by cultured T cell clones in the absence of an APC (28). Obviously the role of the macrophage is complicated, and it is possible that the beneficial role of the liposome includes both delivery of antigens to macrophages and subsequent effects on the processing and presentation of liposome-encapsulated antigens.

Example of a Potential Liposomal Vaccine

Regardless of the mechanism by which liposomes stimulate humoral immunity to proteins, it is evident (Table 41.1) that numerous antigens are now available that might be used as components of potential liposomal vaccines. Experiments by my colleagues and myself at the Walter Reed Army Institute of Research have suggested the possibility that liposomes might serve as useful carriers for a *Plasmodium falciparum* sporozoite vaccine. The feasibility of developing a human malaria sporozoite vaccine was demonstrated previously in a clinical trial by using irradiated sporozoites as antigens (10). Protection against sporozoite infection apparently can be achieved by inducing a high titer of antisporozoite antibodies (10,19–21). The antibodies have only a brief period (a few minutes or hours) to block sporozoite invasion of the liver, and titers in the animal host must therefore be high at the time of transfer of the organism from the mosquito to the host. A major challenge is to induce a high antibody titer that is also long-lived—hopefully with a protective duration of a year or more.

The major sporozoite antigen that is responsible for inducing protective immunity is a protein, the circumsporozoite (CS) protein, that covers the outer surface of the sporozoite. Although large amounts of antigen were not previously available for widespread distribution in a practical malaria vaccine, the problem of antigen availability might be close to being solved because of developments in molecular biology. The gene encoding the CS protein of the human malaria parasite *Plasmodium falciparum* has been cloned and sequenced (11). The middle of the CS protein has a novel antigenic region consisting of 37 Asn-Ala-Asn-Pro tetrapeptide repeats interspersed with four Asn-Val-Asp-Pro tetrapeptides (11). A cloned synthetic protein containing repeated tetrapeptide units has been developed. This antigen ($R32tet_{32}$) is antigenic, and antibodies to it react with intact sporozoites and block invasion of hepatocytes. The $R32tet_{32}$ antigen has been used as the antigen in a vaccine containing aluminum hydroxide (alum) as an adjuvant. This vaccine has been employed in an ongoing clinical trial under an Investigational New Drug protocol that was started in March 1986, and results have indicated that protective immunity can be achieved in humans (7). Although protection with $R32tet_{32}$ can be demonstrated, large amounts of antigen are required for induction of high titers. Experiments that are still ongoing have indicated that liposomes can substantially increase antibody titers to $R32tet_{32}$ in experimental animals.

An alternative method to present an antigen derived from the CS protein is to manufacture a vaccine containing peptides. Sequences consisting of varying numbers of tetrapeptide repeats have been synthesized and conjugated either to bovine serum albumin (BSA), thyroglobulin, or tetanus toxoid as a carrier protein (5,6,31,32). Unfortunately, the peptide–protein conjugates were immunogenic only when emulsified with Freund's adjuvant (5). The antibodies produced in this manner did recognize native CS protein, and they blocked sporozoite invasion of hepatoma cells (6). Research has now indicated that the synthetic peptide–protein conjugate might be rendered highly antigenic by incorporation into liposomes (5).

As shown in Figure 41.2, a single injection of liposome-encapsulated peptide–BSA conjugate induced antibodies in rabbits that were easily detected by ELISA. The antibody titer was greatly enhanced by including lipid A as an adjuvant. The antibodies that were generated reacted only with the synthetic peptide and not with BSA. This finding was determined by the fact that binding occurred to a peptide–thyroglobulin conjugate,

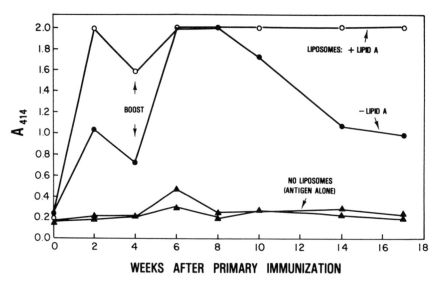

WEEKS AFTER PRIMARY IMMUNIZATION

Fig. 41.2. Immune response against malaria sporozoite peptide. Each line represents the activity of serum (1:100 dilution) from a rabbit immunized intravenously with a 16-amino-acid synthetic peptide conjugated to bovine serum albumin. The peptide was derived from *Plasmodium falciparum* circumsporozoite protein. The antigen was injected alone, in liposomes, or in liposomes containing lipid A. The albumin-conjugated peptide was encapsulated within liposomes, and the liposomes were washed to remove unencapsulated antigen (5). Antibody activity was measured spectrophotometrically against a 16-amino-acid peptide–thyroglobulin conjugate by ELISA. The maximum measurable absorbance (A_{414}) in the assay was 2.0. [From Alving (3), with permission.]

and also by the fact that the antibodies did not bind to BSA, which was used as a blocking agent for nonspecific binding in the solid-phase ELISA. The antipeptide antibodies reacted in fluorescence assays with both living and killed intact sporozoite organisms (5).

Summary

It is often found, as in the case of the synthetic sporozoite peptide, that synthetic antigens that have exciting potential properties as candidates for vaccines have poor or absent immunogenicity. We have now often observed that liposomes, or liposomes containing adjuvants, can greatly stimulate the immune response to synthetic antigens. We presume that the liposome has a variety of effects on macrophages that either influence macrophage functions or interactions with antigen, or carry the antigen through the macrophage in such a manner that the degradation or presentation of the antigen is altered. The future of this line of research will undoubtedly at least partly lie in the direction of elucidating the fate and trafficking patterns of liposome-encapsulated antigens within macrophages. Liposomal adjuvants, such as lipid A, probably have substantial effects on the metabolic processes within macrophages that also would be expected to influence the humoral immune response. Future research may improve on the predictability of induction of immune responses to a variety of liposomal antigens. However, it is likely that even at the present stage of development liposomes could already serve as a useful carrier for a practical vaccine in humans, and among other possibilities a *P. falciparum* sporozoite vaccine might make a logical and interesting candidate in which to develop the potential value of liposomes.

References

1. Alving CR: Therapeutic potential of liposomes as carriers in leishmaniasis, malaria and vaccines, in Gregoriadis G, Senior J, Trouet A (eds): *Targeting of Drugs*. New York, Plenum, 1982, pp 337–353.
2. Alving CR: Delivery of liposome-encapsulated drugs to macrophages. *Pharmacol Ther* 1983;22:407–424.
3. Alving CR: Liposomes as carriers for vaccines, in Ostro M (ed): *Liposomes: From Biophysics to Therapeutics*. New York, Marcel Dekker, 1987, pp 195–218.
4. Alving CR, Richards RL: Immunologic aspects of liposomes, in Ostro M (ed): *Liposomes*. New York, Marcel Dekker, 1983, pp 209–287.
5. Alving CR, Richards RL, Moss J, et al: Effectiveness of liposomes as potential carriers of vaccines: applications to cholera toxin and human malaria sporozoite antigen. *Vaccine* 1986;4:166–172.
6. Ballou WR, Rothbard J, Wirtz RA, et al: Immunogenicity of synthetic peptides from circumsporozoite protein of Plasmodium falciparum. *Science* 1985;228:996–999.
7. Ballou WR, Hoffman SL, Sherwood JA, et al: Safety and efficacy of a recombinant DNA Plasmodium falciparum sporozoite vaccine. *Lancet* 1987;1:1277–1281.
8. Bangham AD, Standish MM, Watkins JC: Diffusion of univalent ions across the lamellae of swollen phospholipids. *J Mol Biol* 1965;13:238–252.
9. Beatty JD, Beatty BG, Paraskevas F, Froese E: Liposomes as immune adjuvants: T cell dependence. *Surgery* 1984;96:345–351.
10. Clyde DF, McCarthy VC, Miller RM, Woodward WE: Immunization of man against falciparum and vivax malaria by use of attenuated sporozoites. *Am J Trop Med Hyg* 1975;24:397–401.
11. Dame JB, Williams JL, McCutchan TF, et al: Structure of the gene encoding the immunodominant surface antigen on the sporozoite of the human malaria parasite Plasmodium falciparum. *Science* 1984;225:593–599.
12. Geiger B, Gitler C, Calef E, Arnon R: Dynamics of antibody- and lectin-mediated endocytosis of hapten-containing liposomes by murine macrophages. *Eur J Immunol* 1981;11:710–716.

13. Hafeman DG, Lewis JT, McConnell HM: Triggering of the macrophage and neutrophil respiratory burst by antibody bound to a spin-label phospholipid hapten in model lipid bilayer membranes. *Biochemistry* 1980;19:5387–5394.

14. Hsu MJ, Juliano RL: Interactions of liposomes with the reticuloendothelial system. II. Nonspecific and receptor-mediated uptake of liposomes by mouse peritoneal macrophages. *Biochim Biophys Acta* 1982;720:411–419.

15. Ivanov VO, Preobrazhensky SN, Tsibulsky VP, et al: Liposome uptake by cultured macrophages mediated by low-density lipoproteins. *Biochim Biophys Acta* 1985;846:76–84.

16. Juliano RL, Lin G: The interaction of plasma proteins with liposomes: protein binding and effects on the clotting systems, in Tom BH, Six HR (eds): *Liposomes and Immunobiology*. Amsterdam, Elsevier/North Holland, 1980, pp 49–66.

17. Leserman LD, Weinstein JN: Receptor-mediated binding and endocytosis of drug-containing liposomes by tumor cells, in Tom BH, Six HR (eds): *Liposomes and Immunobiology*. Amsterdam, Elsevier/North Holland, 1980, pp 241–251.

18. Lewis JT, Hafeman DG, McConnell HM: Kinetics of antibody-dependent binding of haptenated phospholipid vesicles to a macrophage-related cell line. *Biochemistry* 1980;19:5376–5386.

19. Nussenzweig R, Vanderberg J, Most H: Protective immunity produced by the injection of x-irradiated sporozoites of Plasmodium berghei. IV. Dose response, specificity and humoral immunity. *Milit Med* 1969;134:1176–1182.

20. Potocnjak P, Yoshida N, Nussenzweig RS, Nussenzweig V: Monovalent fragments (Fab) of monoclonal antibodies to a sporozoite surface antigen (Pb44) protect mice against malarial infection. *J Exp Med* 1980;151:1504–1513.

21. Rieckmann KH, Beaudoin RL, Cassells JS, Sell KW: Use of attenuated sporozoites in the immunization of human volunteers against falciparum malaria. *Bull WHO* 1979;57(suppl 1):261–265.

22. Roerdink F, Wassef NM, Richardson EC, Alving CR: Phagocytosis of liposomes opsonized by complement: effects of negatively charged lipids. *Biochim Biophys Acta* 1983;734:33–39.

23. Senior JH: Fate and behavior of liposomes in vivo: a review of controlling factors. *CRC Crit Rev Ther Drug Carrier Syst* 1987;3:123–193.

24. Shek PN, Lukovich S: The role of macrophages in promoting the antibody response mediated by liposome-associated protein antigens. *Immunol Lett* 1982;5:305–309.

25. Shepard EG, Joubert JR, Finkelstein MC, Kühn SH: Phagocytosis of liposomes by human alveolar macrophages. *Life Sci* 1981;29:2691–2698.

26. Unanue ER: Antigen-presenting function of the macrophage. *Annu Rev Immunol* 1984;2:395–428.

27. Unanue ER, Allen PM: The basis for the immunoregulatory role of macrophages and other accessory cells. *Science* 1987;236:551–557.

28. Walden P, Nagy ZA, Klein J: Major histocompatibility complex-restricted and unrestricted activation of helper T cell lines by liposome-bound antigens. *J Mol Cell Immunol* 1986;2:191–197.

29. Wassef NM, Alving CR: Complement-dependent phagocytosis of liposomes by macrophages. *Methods Enzymol* 1987;149:124–134.

30. Wassef NM, Roerdink F, Richardson EC, Alving CR: Suppression of phagocytic function and phospholipid metabolism in macrophages by phosphatidylinositol liposomes. *Proc Natl Acad Sci USA* 1984;81:2655–2659.

31. Young JF, Hockmeyer WT, Gross M, et al: Expression of Plasmodium falciparum circumsporozoite proteins in Escherichia coli for potential use in a human malaria vaccine. *Science* 1985;228:958–962.
32. Zavala F, Tam JP, Hollingdale MR, et al: Rationale for development of a synthetic vaccine against Plasmodium falciparum malaria. *Science* 1985;228:1436–1440.

CHAPTER 42

A Short Primer on Vaccine Design: Focus on the Interrelatedness of Antigenic Determinants Addressing Various Lymphocyte Subpopulations

Eli E. Sercarz

In the past effective vaccines directed against disease-causing agents have employed attenuated versions of the whole immunogen. More recently, subunit vaccines, where only a small portion of the antigenic armamentarium of the organism is employed to induce protection, have been developed, introducing both new worries and new opportunities. For example, in a peptide vaccine a troublesome question is the universality of a T-helper (Th)-cell-inducing response. On the other hand, the interesting possibility arises that T-suppressor (Ts)-cell-inducing determinants may possibly be avoided in arriving at the final vaccine.

Features of T-B and T-T interaction, which must be considered carefully in vaccine design especially in relation to peptide and idiotypic immunogens, are discussed here. Most of the considerations mentioned have arisen from our own work, so that in this brief chapter is described an overall perspective; published accounts are cited for the experimental details.

There are several important criteria that influence the design of the new vaccines. The first is the realization that different determinants address different subpopulations of lymphocytes. A second crucial issue is the history of the organism's response to each particular determinant on the vaccine and how that history can influence the type of overall response obtained. A final consideration is the topographic relations among the determinants on the vaccine. Let us consider here an infectious agent with a set of surface antigenic determinants recognized by B cells (BD) that can be either continuous or discontinuous determinants. A variety of Th-cell-inducing determinants (HD) are also present along with Ts-cell-inducing determinants (SD) and determinants inducing cytolytic T lymphocytes (CD). Evidence exists that HD and SD are distinct (15), and the same conclusion has been reached for HD and BD. Peptides recognized by cytotoxic T lymphocytes are just now being described, but because these determinants are recognized in association with class I MHC molecules they would be unlikely to be the same as HD recognized in association with class II molecules.

HD and BD: Induction of "Initial Help" and "Memory Help"

Although in many cases it is necessary to depend on preformed antibodies for protection against the infectious agent (see chap. 43), here only situations in which a memory response suffices for protection are considered. The problems might profitably be considered in an order of increasing complexity, starting with the first world of lymphocytes, the B cells and their progeny. Can a single small peptide induce an antibody response? Although it has been occasionally argued that a small peptide, given in the appropriate adjuvant, is able to induce a complete antibody response (1), it has been far from the usual finding. Even if the HD and BD were to occasionally coincide, what is the likelihood of selecting a peptide, the antibodies to which cross-react with the surface of molecules available on the invasive organism? To examine this question, tests for immunogenicity have involved coupling the peptide to a macromolecular carrier protein such as KLH, which is used to provide an excess of helper cell determinants, and then determining whether the resultant antibodies cross-react with the peptide or the organism. Although a necessary first step in choosing the B cell determinants on the peptide, this procedure alone does not suffice for selecting the eventual peptide vaccine. Establishing B cell memory to the determinant in question may occur, but when the individual is challenged in the field by the infectious agent, the requirement of providing necessary T cell memory for optimal interaction will not have been accomplished; i.e., KLH-like determinants are unlikely to be present on the infectious agent.

It is vital to separate the requirements for *initial help,* needed to produce antibodies specific for the BD, from the requirements for *memory help,* which represents those Th enlisted at the time of confrontation with the infectious agent. Another example of the problem of ensuring both these types of help is seen with the idiotype vaccines. Here the immunogen is either a monoclonal or a polyclonal antiidiotypic antibody, selected because it mimics epitope(s) on the eliciting antigen, serving as an "internal image" of the external epitope. The constant regions of self-immunoglobulin and shared V region frameworks probably cannot serve as HD owing to tolerance induction. If initial T helper cell activation depends on an idiotypic V region HD, it may be quite weak. However, the use of xenogeneic idiotypic immunoglobulin may bring about premature disappearance of the desirable determinant. Even if the initial help were successful, or if coupling to an immunogenic carrier were employed, a protective response to the infectious organism would be unlikely for want of relevant "memory help." Actually, a major deficit in experimental knowledge, especially with peptides, concerns the level of protection conferred by primed T cells alone: it may even be the most essential ingredient in the vaccine recipe.

When deciding on the most efficient way to induce B cell memory to a desirable (set of) epitope(s), two possibilities appear to be most interesting. (a) Geysen and his colleagues (8) have prepared "three-dimensional peptides" that mimic sites found on the surface of macromolecules. The process of doing so requires trial-and-error, reiterative synthesis of peptides with an increasingly better fit to a particular monoclonal antibody (mAb). By the use of unusual component residues, turns and bends can be designed, and extremely high affinities of interaction of the "three-dimensional" designer peptide and the given mAb can be obtained. It would be advantageous to know that the designer B cell determinant is a dominant one that is normally activated during an immune response to the infectious agent (see below). (b) Coupling of the immunogen to lipopolysaccharide has been utilized to directly address B cells of the desired specificity, even in the absence of strong T cell help (6). However, the establishment of *memory* by this procedure remains to be tested.

MHC Determination of Restrictions in the T Cell Repertoire

When it comes to designating an appropriate T cell determinant, the important dictum is that there probably is no single universal Th-cell-inducing determinant that works (a) in all haplotypes and (b) on all molecules. Class II molecules of the major histocompatibility complex (MHC) are the primary arbiters of whether a response can take place to a particular determinant. It has been the experience of many workers in different protein antigen systems in the mouse that animals of a certain MHC haplotype display a characteristic pattern of responsiveness to peptides comprising the native molecule. What appears to be the controlling factor is the existence of sites on the peptide (agretopes) that either can or cannot interact with complementary polymorphic sites (desetopes) on the class II molecules: desetopes appear to be unique in each MHC. Both an agretope and an epitope must be available on a continuous peptide from the original primary structure of the immunogen. When a protein antigen is taken up by a cell capable of antigen presentation, processing occurs preceding the agretope–desetope interaction and can be assumed to be required for opening up of the molecule to reveal internal determinants (9). Whether the molecule simply needs to be unfurled or requires actual breakage of disulfide and peptide bonds to reveal the agretopes and epitopes of the HD will probably be a function of the particular antigen (18). A second reason for differences between haplotypes is that a particular epitope in conjunction with the Ia molecule may conceivably represent a hole in the repertoire for that haplotype owing to self-tolerance against a related self-epitope. What may be one haplotype's gap in the repertoire may be another's dominant epitope.

To summarize our experience in studies on T cell responsiveness to chicken lysozyme in three mouse haplotypes (7,21), as well as the experience of others in a fourth, it is not yet possible to predict the breadth and nature of the repertoire directed against such a protein. Portions of lysozyme that are β-pleated sheets or others that are in α-helices are each usable in particular MHC haplotypes.

Although the problem of selecting the right T cell helper determinant may be less stringent in the heterozygous human (with a more complex class II population) than the inbred mouse, occasions arise frequently where lack of response to a vaccine can be attributed to a lack of matching between agretopes and MHC desetopes, and a consequent inability to utilize the vaccine peptide to produce both T cell help and B cell responsiveness (4).

Dominance of Certain Th-Cell-Inducing Determinants

The second feature, hierarchical patterns of peptide dominance, has not yet generally been appreciated, so some space is devoted here to its ramifications. We can consider an antigen with potential HD 1-6: 1, 3, and 5 may be immunogenic in strains of MHC a, and determinants 2, 4, and 6 may be immunogenic in strains with a second MHC b, if the determinants are presented in the form of peptides or small fragments of the native molecule. However, when the native molecule is used as an immunogen in strain a, possibly only determinant 3 will ever have a chance to address the available Th in the individual, even though the strain could potentially respond to determinants 1, 3, and 5. Of course, from the point of view of the vaccinologist, it would be essential to select determinant 3 in designing a protective immunogen. With memory Th raised to HD 3, when the infectious agent is met in the field, Th memory for this determinant would be utilized, whereas memory Th raised to the subdominant HD 5 might be superfluous and unable to be productively engaged in response to the whole molecule.

An important feature of this hierarchy of determinant utilization is that *it depends on the form of the antigen* in which the determinant is presented. We can cite two examples from our work: First, in the induction of responses to determinants within peptide 74–96 of lysozyme, when a large fragment (13–105) was the immunogen, only clones of specificity for 81–96 were obtained, whereas with native lysozyme as the immunogen only clones of specificity 74–90 were found. The search included more than 50 clones, and no exception was found. Thus the molecular context of a determinant governs whether and how it will serve as an immunogen. A second example illustrates how residues distant from the site of T cell recognition can play a profound role in the consequences of an encounter with the immunogen (17). Ring-necked pheasant lysozyme (REL) is 100 times as efficient as HEL in activating all H-2^b clones, e.g., those directed against amino acid residues 74–90 or 81–96. However, when the amino and carboxyl ends of the molecule are removed by cyanogen bromide, leaving 13–105 (HEL) and 16–105 (REL), these large central fragments become equally effective immunogens. Interestingly, REL and HEL are identical between residues 81 and 96. Thus the N and C regions of the lysozymes must play an essential role in the early molecular interactions (i.e., processing) within the cell. The scaffolding surrounding (and protecting) the determinant region can apparently play a decisive role in determinant choice.

What are the possible reasons for the dominance of certain determinants? One underlying force is clearly the affinity of the desetope–agretope interaction for the dominant determinant. Alternatively, it could be due to regulatory suppression of Th cells directed against certain nondominant determinants. Most likely, dominance will also derive from the processing pathways available from the native molecule to its effective products. We have argued elsewhere (16) that the details of antigen processing of the residues surrounding a determinant lead to differing pathways of immunogen breakdown. Depending on such differences as the amino-terminal amino acid (2), the availability of a nearby enzyme cleavage site, or the availability of initially disclosed agretopes, a particular determinant becomes more or less utilizable for antigen presentation by the MHC. We have postulated that just as the precise specificity of the B cell receptor can be implicated in protecting certain residues of the antigen by residues in the immunoglobulin active site, interaction of Ia molecules with sites on the antigen extrinsic to that of T cell recognition might "guide" determinant choice by enhancing or protecting residues from proteolytic attack. The end result would be the favoring of certain potentialities inherent in the system, which would then appear to be dominant.

One method of assessing dominance would be to examine the pattern of T cell proliferation to candidate peptides from the infectious agent in convalescents. This examination may provide a list of dominant peptides for each MHC haplotype, which should be important in vaccine design.

Th Specific for Certain HD and B Specific Cells for Certain BD: Preferential Collaboration

Not only is the genetic potential to act as an HD important, as well as its dominance characteristics, but each HD does not provide equivalent help for B cells specific for all B cell determinants on the molecule. There are preferential partners in Th–B cell interactions, at least where large antigens are concerned (12). Helper cells raised against certain peptides on β-galactosidase show marked favoritism for B cells directed against certain other sites on the native enzyme. This area is just one of many that must be eventually sorted out in experiments designed to investigate relations among the subpopulations of lymphocytes as to the extent and specificity of memory in each compartment. At the present time, it still is not clear how well T cell memory or B cell memory alone would suffice in infectious situations, although the premise in this chapter is that both expanded T and B cell memory must be available for optimal effect in cases where antibody production will be protective.

In the collaboration to produce antibody, preference of collaborating cells specific for certain HD and BD may be a matter of proximity. In its activity as an antigen-presenting cell, the B cell performs a needed service for the T cell and, in turn, receives the just return of a required T cell stimulus—a perfect marriage (3). The BD buried in the B cell receptors on one cell is protected by immunoglobulin residues and surrounded by some HD, which may be afforded different levels of protection (11,12). One presumption is that class II molecule attachment will occur more effectively with the HD that have been most close to the BD and presumably most protected. The closest HD might be carried along, protected during the drama of endocytosis and processing, until the time for interaction with Ia, followed by appearance of the complex on the surface of the B cell. Figure 42.1 depicts one B cell specific for the triangular epitope and another for the rectangular epitope. Each B cell dominantly displays a *different* class II–HD complex on its surface. This finding is in accord with our previous results with β-galactosidase, suggesting that only certain HDs are displayed by B cells of a particular fine specificity (12).

This whole issue of preferential pairing may not be important when a multideterminant structure is used as the vaccine. However, when only a single B cell determinant is incorporated into the vaccine, preferential pairing rules suggest that increased care should be taken in the correct choice of T cell determinants. One approach to this problem is to not rely on either a single BD or a single HD. This approach ameliorates several problems at once, including that of the nonuniversality of HD in persons of different MHC.

Inducing Cytotoxic T Lymphocytes with Peptides

A revolution in thinking about Tc specificity has been under way over the past few years, spearheaded by the work by Townsend and his colleagues on influenza virus (19). They have identified a peptide necessary for response to the core protein of influenza. Whereas it was once thought that a viral subunit would interact directly with a class I MHC glycoprotein before recognition by a cytotoxic T cell (CTL) precursor, this more recent influenza work indicates that the pathways of processing of HD and CD may be different but parallel: A peptide from an internal protein of the virus can become immunogenic through its display on the cell surface together with class I molecules.

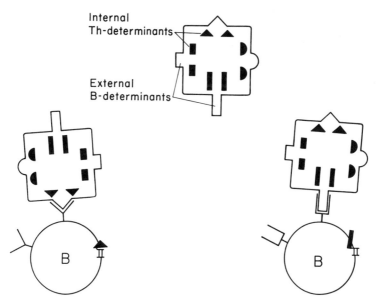

Fig. 42.1. Special partnerships between Th and B cells with certain specificities. The structure at the top center of the figure is a multideterminant antigen molecule displaying external conformational determinants recognized by B cell receptors. Buried within it are internal Th-cell-inducing determinants (HD) that become revealed to the T cell through processing that occurs in the B cell. Because of presumed proximity relations discussed further in the text, only certain HDs, in conjunction with class II molecules, are preferentially displayed by each B cell (depending on its specificity) to ambient T cells.

These findings also open up the possibility that when CTLs play an important role in disease resistance, a CD peptide might be included in the vaccine as well. Indeed, cytotoxic T cells may be the prime movers in the defense against a virus; and, in fact, antibody may play no role or an antagonistic one (e.g., blocking antibody in tumor systems). In such a case, the CD peptide alone could be employed, which underscores the axiom that it should be possible to tailor a vaccine to the exact requirements of the disease situation.

Avoiding Suppression by "Amputation" of Suppressor Determinants

One potential advantage of subunit/peptide vaccines is that it might be possible to avoid inclusion of an SD that could vitiate the immunization. Experiments in our laboratory with lysozymes and β-galactosidase continue to indicate that SD and HD are generally distinct and nonoverlapping (11,15). Nevertheless, in some systems SD and HD do overlap, although they are never identical. Therefore by having a minimal immunogenic (HD) determinant, it is unlikely that an SD will also be included. Here the universality of SD is at issue, for it is conceivable that one strain's HD is another's SD. However, in the lysozyme system both H-2[a] (13) and H-2[b] (20) strains seem to recognize the amino-terminus of chicken lysozyme (HEL) as an SD. "Amputation" experiments in which the three amino-terminal residues were removed from HEL have converted a nonimmunogenic molecule into a immunogen in genetically nonresponsive H-2[b] strains. (Not only has the amputation indicated the position of an SD, but it has revealed the existence of otherwise silent HDs.) Presumably, a limited number (probably one) of Ts-inducing determinants are present on HEL for H-2[b] mice. Even if potential, subdominant suppressogenic determinants actually exist on the amputated HEL, they are de facto ineffective. Although the question of universality of SDs in different strains deserves detailed further study, the argument has already been made (9) that SDs are superficial and include hydrophilic residues as essential components, whereas HDs are internal and include critical hydrophobic residues (5,9,14). One can be optimistic in predicting that because of their physicochemical distinctness exclusions of SD should be achievable from an HD–CD–BD vaccine. Techniques of site-directed mutagenesis may also be successful in removing SD from recombinant DNA constructs.

We should discuss another point brought forth lately by the experiments of Herzenberg and her colleagues (10). Previous experience with a carrier antigen may lead to suppression of future antibody responses directed to haptens or other epitopes attached to the carrier. The suppression might in any case be temporary, as we have demonstrated earlier in the β-galactosidase (GZ) system in which waves of help, suppression, and help followed priming with the native protein, GZ (15). However, with the opportunity of deleting SD from the original immunogen, this "epitope-specific suppression" would no longer be a concern.

Conclusions

It is reassuring to those of us who have worked on model protein antigen systems for a long time that so many useful generalities seem to emerge from the study of one or two such proteins. The following list of caveats were derived from our work with either the lysozyme or the β-galactosidase system.

1. Th-cell-inducing determinants (HD) are not universal; they differ among MHC haplotypes.

2. Ordinarily, the entire potential repertoire is not expressed: rather, acharacteristic hierarchy of epitope dominance is found. It is therefore reasonable to avoid subdominant determinants in vaccine design.
3. HD and B-cell-inducing determinants are likely to be nonidentical, and therefore it is necessary to include a carefully chosen HD(s).
4. A preferential relation exists between certain HDs and other BDs on the native antigen: because a *small number* of BDs are included in any subunit vaccine, it is important to determine that the HD in the vaccine is located appropriately, so that the memory in both the Th and B cell compartments can be effectively engaged.
5. Ts-inducing determinants (SD and HD) are generally distinct and non-overlapping, and never identical. It therefore seems likely that it is possible to avoid including SD in a potential vaccinogen.
6. To reiterate the earlier points, the principles of hierarchical dominance, position and number of determinants in the native structure, and the requirements for memory induction in different subpopulations must be considered in vaccine design. Attenuated vaccines might fulfill certain of these criteria easily because of their close relation to the native infectious agent. Failures of naive versions of subunit/peptide vaccines, however, should not discourage the quest for an effective, "designer peptide" multideterminant vaccine.

The suggestion to include more than one dominant HD and BD in the eventual vaccine has already been made and might answer points 1 and 2 above, especially in the heterozygous, polymorphic human species. Another suggestion that may circumvent the "preferential partner" and "insufficient help" problems would be to separately immunize with the appropriate T-memory- and B-memory-inducing vaccines. T memory might be induced with peptides representing dominant HDs that are correctly positioned on the native molecule to provide help for the designated BD; B memory could be induced with a powerful conjugate containing a previously unused carrier protein, rich in HD (e.g., KLH), coupled to an otherwise poorly immunogenic BD (e.g., a monoclonal, syngeneic, idiotypic vaccine).

Much remains to be learned about the efficacy of separate induction of T and B cell memory, the details of suppressor cell induction, and even the full story of T–B collaboration. In certain disease models it will be necessary to learn which immune subpopulation is the protective one— T helpers, T cells capable of delayed-type hypersensitivity, cytotoxic T cells, B cells—in order to appropriately direct the choice of determinants for the vaccine. These "holes in our repertoire" are being swiftly filled so that soon a truly rational approach to vaccine design will be within grasp.

448   Eli E. Sercarz

Acknowledgments. I would like to thank Drs. Alexander Miller, Norma Kenyon, and Guy Gammon for reviewing the manuscript; Vicky Godoy for preparing it; and Margaret Kowalczyk for drawing the figure. The work herein was supported by grants AI-11183 and CA-24442 from the National Institutes of Health.

References

1. Atassi MZ, Young CR:. Discovery and implications of the immunogenicity of free small synthetic peptides: powerful tools for manipulating the immune system and for production of antibodies and T-cells of preselected submolecular specificities. *Crit Rev Immunol* 1985;5:387–410.
2. Bachmair A, Finley D, Varshavsky A: In vivo half-life of a protein is a function of its amino-terminal residue. *Science* 1986;234:179–186.
3. Berzofsky JA: T–B reciprocity: an Ia-restricted epitope-specific circuit regulating T cell–B cell interaction and antibody specificity. *Surv Immunol Res* 1983;2:223.
4. Del Giudice G, Cooper JA, Merino J, et al: The antibody response in mice to carrier-free synthetic polymers of Plasmodium falciparum circumsporozoite repetitive epitope is I-A^b-restricted: possible implications for malaria vaccines. *J Immunol* 1986;137:2952–2955.
5. DeLisi C, Berzofsky JA: T-cell antigenic sites tend to be amphipathic structures. *Proc Natl Acad Sci USA* 1985;82:7048.
6. Furman A, Sercarz EE: The failure of non-responder mice to develop IgG memory assessed by in vitro culture with an antigen-LPS conjugate. *J Immunol* 1981;126:2430–2435.
7. Gammon G, Shastri N, Cogswell J, et al: The choice of T cell epitopes utilized on a protein antigen depends on factors distant from, as well as at, the determinant site. *Immunol Revs* 1987;98:53–73.
8. Geysen HM:. Antigen–antibody interactions at the molecular level: adventures in peptide synthesis. *Immunol Today* 1985;5:364–369.
9. Goodman JW, Sercarz E: The complexity of structures involved in T cell activation. *Annu Rev Immunol* 1983;1:465–498.
10. Herzenberg LA, Tokuhisa T, Hayakawa K: Epitope-specific regulation. *Annu Rev Immunol* 1983;1:609–632.
11. Krzych U, Sercarz E: The relationships between Ts-inducing and Th and Tp-inducing determinants on a large protein antigen. *Dev Biol Stand* 1986;63:41–51.
12. Manca F, Kunkl A, Fenoglio D, et al: Constraints in T-B cooperation related to epitope topology in E. coli β-galactosidase. I. The fine specificity of T cells dictates the fine specificity of antibodies directed to conformation-dependent determinants. *Eur J Immunol* 1985;15:345–350.
13. Oki A, Sercarz EE: T cell tolerance studied at the level of antigenic determinants. I. Latent reactivity to lysozyme peptides that lack suppressogenic epitopes can be revealed in lysozyme-tolerant mice. *J Exp Med* 1985;161:897–911.
14. Rothbard JB, Taylor WR: A sequence pattern common to T cell epitopes. *EMBO J* 1988;7:93.

15. Sercarz EE, Yowell RL, Turkin D, et al: Different functional specificity repertoires for suppressor and helper T cells. *Immunol Rev* 1978;39:109–137.
16. Sercarz E, Wilbur S, Sadegh-Nasseri S, et al: The molecular context of a determinant influences its dominant expression in a T cell response hierarchy through "fine processing," in *Progress in Immunology VI: Sixth International Congress of Immunology* 1986; pp. 227–237.
17. Shastri N, Miller A, Sercarz EE: Amino acid residues distinct from the determinant region can profoundly affect activation of T cell clones by related antigens. *J Immunol* 1986;136:371–376.
18. Streicher HZ, Berkower IJ, Busch M, et al: Antigen conformation determines processing requirements for T-cell activation. *Proc Natl Acad Sci USA* 1984;81:6831.
19. Townsend ARM, Rothbard J, Gotch FM, et al: The epitopes of influenza nucleoprotein recognized by cytotoxic T lymphocytes can be defined with short synthetic peptides. *Cell* 1986;44:959–968.
20. Wicker LS, Katz M, Sercarz EE, Miller A: Immunodominant protein epitopes. I. Induction of suppression to hen egg white lysozyme is obliterated by removal of the first three N-terminal amino acids. *Eur J Immunol* 1984;14:442–447.
21. Zanetti M, Sercarz EE, Salk J: The immunology of second generation vaccines. *Immunol Today* 1987;8:18–25.

CHAPTER 43

Next Steps in the Evolution of Vaccinology

Jonas Salk and Maurizio Zanetti

Long before viruses or the immune system were recognized, Jenner had the creative insight that ultimately led to the eradication of smallpox less than two centuries later. We have come a long way since the days of Jenner, Pasteur, and the other pioneers who advanced the sciences and technologies related to vaccinology. The creative co-evolution of the relevant sciences and technologies now brings us to the threshold of altering significantly the relationship between humans and their microbial pathogens.

Vaccinology, as an integrating concept and a science, is more easily sensed than defined. It can be regarded as an organized body of knowledge and principles required to induce protective immunity. The latter can be accomplished through an understanding of the fundamental properties of the immune system, of the immunogens needed to induce protective effects, and of the ways by which these factors may be applied in practice.

Great progress has been made in the evolution of vaccinology. The many facets of this integrated discipline provide not only a retrospective view of the past but also implications of what can be anticipated in the future. The thoughts to follow are intended to suggest trends and next steps that need to be considered in the further evolution of vaccinology.

The advances made in viral vaccines in recent years, as seen in Table 43.1, acknowledge two factors: (a) effective and durable immunity can now be achieved with suitably constituted noninfectious vaccines; and (b) the experience of infection is not required for inducing such an effect. The latter is recognized in the creation of a new generation of noninfectious vaccines using (a) recombinant DNA technology, (b) synthesis of peptides, (c) internal image (idiotype) antibodies, and (d) the inactivated intact virion, or its subunits, replicated in continuously propagating cell lines.

We wish here to draw attention to a few of the basic principles and requirements for inducing effective and durable immunity established earlier in the course of studies using noninfectious whole virion vaccines (27). They are applicable to the formulation of the new-generation vaccines now in the process of development.

Table 43.1. Synopsis of development of virus vaccines.

Years	Disease	Substrate used for production of vaccine	State of virus or protective antigen	
			Attenuated	Noninfectious
1700–1900	Smallpox	Calf	+	
	Rabies	Rabbit	+	
1935–1945	Yellow fever	Mouse, chicken, embryo	+	
	Influenza	Chicken embryo		+
	Mumps	Chicken embryo		+
1955–1964	Polio	Primary monkey cells	+	+
	Measles	Chicken embryo	+	+
	Rabies	Duck embryo	+	
1965–1974	Mumps	Chicken embryo	+	+
	Rubella	Duck embryo	+	
	Polio	Human cell lines	+	
	Measles	Human cell lines	+	
	Mumps	Human cell lines	+	
	Rubella	Human cell lines	+	
1975–1984	Hepatitis B	Human plasma		+
	Rabies	Human cell lines		+
	Polio	Monkey cell line		+
1985–	X	Recombinant DNA		+
	Y	Synthetic peptides		+
	Z	Internal-image antibody		+

from Zanetti et al (39), with permission of Elsevier Science Publishers B.V.

Immunology

Mechanisms of Immunity

Because exogenous pathogens induce disease by different pathways, different strategies are required for their interception. Those that initially come into contact with external secretions, or enter the bloodstream in a cell-free form, can be deflected by humoral mechanisms. Humoral factors that prevent the initiation of intracellular parasitism also block transmission of pathogens from infected to uninfected cells. For pathogens that have become established intracellularly, cell-mediated as well as humoral immunity are needed. The mode of action of vaccine-induced humoral immunity in preventing disease differs depending on the length of the incubation period.

Incubation Period

A disease of short incubation period (less than 3 days, as for influenza) requires that protective levels of serum neutralizing antibody be present at the time of exposure to prevent the establishment of infection. The

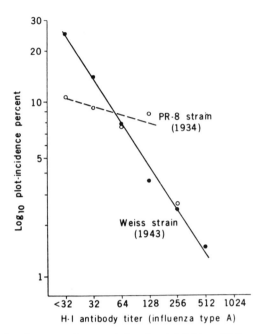

Fig. 43.1. Relation between hemagglutination-inhibiting (H-I) antibody titer and illness rate. [From Salk (23), with permission.]

portal of entry, in this instance, is also the site of pathology in the respiratory tract.

The degree of resistance to influenza virus infection (Fig. 43.1) is directly proportional to the level of specific hemagglutination-inhibition antibody in the serum and to the level of specific protective antibody in the secretions of the respiratory tract. Therefore for the prevention of influenza it is necessary to induce and maintain antibody titers above levels associated with protection against the appropriate type-specific variants. It can be done by re-immunization or by use of a potent immunologic adjuvant, to which reference is made later (26).

As shown in Figure 43.2 for diseases of longer incubation period, e.g., paralytic poliomyelitis (which requires more than 3 days for central nervous system invasion from the primary site of infection in the intestinal tract), it is necessary merely to prime the immune system (to induce immunologic memory) for inducing durable immunity to paralysis.

The degree of memory induced by infection or vaccination is measured by the anamnestic antibody response elicited by a challenge dose of vaccine, as shown in Figure 43.3. Following a single primary dose of a suitably potent vaccine, the degree of anamnestic responsiveness increases progressively over a period of 6 months to a plateau that tends to persist

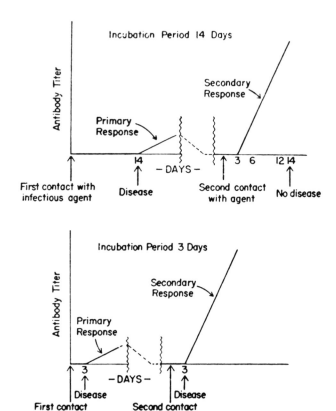

Fig. 43.2. Effect of the length of the incubation period on the development of permanent immunity. Polio is an example of a disease with a long incubation period (top graph) and influenza is an example of a disease with a short incubation period (bottom graph). [From MacLeod (17) Relation of the incubation period, *Journal of Immunology*, vol. 70, 420–425, © by Williams & Wilkins, 1953.]

thereafter, as depicted in Figure 43.4. The degree of anamnestic responsiveness induced by a vaccine is proportional to the quantity of antigen administered for primary immunization, as may be seen in Figure 43.5. It is revealed by the degree of response to a uniform challenge dose administered 1 year after graded primary doses of vaccine.

Because of the long half-life of memory cells (8) the anamnestic response persists, maintaining a state of immunity without need for periodic reinforcement in such diseases as poliomyelitis and hepatitis. It is as if, in the course of evolution, the primary antibody response served to clear the host of the invading organisms and the secondary-type response in order to prevent reinfection or disease.

In summary, an anamnestic response in diseases with incubation periods longer than 3 days results in a rapid outpouring of antibody, even in individuals in whom none was detectable. Thus the distinction may be made

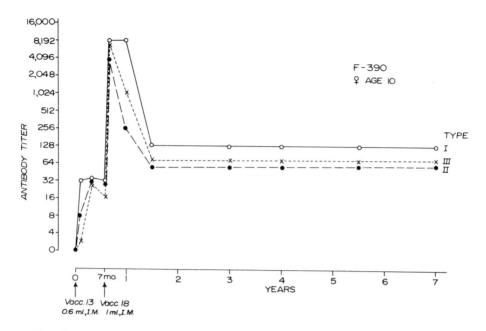

Fig. 43.3. Pattern of antibody response and persistence following a primary and booster immunization in a 10-year-old child followed for 7 years. [From Salk (25). with permission.]

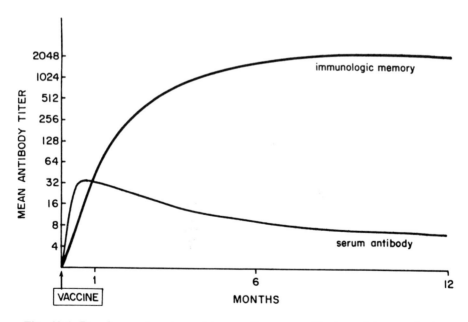

Fig. 43.4. Development and persistence of serum antibody and immunologic memory following one dose of noninfectious poliovirus vaccine. [From Salk (24), with permission. Copyright 1984 The University of Chicago Press.]

Primary Dose (Vaccine A)	Challenge Dose (Vaccine J)	Number of Subjects
2 ml	1 ml	24
1	1	21
1/2	1	26
1/4	1	27
1/8	1	30
1/16	1	26
Control	1	33

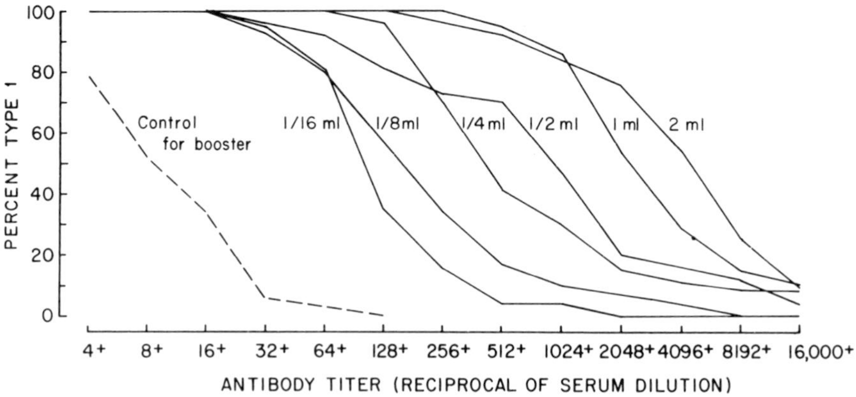

Fig. 43.5. Degree of immunologic memory induced by vaccination. Data are expressed in terms of percentage of individuals with titers of type 1 antibody at or above the indicated levels 2 weeks after a uniform challenge dose of vaccine J given 1 year after a two-dose primary series (2-week interval) in groups given different quantities of reference vaccine A for primary immunization. [From Salk et al (28), with permission of S. Karger AG, Basel.]

between *memory-dependent immunity* compared to *antibody-dependent immunity* in which antibody must be present at the time of primary infection for diseases with incubation periods shorter than 3 days.

Antigenic Complexity

Microbial pathogens possess a complex mosaic of epitopes, not all of which induce protective immunity, and may even change in specificity in the course of infection or spread. Moreover, the induction of the necessary cooperative immune responses at the level of B, T, and antigen-presenting cells requires attention to sufficiency of antigenic complexity for inducing a protective response. For the generation of vaccines now emerging, the omission or deletion of any of the essential functional epitopes could limit their effectiveness. For this reason it will be necessary to use adjuvants and develop immunopotentiators to produce the desired effects with such vaccines.

When formulating and using new as well as conventional vaccines, the foregoing factors need to be taken into consideration for those pathogens that undergo antigenic drift and antigenic shift as is seen with influenza and other infectious or parasitic diseases. The success or failure of vaccines and vaccination programs depends on attention to all of the relevant elements involved in optimizing the immune response required for the prevention of natural infection and disease.

Technology

Recombinant DNA Vaccines

The expression of specific proteins of a pathogen's genome encoding for antigenic regions of potential protective value has already been achieved in several instances, as indicated in Table 43.2. Animals immunized with viral recombinant DNA vaccines, administered in adjuvant, that produce neutralizing antibody (3,11) are protected against challenge infection (3,18). However, the ability of these recombinant vaccines to induce cell-mediated immunity remains to be fully explored. Several theoretical questions need to be answered that concern differences between the native protein of the virus and that produced by recombinant DNA technology.

In addition to the use of recombinant DNA technology to produce non-infectious immunogens, progress has also been made in constructing chimeric vaccinia viruses expressing viral proteins. Genetically engineered

Table 43.2. Systems in which idiotype vaccines have been tested.

Infectious agent	Nature of idiotype vaccines[a]	Species tested	Adjuvant	Protection
Viruses				
Hepatitis B	P/M	Mice	+	+
Rabies	P	Mice	+	N.D.
Tobacco mosaic	P	Mice	+	N.D.
Polio type II	M	Mice	−	−
Venezuelan equine	P	Mice	+	N.D.
encephalomyelitis	M	Mice	−	N.D.
Reovirus	M	Mice	−	+
Sendai virus				
Bacteria				
Streptococcus pneumoniae	M	Mice	−	+
Escherichia coli	M	Mice	+	+
Listeria monocytogenes	M	Mice	+	+
Parasites				
Trypanosoma rhodesiense	M	Mice	−	+
Schistosoma mansoni	M	Rats	−	+
Trypanosoma cruzi	P	Mice	+	N.D.

[a]N.D. = not done. P = polyclonal antibody; M = monoclonal antibody.
From Zanetti et al (39), with permission of Elsevier Science Publishers B.V.

vaccinia viruses rely on their ability to express immunogens of foreign cloned genes, as shown in Table 43.3.

Laboratory animals vaccinated with live recombinant vaccinia viruses produce neutralizing antibody (4,10,21,22). This type of vaccine may also prime and stimulate a specific cytolytic T lymphocyte (CTL) response; however, the demonstration of such a CTL response thus far requires secondary in vitro immunization of T cells with the native virus (37).

With respect to the use of vaccinia virus for human vaccination, two important considerations need to be taken into account. One is that, in 1968 in the United States 572 complications including nine deaths were reported in approximately 14 million vaccinees (12). The other is that the immunity induced to the vaccinia virus may limit its practical usefulness for repeated administration in this way.

Synthetic Peptide Vaccines

The synthesis of peptides, corresponding in sequence to the primary structure of certain antigenic regions of a pathogen, represents another way of constructing noninfectious surrogate vaccines (13). Because of the progress in gene cloning of structural proteins, primary amino acid sequences predicted from their nucleotide sequences are readily obtained, and synthetic peptides can be constructed accordingly. The effectiveness of a synthetic peptide vaccine would be reflected in its ability to elicit the formation of neutralizing antibody and/or immunologic memory.

Vaccines consisting of specific synthetic peptides coupled to carrier

Table 43.3. Recombinant DNA vaccines based on use of vaccinia virus.

Virus	Reference
Infectious	
Hepatitis B virus	Paoletti et al (21)
Herpes virus type 1	Cremer et al (4)
Influenza virus	Smith et al (33)
Rabies virus	Kieny et al (10)
Vesicular stomatitis virus	Mackett et al (16)
Epstein-Barr virus	Mackett and Arrand (15)
Plasmodium knowlesi	Smith et al (31)
Noninfectious	
Foot-and-mouth disease	Kleid et al (11)
Cholera toxin	Mekalanos et al (20)
Herpes virus type 1	Berman et al (3)
Plasmodium falciparum	Enea et al (6)
Mycobacterium leprae	Young et al (38)
Hepatitis B virus	McAleer et al (18)
HIV (experimental)	—

protein in adjuvant have been shown to confer a degree of protection against influenza and polio type 1 in laboratory animals (5,29).

Synthetic peptides can be made that are recognized by T cells, whereas others can be constructed to mimic B cell epitopes. The essential problem until now has been in determining which of the numerous peptides are needed to trigger protective immunity. Although some native B cell epitopes are recognized by their linear sequence, most protein antigenic determinants recognized by B cells are conformational. Criteria such as hydrophilicity (36), mobility (35), and surface availability have been employed to attempt to better localize immunogenic B cell sites. Criteria for predicting T cell sites are still in the exploratory stage.

In some instances, where protection is dependent on the presence of antibody rather than on memory alone, as in the case of the circumsporozoite (CS) antigen of human malaria, a peptide vaccine preparation has to include an appropriate carrier molecule necessary for generating high levels of antibody. In this system, the immunodominant epitope, a 23 times tandem repetition of four amino acids (40), is recognized by neutralizing monoclonal antibodies, as well as by human sera from endemic areas of malaria. Rabbits immunized with a synthetic CS dodecapeptide (three repeats) in adjuvant produced antibodies that neutralized *P. falciparum* sporozoites in vitro (40).

The essential problems for the development of synthetic peptide vaccines are to determine which of the numerous peptides might trigger protective immunity and if peptides that mimic only the primary structure of the antigen are good immunogens at all.

Idiotype Vaccines

It has become evident that antibodies can be used to induce immunity of predetermined specificity. Antigenic determinants of immunoglobulins, idiotypes, may substitute for protein or even carbohydrate determinants on conventional antigens and trigger immune responsiveness at the level of B and/or T lymphocytes. This phenomenon, predicted by Lindeman (14) and theorized by Jerne (9), reflects the fact that the B cell repertoire in its vast spectrum of specificities can also form antibodies that are the internal image of virtually any antigen.

As shown in Table 44.3, antibody idiotypes have been used as surrogate vaccines to induce specific immunity against viruses, bacteria, and parasites. Idiotypic vaccines have been shown to induce neutralizing antibodies for the corresponding pathogen and, in some instances, to protect from an otherwise lethal challenge with the infectious agent.

In this volume others have discussed idiotype vaccines and the idiotype network theory from which the internal image concept originates. We intend simply to comment on a few points of theoretical and practical interest.

One is that idiotype vaccines can elicit specific humoral and/or cellular immunity in a non-MHC restricted fashion; that is, they may be immunogenic in the largest possible assortment of MHC haplotypes (7,30). Another is that idiotype vaccines can substitute for nonprotein antigens such as polysaccharides (34). It is known that infants are unresponsive to most polysaccharide antigens (hence their failure to develop immunity to *Hemophilus influenzae, Neisseria meningitidis,* and *Streptococcus pneumoniae* infections), and adults respond poorly to vacccines whose immunogenic moiety is a sugar unless conjugated to a protein. Recombinant DNA and synthetic peptide vaccines lack, a priori, the capacity to induce immunity efficiently against carbohydrate-composed epitopes on pathogens. Thus idiotype vaccines may be employed to avoid the inconvenience in (a) the delay in immune responsiveness to polysaccharides during the neonatal period and (b) the poor immunogenicity of polysaccharides in adults.

Immunological Adjuvants

As pointed out above, some of the new-generation antigens may possess the specificity but not necessarily the immunogenic potency of the antigenic architecture of the organism from which they are derived. For this reason, adjuvants and immunopotentiators will be needed to express the full immunogenic potential of such antigens. Adjuvants would also be of value for enhancing the potency of present vaccines. Although many such reagents exist, the most potent have not yet been adopted for use in humans.

Of this group, incomplete Freund's adjuvant (IFA) is one of the more powerful. The occurrence of objectional nodules or cysts at the inoculation site are now avoidable by use of purified antigens and reaction-free reagents. As for possible long-term side effects such as autoimmunity, immune complex disease, and neoplasia, a prospective study (1,2) has been under way since 1951–1953 to answer these questions in a group of approximately 18,000 military personnel who had received influenza virus vaccine in IFA (Arlacel–mineral oil). Data from the 10- and 18-year follow-up are to be supplemented with a 33- to 35-year follow-up to provide further reassurance of long-term safety beyond that indicated by the 10- and 18-year observation periods (1,2).

The enhancing effect of IFA on immunogenicity of influenza vaccine in humans is illustrated in Figure 43.6. The potentiating effect of IFA is also evident in the results of an antigenic extinction titration in humans as shown in Figure 43.7. These results suggest that it is possible to reduce the amount of antigen required for an effective response, which in turn would allow preparation of multiple-disease vaccines as well as broad-spectrum vaccines needed to deal with strain variation due to antigenic drift and antigenic shift. The striking protective effect in humans (Fig.

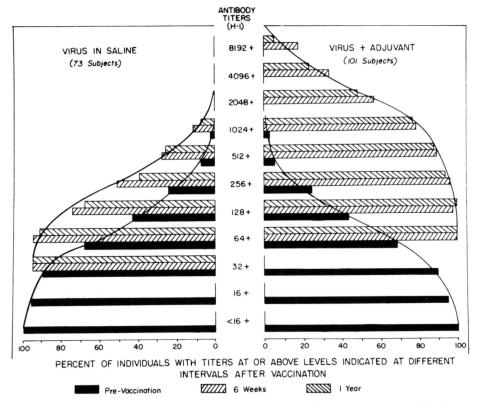

Fig. 43.6. Comparison of antibody response and persistence in human subjects inoculated with influenza virus in aqueous (saline) or emulsified (adjuvant) vaccines. [From Salk et al (26). The Journal of the American Medical Association, vol. 151(14), 1168–75. Copyright 1953, American Medical Association.]

43.8) of a vaccine containing six strains of influenza and two of adenovirus in IFA reveals the usefulness of such an adjuvant for this purpose (19).

A metabolizable oil of animal origin, i.e., squalene or shark oil, is now available for use in an IFA type of emulsion. A comparative study of adjuvants and immunopotentiators will provide data for choosing appropriate enhancers for weak antigens. Such a study is now required to derive the full potential of the newly emerging repertoire of antigens from which multiple disease vaccines can be made. It will be necessary to take into consideration the nonspecific effect of immunization that is needed in the balanced activation of the correct T and B cell compartments.

The Future

The theoretical and experimental considerations discussed in this communication suggest that it should be possible to design vaccines with the immunogenic potential of natural pathogens. Considerable progress in the

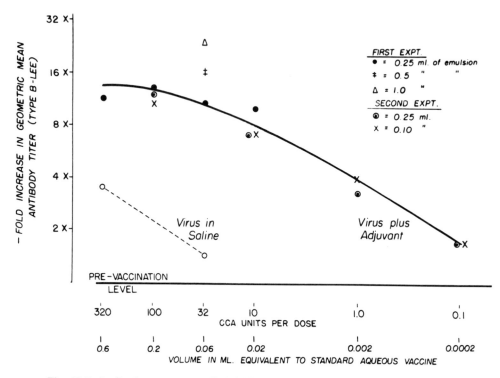

Fig. 43.7. Antibody response to diminishing quantities of influenza virus in aqueous (saline) or emulsified (adjuvant) vaccines. Each point represents the mean value of type B Lee antibody titer rise for a group of about 50 young adults. The vaccines used contained equal parts of PR-8 (type A), Cuppett (type A₁), and Lee (type B) strains. The standard aqueous vaccine then in use contained 500 CCA units per 1.0-ml dose. [From Salk et al (26). The Journal of the American Medical Association, vol. 151(14), 1168–75. Copyright 1953, American Medical Association.]

technology for the development of a new generation of vaccines has been made; however, the science for optimizing vaccine composition and immunogenicity is still in development.

It is not yet possible to predict which among the various means for formulating new-generation vaccines will be most advantageous. Small subunit vaccines obtained by either recombinant DNA technology or chemical synthesis pose limitations in their capacity to express the immunogenic quality and potency of the natural pathogen. It is possible, by choosing a combination of peptides with the appropriate effect on T and B cells, that this obstacle may be overcome. Conjugation to adjuvants may be necessary to potentiate such antigens. As for idiotype vaccines, it will be necessary to improve ways to analyze and modify appropriately the variable region of antibody molecules of specific interest, e.g., by site-directed mutagenesis.

The effect to be achieved when designing new vaccines, whatever their

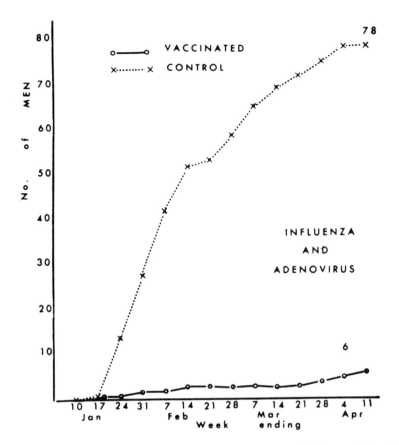

Fig. 43.8. Number of men with influenza or adenovirus infections admitted from vaccinated and control groups. [From Meiklejohn (19). The Journal of the American Medical Association, vol. 179, 594–97. Copyright 1962, American Medical Association.]

formulation may be, is the full expression of both their immunogenic and anamnestic potential with the fewest number of administrations. It is likely that it will be achieved with the help of the broadening spectrum of immunologic adjuvants that will become available for human use. Sophistication in the design and preparation of vaccines will be by optimizing epitope quality, density, and position.

At this time of continued technologic advancement, the science of vaccinology needs to be critically reevaluated for the possibilities inherent in the new generation of vaccines now being developed as well as of vaccines that presently exist. The theoretical, experimental, and empirical approach used in recent decades to study human vaccination with non-infectious and infectious vaccines have established the parameters for the further development of even more efficient means for the control of diseases preventable by vaccination. In the years to come, a significant change in the relationship between humans and their microbial pathogens can be anticipated.

References

1. Beebe GW, Simon AH, Vivona S: Follow-up study on army personnel who received adjuvant influenza virus vaccine 1951–1953. *Am J Med Sci* 1964;247:385–406.
2. Beebe GW, Simon AH, Vivona S: Long-term mortality follow-up of army recruits who received adjuvant influenza virus vaccine in 1951–1953. *Am J Epidemiol* 1972;96:337–346.
3. Berman PW, Gregory T, Crase D, Lasky LA: Protection from genital herpex simplex virus type 2 infection by vaccination with cloned type 1 glycoprotein D. *Science* 1984;227:1490–1492.
4. Cremer KJ, Mackett M, Wohlenberg C, et al; Vaccinia virus recombinant expressing herpes simplex virus type 1 glycoprotein D prevents latent herpes in mice. *Science* 1985;228:737–740.
5. Emini EA, Jameson BA, Wimmer E: Priming for and induction of anti-poliovirus neutralizing antibodies by synthetic peptides. *Nature* 1983;304:699–703.
6. Enea V, Ellis J, Zavala F, et al: DNA cloning of Plasmodium falciparum circumsporozoite gene: amino acid sequences of repetitive epitope. *Science* 1984;225:268–230.
7. Ertl HCJ, Homans E, Tournas S, Finberg RW: Sendai virus-specific T cell clones. V. Induction of a virus-specific response by antiidiotypic antibodies directed against a T helper cell clone. *J Exp Med* 1984;159:1778–1783.
8. Jerne NK: Idiotypic networks and other preconceived ideas. *Immunol Rev* 1984;79:5–24.
9. Jerne NK: Towards a network theory of the immune system. *Ann Immunol (Paris)* 1974;125:373–389.
10. Kieny MP, Lathe R, Drillien R, et al: Expression of rabies virus glycoprotein from a recombinant vaccinia virus. *Nature* 1984;312:163–166.
11. Kleid DG, Yansura D, Small B, et al: Cloned viral protein vaccine for foot-and-mouth disease: responses in cattle and swine. *Science* 1981;214:1125–1129.
12. Lane JM, Ruben FL, Neff JM, Millar JD: Complications of smallpox vaccination, 1968. *N Engl J Med* 1968;281:1201–1208.
13. Lerner RA: Tapping the immunological repertoire to produce antibodies to predetermined specificity. *Nature* 1982;299:593–596.
14. Lindeman J: Speculations on idiotypes and homobodies. *Ann Immunol (Paris)* 1973;124:171–184.
15. Mackett M, Arrand JR: Recombinant vaccinia virus induces neutralising antibodies in rabbits against Epstein-Barr virus membrane antigen gp340. *EMBO J* 1985;4:3229–3234
16. Mackett M, Yilma TY, Rose JA, Moss B: Vaccinia virus recombinants: expression of VSV genes and protective immunization of mice and cattle. *Science* 1985;227:433–435.
17. MacLeod CM: Relation of the incubation period and the secondary immune response to lasting immunity to infectious diseases. *J Immunol* 1953;70:421–425.

18. McAleer WJ, Buynak EB, Maigetter RZ, et al: Human hepatitis B vaccine from recombinant yeast. *Nature* 1984;307:178–180.
19. Meiklejohn G: Adjuvant influenza adenovirus vaccine. *JAMA* 1962;179:594–597.
20. Mekalanos JJ, Swartz DJ, Pearson GDN, et al: Cholera toxin genes: nucleotide sequence, deletion analysis and vaccine development. *Nature* 1983;306:551–557.
21. Paoletti E, Lipinskas BR, Samsonoff C, et al: Construction of live vaccines using genetically engineered poxviruses: biological activity of vaccinia virus recombinants expressing the hepatitis B virus surface antigen and the herpes simplex virus glycoprotein D. *Proc Nat Acad Sci* USA 1984;81:193–197.
22. Perkus ME, Piccini A, Lipinskas BR, Paoletti E: Recombinant vaccinia virus: immunization against multiple pathogens. *Science* 1985;229:981–984.
23. Salk J: Mechanisms of immunity in virus infections, in: *Recent Progress in Microbiology*. Toronto, University of Toronto Press, 1963, pp 388–398.
24. Salk J: One-dose immunization against paralytic poliomyelitis using a non-infectious vaccine. *Rev Infect Dis* 1984;6:S444–S450.
25. Salk J: Persistence of immunity after administration of formalin-treated polio-virus vaccine. *Lancet* 1960;2:715–723.
26. Salk J, Contakos M, Laurent AM, et al: Use of adjuvants in studies on influenza immunization. 3. Degree of persistence of antibody in human subjects two years after vaccination. *JAMA* 1953;151:1169–1175.
27. Salk J, Salk D: Control of influenza and poliomyelitis with killed virus vaccines. *Science* 1977;195:834–847.
28. Salk J, Van Wezel AL, Stoeckel P, et al: Theoretical and practical considerations in the application of killed poliovirus vaccine for the control of paralytic poliomyelitis. *Dev Bio Stand* 1981;47:181–198.
29. Shapira M, Jibson M, Muller G, Arnon R: Immunity and protection against influenza virus by synthetic peptide corresponding to antigenic sites of hem-agglutinin. *Proc Nat Acad Sci USA* 1984;81:2461–2465.
30. Sharpe AH, Gaulton GN, McDade KK, et al: Syngeneic monoclonal antiidi-otype can induce cellular immunity to reovirus. *J Exp Med* 1984;160:1195–1205.
31. Smith GL, Godson EN, Nussenzweig V, et al: Plasmodium knowlesi sporozoite antigen: expression by an infection recombinant vaccinia virus. *Science* 1984;224:397–399.
32. Smith GL, Moss B: Infectious poxvirus vectors have capacity for at least 25,000 base pairs of foreign DNA. *Gene* 1983;25:21–28.
33. Smith GL, Murphy BR, Moss B: Construction and characterization of an in-fectious vaccinia virus recombinant that expresses the influenza hemagglutinin gene and induces resistance to influenza infection in hamsters. *Proc Nat Acad Sci* USA 1983;90:7155–7159.
34. Stein KE, Soderstrom T: Neonatal administration of idiotype or antiidiotype primes for protection against Escherichia coli K13 infection in mice. *J Exp Med* 1984;160:1001–1011.
35. Tainer JA, Getzoff ED, Alexander H, et al: The reactivity of anti-peptide antibodies is a function of the atomic mobility of sites in a protein. *Nature* 1984;312:127–134.

36. Westhof E, Altshuh D, Moras D, et al: Correlation between segmental mobility and the location of antigenic determinants in proteins. *Nature* 1984;311:123–126.
37. Wiktor TJ, Mac Farlan RI, Reagan KJ, et al: Protection from rabies by a vaccinia virus recombinant containing the rabies virus glycoprotein gene. *Proc Nat Acad Sci* USA 1984;81:7194–7198.
38. Young RA, Mehra V, Sweetser D, et al: Genes for the major protein antigens of the leprosy parasite Mycobacterium leprae. *Nature* 1985;316:450–452.
39. Zanetti M, Sercarz E, Salk J: Immunology of new generation vaccines. *Immunol Today* 1987;8:18–25.
40. Zavala F, Tam JP, Hollingdale MR, et al: Rationale for development of a synthetic vaccine against Plasmodium falciparum malaria. *Science* 1985;228:1436–1440.

Index